Advances in Mathematical Fluid Mechanics

Advances in Mathematical Fluid Mechanics is a forum for the publication of high quality monographs, or collections of works, on the mathematical theory of fluid mechanics, with special regards to the Navier-Stokes equations. Its mathematical aims and scope are similar to those of the *Journal of Mathematical Fluid Mechanics*. In particular, mathematical aspects of computational methods and of applications to science and engineering are welcome as an important part of the theory. So also are works in related areas of mathematics that have a direct bearing on fluid mechanics.

The monographs and collections of works published here may be written in a more expository style than is usual for research journals, with the intention of reaching a wide audience. Collections of review articles will also be sought from time to time.

Fundamental Directions in Mathematical Fluid Mechanics

Giovanni P. Galdi
John G. Heywood
Rolf Rannacher
Editors

Birkhäuser Verlag
Basel · Boston · Berlin

Editors' addresses:

Giovanni P. Galdi
School of Engineering
Department of Mechanical
Engineering
University of Pittsburgh
3700 O'Hara Street
Pittsburgh, PA 15261
USA
e-mail: galdi@engrng.pitt.edu

John G. Heywood
Department of Mathematics
University of British Columbia
Vancouver BC
Canada V6T 1Y4
e-mail: heywood@math.ubc.ca

Rolf Rannacher
Institut für Angewandte Mathematik
Universität Heidelberg
Im Neuenheimer Feld 293/294
Germany
e-mail: rannacher@iwr.uni-heidelberg.de

2000 Mathematical Subject Classification 76D05, 35Q30; 76M10, 76N10, 35B32

A CIP catalogue record for this book is available from the
Library of Congress, Washington D.C., USA

Deutsche Bibliothek Cataloging-in-Publication Data
Fundamental directions in mathematical fluid mechanics / Giovanni P. Galdi ... ed.. - Basel ; Boston ;
Berlin : Birkhäuser 2000
(Advances in mathematical fluid mechanics)
ISBN 3-7643-6414-9

ISBN 3-7643-6414-9 Birkhäuser Verlag, Basel – Boston – Berlin

Printed on acid-free paper produced from chlorine-free pulp. TCF ∞
Printed in Germany
ISBN 3-7643-6414-9

9 8 7 6 5 4 3 2 1

Contents

Preface

This volume consists of six articles, each treating an important topic in the theory of the Navier-Stokes equations, at the research level. Some of the articles are mainly expository, putting together, in a unified setting, the results of recent research papers and conference lectures. Several other articles are devoted mainly to new results, but present them within a wider context and with a fuller exposition than is usual for journals.

The plan to publish these articles as a book began with the lecture notes for the short courses of G.P. Galdi and R. Rannacher, given at the beginning of the International Workshop on Theoretical and Numerical Fluid Dynamics, held in Vancouver, Canada, July 27 to August 2, 1996. A renewed energy for this project came with the founding of the Journal of Mathematical Fluid Mechanics, by G.P. Galdi, J. Heywood, and R. Rannacher, in 1998. At that time it was decided that this volume should be published in association with the journal, and expanded to include articles by J. Heywood and W. Nagata, J. Heywood and M. Padula, and P. Gervasio, A. Quarteroni and F. Saleri. The original lecture notes were also revised and updated.

The topics of the articles here are from that part of fluid mechanics which seeks its foundation in the rigorous mathematical treatment of the Navier-Stokes equations. As is well known, the most famous problem in this area is to go beyond the presently known global existence of weak solutions, to the global existence of smooth solutions, for which uniqueness and continuous dependence on the data can be proven. The article of Galdi gives an overview of this topic.

The finite element Galerkin method shares the same variational setting as the mathematical theory of the equations, which provides it with a rigorous basis that has recently led to new codes of high reliability and flexibility, and great promise for further development. The article by Rannacher takes up both the theory and implementation of the finite element method, with an emphasis on a priori and a posteriori error estimation and adaptive mesh refinement.

Spectral Galerkin methods also share the same variational setting as the mathematical theory, and in modern practical implementations may be considered as a special class of finite element methods utilizing high order polynomials. The article by Gervasio, Quarteroni and Saleri addresses, particularly, the extension of these methods to domains with complicated geometries, by employing the techniques of domain decomposition.

The rigorous explanation of bifurcation phenomena in fluids has long been a central topic in the theory of the Navier- Stokes equations. The article by Heywood and Nagata introduces bifurcation theory in a general setting that is particularly convenient for application to the Navier-Stokes equations.

The extension of Navier-Stokes theory to the case of compressible viscous flow presents a fascinating array of new problems for theoretical and numerical investigation. The two articles of Heywood and Padula offer a simplified approach to the theory of steady compressible viscous flow. The second of these papers, on the steady transport equation, provides simplified proofs of results needed in the first paper.

Giovanni P. Galdi John G. Heywood Rolf Rannacher

An Introduction to the Navier-Stokes Initial-Boundary Value Problem

Giovanni P. Galdi

Rechts auf zwei hohen Felsen befinden sich Schlösser,
unten breitet sich die Stadt

J.W. GOETHE

0. Introduction

The equations of motion of an incompressible, Newtonian fluid – usually called *Navier-Stokes equations* – have been written almost one hundred eighty years ago. In fact, they were proposed in 1822 by the French engineer C.M.L.H. Navier upon the basis of a suitable molecular model. It is interesting to observe, however, that the law of interaction between the molecules postulated by Navier were shortly recognized to be totally inconsistent from the physical point of view for several materials and, in particular, for liquids. It was only more than twenty years later that the same equations were rederived by the twenty-six year old G. H. Stokes (1845) in a quite general way, by means of the theory of continua.

In the case where the fluid is subject to the action of a body force $\boldsymbol{f}$, the Navier-Stokes equations can be written as follows

$$\begin{aligned} &\frac{\partial \boldsymbol{v}}{\partial t} + \boldsymbol{v} \cdot \nabla \boldsymbol{v} = \nu \Delta \boldsymbol{v} + \nabla p + \boldsymbol{f} \\ &\operatorname{div} \boldsymbol{v} = 0 \end{aligned} \tag{0.1}$$

where $\boldsymbol{v} = \boldsymbol{v}(x,t)$ is the velocity field evaluated at the point $x \in \Omega$ and at time $t \in [0,T]$, ρp is the pressure field, ρ is the constant density of the fluid, and ν (> 0) is the coefficient of kinematical viscosity. Finally, Ω denotes the relevant geometrical domain where the spatial variables are ranging. Therefore, it will coincide with the region of flow for three-dimensional motions (*i.e.*, $\Omega \subset \mathbb{R}^3$), while it will coincide with a two-dimensional region, in case of plane flows ($\Omega \subset \mathbb{R}^2$).

To the equations (0.1) we append the *initial condition*: [1]

$$\boldsymbol{v}(x,0) = \boldsymbol{v}_0, \quad x \in \Omega \tag{0.2}$$

and the *boundary condition*

$$\boldsymbol{v}(y,t) = 0, \quad y \in \partial\Omega, \ t > 0 \ ^2. \tag{0.3}$$

[1]Without loss of generality, we can take 0 as initial time.

In the case where Ω extends to infinity, we should impose also convergence conditions on $\boldsymbol{v}(x,t)$ (and/or, possibly, on $p(x,t)$) when $|x| \to \infty$.

Several mathematical properties for system (0.1) have been deeply investigated over the years and are still the object of profound researches. However, after more than one hundred seventy years from their formulation, the *Fundamental Problem* (FP)related to them remains still unsolved, that is:

Given the body force $\boldsymbol{f}$ *and the initial distribution of velocity* $\boldsymbol{v}_0$ *(no matter how smooth), to determine a corresponding unique regular solution* $\boldsymbol{v}(x,t), p(x,t)$ *to* (0.1)–(0.3) *for all times* $t > 0$.

So far, this problem is only partially solved, despite numerous efforts by mathematicians and despite being viewed as an "obvious truth" by engineers. All this adds more weight to the following profound consideration due to Sir Cyril Hinshelwood, see Lighthill (1956, p. 343)

> Fluid dynamicists were divided into hydraulic engineers who observe what cannot be explained and mathematicians who explain things that cannot be observed

One of the aims of this article is to furnish an elementary presentation of some of the basic results so far known for (FP). In Section 1, we shall discuss the main features of system (0.1) and describe the main difficulties related to it. Successively, following the classical methods of Leray (1934a, 1934b) and Hopf (1951/1952), we introduce the definition of *weak solution* to (0.1)–(0.3) and study some of the related properties (Section 2). These solutions play a major role in the mathematical theory of Navier-Stokes equations, in that they are the only solutions, so far known, which exist for *all times* and *without restrictions on the size of the data.* In Section 3 we shall show the existence of a *weak solution* for all times $t > 0$. Uniqueness and regularity of Leray-Hopf solutions will be presented in Sections 4 and 5, respectively. Due to the particular form of the nonlinearity involved in the Navier-Stokes equations, this study will naturally lead to the functional class $L^{s,r} \equiv L^r(0,T;\boldsymbol{L}^s(\Omega))$, $n/s + 2/r = 1$, $s > n$, [3] such that any weak solution belonging to $L^{s,r}$ is unique and regular. In view of this result, we shall see that every weak solution in dimension two is unique within its class, and that it possesses as much space-time regularity as allowed by the data. Since it is not known if a weak solution in dimension three is in $L^{s,r}$, it is not known if these properties continue to hold for three-dimensional flows. However, "partial regularity" results are available. To show some of these latter, we begin to prove the existence of more regular solutions in Sections 6. This existence theory will lead to the celebrated "théorème de structure" of Leray, which, roughly speaking, states that every weak solution is regular in space and time, with the possible exception of a set of times I of zero 1/2-dimensional Hausdorff measure. Moreover, defining a finite time $t_1 \in I$ an *epoch of irregularity* for a weak solution $\boldsymbol{v}$, if $\boldsymbol{v}$ is regular in a left-neighborhood of t_1 but it can not be extended to a regular solution after t_1,

[3] n denotes the space dimension.

we shall give blow-up estimates for the Dirichlet norm of $\boldsymbol{v}$ at any (possible) epoch of irregularity. In view of the relevance of the functional class $L^{s,r}$, in Section 7 we will investigate the existence of weak solutions in such a class. Specifically, we shall prove the existence of weak solutions in $L^{s,r}$, at least for small times, provided the initial data are given in Lebesgue spaces L^q, for a suitable q. To avoid technical difficulties, this study will be performed for the case $\Omega = \mathbb{R}^n$ (Cauchy problem). As a consequence of these results, we shall enlarge the class of uniqueness of weak solutions, to include the case $s = n$. In addition, we shall give partial regularity results of a weak solution belonging to $L^{n,\infty}$. The important question of whether a weak solution in $L^{n,\infty}$ is regular, is left open.

1. Some considerations on the structure of the Navier-Stokes equations

Before getting involved with weak solutions à la Leray-Hopf and with their regularity, we wish to emphasize the main mathematical difficulties relating to (FP). First of all, we should notice that the unknowns $\boldsymbol{v}, p$ do not appear in (0.1) in a "symmetric way". In other words, the equation of conservation of mass is *not* of the following form

$$\frac{\partial p}{\partial t} = G(p, \boldsymbol{v}).$$

This is due to the fact that, from the mechanical point of view, the pressure plays the role of *reaction force* (Lagrange multiplier) associated with the isochoricity constraint $\operatorname{div} \boldsymbol{v} = 0$. In these regards, it is worth noticing that, in a perfect analogy with problems of motion of constrained rigid bodies, the pressure field must be generally deduced in terms of the velocity field, once this latter has been determined. In particular, we recall that the field $p(x,t)$ can be formally obtained – by operating with "div" on both sides of (0.1_1) – as a solution of the following Neumann problem

$$\begin{aligned} \Delta p &= \operatorname{div}(\boldsymbol{v}\cdot\nabla\boldsymbol{v} - \boldsymbol{f}) \ \text{ in } \Omega \\ \frac{\partial p}{\partial n} &= -(\nu\Delta\boldsymbol{v} + \boldsymbol{f})\cdot\boldsymbol{n} \ \text{ at } \partial\Omega \end{aligned} \tag{1.1}$$

where $\boldsymbol{n}$ denotes the unit outer normal to $\partial\Omega$ [4].

Because of the mentioned lack of "symmetry" in $\boldsymbol{v}$ e p, the system (0.1) does not fall in any of the classical categories of equations, even though, in a sense, it could be considered close to a quasi-linear parabolic system. Nevertheless, the basic difficulty related to problem (0.1)–(0.3) does not arise from the lack of such a symmetry but, rather, from the *coupled* effect of the *lack* of symmetry and of

[4] From this it is clear that to prescribe the values of the pressure at the bounding walls or at the initial time *independently* of $\boldsymbol{v}$, could be incompatible with (1.1) and, therefore, could render the problem ill-posed.

the *presence* of the nonlinear term. In fact, the (FP) formulated for any of the following systems

$$\begin{aligned} &\frac{\partial \boldsymbol{v}}{\partial t} = \nu \Delta \boldsymbol{v} + \nabla p + \boldsymbol{f} \\ &\operatorname{div} \boldsymbol{v} = 0 \end{aligned} \tag{0.1$'$}$$

$$\frac{\partial \boldsymbol{v}}{\partial t} + \boldsymbol{v} \cdot \nabla \boldsymbol{v} = \nu \Delta \boldsymbol{v} + \boldsymbol{f} \tag{0.1$''$}$$

obtained from (0.1) by disregarding either the nonlinear term $[(0.1')]$ or the isochoricity condition $[(0.1'')]$ can be completely solved. While for $(0.1')$ this solvability will be clear when we shall consider the solvability of (FP) for (0.1), the solvability of $(0.1'')$ is a consequence of an interesting *a priori* estimate discovered by Kiselev and Ladyzhenskaya (1957) and based on a maximum principle that we would like to mention here. Setting

$$\boldsymbol{u}(x,t) = \boldsymbol{v}(x,t) e^{-\alpha t} \quad \alpha > 0$$

from $(0.1'')$ we obtain

$$\tfrac{1}{2} \frac{\partial \boldsymbol{u}^2}{\partial t} + \tfrac{1}{2} \boldsymbol{v} \cdot \nabla \boldsymbol{u}^2 + \alpha \boldsymbol{u}^2 = \nu \Delta \boldsymbol{u} \cdot \boldsymbol{u} + \boldsymbol{f} \cdot \boldsymbol{u} e^{-\alpha t}. \tag{1.2}$$

Consider a point $P = (\widetilde{x}, \widetilde{t})$ of the cylinder $\overline{\Omega} \times [0,T]$ where $\boldsymbol{u}^2$ assumes its maximum. If such a point lies either on the bottom face of the cylinder (*i.e.*, at $\widetilde{t} = 0$) or on its lateral surface (*i.e.*, at $\widetilde{x} \in \partial\Omega$) we have

$$\max_{x,t} \boldsymbol{u}^2(x,t) \le \boldsymbol{u}^2(\widetilde{x}, 0) \le \max_x \boldsymbol{v}_0^2(x). \tag{1.3}$$

If, on the contrary, P is an interior point of the cylinder or lies on its top face we find

$$\left. \begin{aligned} &\frac{\partial \boldsymbol{u}^2}{\partial t} \ge 0, \quad \nabla \boldsymbol{u}^2 = 0, \\ &-\boldsymbol{u} \cdot \Delta \boldsymbol{u} = |\nabla \boldsymbol{u}|^2 - \tfrac{1}{2} \Delta \boldsymbol{u}^2 \ge 0 \end{aligned} \right\} \quad \text{evaluated at} \quad (\widetilde{x}, \widetilde{t}) = P.$$

Therefore, from (1.2) we deduce

$$\alpha \boldsymbol{u}^2(\widetilde{x}, \widetilde{t}) \le \boldsymbol{f}(\widetilde{x}, \widetilde{t}) \cdot \boldsymbol{u}(\widetilde{x}, \widetilde{t}) e^{-\alpha \widetilde{t}}. \tag{1.4}$$

As a consequence, from (1.3), (1.4) we prove the following *a priori* estimate holding for all (sufficiently regular) solutions to system $(0.1)''$

$$|\boldsymbol{v}(x,t)| \le e^{\alpha t} \left\{ \frac{1}{\alpha} \max_{x,t} [e^{-\alpha t} |\boldsymbol{f}(x,t)|] + \max_x |\boldsymbol{v}_0(x)|^2 \right\}. \tag{1.5}$$

Notice that (1.5) is *independent* of the spatial dimension. Unfortunately, nothing similar to (1.5) is so far known for system (0.1) in dimension 3. Nevertheless, as we shall show later on, in dimension 2 the global (*i.e.*, for all times) estimates that we are able to derive will suffice to ensure the existence and uniqueness of a regular solution for (0.1).

2. The Leray-Hopf weak solutions and related properties

We shall begin by giving the definition of weak solution in the sense of Leray-Hopf. To this end, we need to introduce some notation. By $L^q(\Omega)$ and $W^{m,q}(\Omega)$, $1 \le q \le \infty$, $m = 0, 1, \ldots$, we denote the usual Lebesgue and Sobolev spaces, respectively. The norm in $W^{m,q}$ is indicated by $\|\cdot\|_{m,q}$. For $m = 0$, it is $W^{0,q} \equiv L^q$ and we set $\|\cdot\|_{0,q} \equiv \|\cdot\|_q$. Whenever we need to specify the domain D on which these norms are evaluated, we shall write $\|\cdot\|_{m,q,D}$. We denote by $W_0^{m,q}(\Omega)$ the completion in the norm $\|\cdot\|_{m,q,\Omega}$ of the space $C_0^\infty(\Omega)$ constituted by all infinitely differentiable functions with compact support in Ω. The dual space of $W_0^{m,q}$ will be denoted by $W^{-m,q'}$.

Let [5]

$$\mathcal{D}(\Omega) = \{\boldsymbol{\psi} \in \boldsymbol{C}_0^\infty(\Omega) : \ \operatorname{div} \boldsymbol{\psi} = 0 \text{ in } \Omega\}.$$

We define $H_q = H_q(\Omega)$ as the completion of $\mathcal{D}(\Omega)$ in the Lebesgue space $\boldsymbol{L}^q(\Omega)$. Moreover, we denote by $H_q^1(\Omega)$ the completion of $\mathcal{D}(\Omega)$ in the Sobolev space $\boldsymbol{W}^{1,q}(\Omega)$. For $q = 2$, we shall simply write H and H^1, respectively. It is well known, see Galdi (1994, Chapter III), that if Ω has a bounded boundary which is locally lipschitzian, or if Ω is a half-space, the following characterizations for H_q and H_q^1 hold, for $1 < q < \infty$:

$$H_q(\Omega) = \{\boldsymbol{u} \in \boldsymbol{L}^q(\Omega) : \ \operatorname{div} \boldsymbol{u} = 0, \ \boldsymbol{u} \cdot \boldsymbol{n}|_{\partial\Omega} = 0\}$$

$$H_q^1(\Omega) = \{\boldsymbol{u} \in \boldsymbol{W}^{1,q}(\Omega) : \ \operatorname{div} \boldsymbol{u} = 0, \ \boldsymbol{u}|_{\partial\Omega} = 0\}$$

where the values at the boundary have to be understood in the trace sense. Furthermore, if Ω is of class C^1, the following *Helmholtz-Weyl decomposition* holds

$$\boldsymbol{L}^q(\Omega) = H_q(\Omega) \oplus G_q(\Omega) \tag{2.1}$$

where

$$G_q(\Omega) = \{\boldsymbol{u} \in \boldsymbol{L}^q(\Omega) : \ \boldsymbol{u} = \nabla p, \ \text{for some } p \in L^1_{loc}(\Omega) \ \text{with } \nabla p \in \boldsymbol{L}^q(\Omega)\}$$

(we set $G = G_2$). The projection of $\boldsymbol{L}^q$ onto H_q is denoted by P_q ($\equiv P$, for $q = 2$). In the case $q = 2$, H and G are orthogonal subspaces of $\boldsymbol{L}^2$, and (2.1) holds for any open set Ω.

For $T \in (0, \infty]$ we set $\Omega_T = \Omega \times [0, T)$ and define

$$\mathcal{D}_T = \{\boldsymbol{\varphi} \in \boldsymbol{C}_0^\infty(\Omega_T) : \ \operatorname{div} \boldsymbol{\varphi}(x,t) = 0 \text{ in } \Omega_T\}$$

Notice that for $\boldsymbol{\varphi} \in \mathcal{D}_T$, $\boldsymbol{\varphi}(x, 0)$ need not be zero. For $\boldsymbol{a}, \boldsymbol{b}$ vector functions in Ω we put

$$(\boldsymbol{a}, \boldsymbol{b}) \equiv \int_\Omega \boldsymbol{a} \cdot \boldsymbol{b}, \quad \|\boldsymbol{a}\|_2 = (\boldsymbol{a}, \boldsymbol{a})^{1/2}$$

$$(\nabla \boldsymbol{a}, \nabla \boldsymbol{b}) \equiv \sum_{i,j=1}^n \int_\Omega \frac{\partial a_i}{\partial x_j} \frac{\partial b_i}{\partial x_j}, \quad \|\nabla \boldsymbol{a}\|_2 = (\nabla \boldsymbol{a}, \nabla \boldsymbol{a})^{1/2}.$$

---

[5] If X is a space of scalar functions, we denote by $\mathbf{X}$ the space constituted by vector or tensor valued functions having components in X.

If we need to specify the domain D on which these quantities are evaluated, we shall write

$$(\cdot,\cdot)_D, \quad \|\cdot\|_{2,D}.$$

Moreover, for a given Banach space X, with associated norm $\|\cdot\|_X$, and a real interval (a,b), we denote by $L^q(a,b;X)$ the linear space of (equivalence classes of) functions $f:(a,b)\to X$ such that the functional

$$\|f\|_{L^q(a,b;X)} \equiv \begin{cases} \left(\displaystyle\int_a^b \|f(t)\|_X^q dt\right)^{1/q} & \text{if } 1\le q<\infty \\ \operatorname*{ess\,sup}_{t\in(a,b)} \|f(t)\|_X & \text{if } q=\infty \end{cases}$$

is finite. It is known that this functional defines a norm with respect to which $L^q(a,b;X)$ becomes a Banach space (Hille and Phillips 1957, Chapter III). Likewise, for r a non-negative integer and I a real interval, we denote by $C^r(I;X)$ the class of continuous functions from I to X, which are differentiable in I up to the order r included. Finally, if I is open or semi-open, by $BC(I;X)$, we denote the subspace of $C(I;X)$ such that $\sup_{t\in I}\|u(t)\|_X<\infty$. Depending on X, these spaces might share several properties with the "usual" Lebesgue spaces $L^q(a,b)$ and spaces $C^r(I)$, and we refer to the monograph of Hille and Phillips for further information.

Assume now $\boldsymbol{v}(x,t), p(x,t)$ is a classical solution to (0.1)–(0.3). [6] Then, multiplying $(0.1)_1$ by $\boldsymbol{\varphi}\in\mathcal{D}_T$ and integrating over Ω_∞ we find

$$\begin{aligned}\int_0^\infty \left\{\left(\boldsymbol{v},\frac{\partial\boldsymbol{\varphi}}{\partial t}\right) - \nu(\nabla\boldsymbol{v},\nabla\boldsymbol{\varphi}) - (\boldsymbol{v}\cdot\nabla\boldsymbol{v},\boldsymbol{\varphi})\right\}dt \\ = -\int_0^\infty (\boldsymbol{f},\boldsymbol{\varphi})dt - (\boldsymbol{v}_0,\boldsymbol{\varphi}(0)), \\ \text{for all } \boldsymbol{\varphi}\in\mathcal{D}_T.\end{aligned} \tag{2.2}$$

Conversely, if $\boldsymbol{v}(x,t)$ is a vector field satisfying (2.2), and having enough smoothness as to allow for integration by parts over Ω_∞ in some sense, [7] we easily obtain

$$\int_0^\infty \left(\frac{\partial\boldsymbol{v}}{\partial t} + \boldsymbol{v}\cdot\nabla\boldsymbol{v} - \nu\Delta\boldsymbol{v} - \boldsymbol{f}, h(t)\boldsymbol{\psi}\right)dt = 0$$

for all $h\in C_0^\infty((0,T))$ and $\boldsymbol{\psi}\in\mathcal{D}(\Omega)$. Therefore, for every such $\boldsymbol{\psi}$

$$\left(\frac{\partial\boldsymbol{v}}{\partial t} + \boldsymbol{v}\cdot\nabla\boldsymbol{v} - \nu\Delta\boldsymbol{v} - \boldsymbol{f}, \boldsymbol{\psi}\right) = 0,$$

[6] For instance, $\boldsymbol{v}$ is one time differentiable in t and twice differentiable in x, while p is one time differentiable in x. Moreover, $\boldsymbol{v}$ assumes continuously the initial and boundary data.

[7] For instance, in the sense of generalized differentiation.

and by a well known result of Hopf (1950/1951), see Galdi (1994, Lemma III.1.1), we conclude the validity of (0.1) for some pressure field $p(x,t)$. However, it is clear that if $\boldsymbol{v}(x,t)$ is a solenoidal vector field that satisfies (2.2) but is *not* sufficiently differentiable, we can *not* go from (2.2) to $(0.1)_1$ and it is precisely in this sense that (2.2) has to be considered as the *weak formulation* of $(0.1)_1$.

Remark 2.1 It is simple to give examples of solenoidal vector fields which satisfy (2.2) but which do not have enough smoothness as to verify $(0.1)_1$, no matter how smooth $\boldsymbol{f}$ and $\boldsymbol{v}_0$ are. Take, for instance

$$\overline{\boldsymbol{v}}(x,t) = a(t)\nabla\sigma(x), \quad \Delta\sigma = 0 \text{ in } \Omega \tag{2.3}$$

where $a(t)$ has no more regularity than the local integrability in $(0,T)$ with $a(0)$ finite. Since

$$\overline{\boldsymbol{v}}\cdot\nabla\overline{\boldsymbol{v}} = \tfrac{1}{2}\nabla(\nabla\sigma)^2$$

and $\operatorname{div}\overline{\boldsymbol{v}} = 0$, we deduce that $\overline{\boldsymbol{v}}(x,t)$ is a non-smooth solenoidal vector field satisfying (2.2) with $\boldsymbol{f} = 0$ and $\boldsymbol{v}_0 = a(0)\nabla\psi$. Notice that, since σ is harmonic, $\boldsymbol{v}_0$ is analytic.

We wish to give a generalized meaning to the solenoidality condition $(0.1)_2$ and to the boundary condition (0.4). This will be accomplished, for instance, if we require that, for almost all times $t \in [0,T]$, $\boldsymbol{v}(\cdot,t)$ belongs to $H^1(\Omega)$. Moreover, to ensure that all integrals in (2.2) are meaningful, we may require $\boldsymbol{v} \in L^2(0,T;H^1)$.

These considerations then lead to the following definition of weak solution, due to Leray (1934a, 1934b) and Hopf (1951/1952).

Definition 2.1 Let $\boldsymbol{v}_0 \in H(\Omega)$, $\boldsymbol{f} \in \boldsymbol{L}^2(\Omega_T)$. A measurable function $\boldsymbol{v} : \Omega_T \to \mathbb{R}^n$, $n = 2,3$, [8] is said a *weak solution* of the problem (0.1)–(0.4) in Ω_T if

a) $\boldsymbol{v} \in V_T \equiv L^2(0,T;H^1) \cap L^\infty(0,T;H)$;

b) $\boldsymbol{v}$ verifies (2.2).

If $\boldsymbol{f} \in \boldsymbol{L}^2(\Omega_T)$ for all $T > 0$, $\boldsymbol{v}$ will be called a *global weak solution* if it is a weak solution in Ω_T for all $T > 0$.

Remark 2.2 In a) we have included the condition that $\boldsymbol{v} \in L^\infty(0,T;H)$ which, *a priori*, does not seem to be strictly necessary. However, on one hand, this condition ensures that the kinetic energy of a weak solution is essentially bounded in the time interval $[0,T]$, and this is a natural request from the physical point of view. On the other hand, excluding such a condition would result in a definition of weak solution too poor to allow for the development of any further relevant property. And last, but not least, we shall prove that the class of weak solutions is not empty, see Theorem 3.1.

Remark 2.3 Definition 2.1 is apparently silent about the pressure field. Later on (Theorem 2.1) we shall see that to every weak solution we can always associate a corresponding pressure field.

---

[8] Of course, the definition of weak solution can be given for any spatial dimension $n \geq 2$, but we shall be mainly interested in the physical interesting cases of two and three dimensions.

Our next objective is to collect a certain number of properties of weak solutions which will eventually lead, among other things, to a definition equivalent to Definition 1.1. The following result is due to Hopf (1951/1952, Satz 2.1); see also Prodi (1959, Lemma 1) and Serrin (1963, Theorem 4).

Lemma 2.1 *Let $\boldsymbol{v}$ be a weak solution in Ω_T. Then $\boldsymbol{v}$ can be redefined on a set of zero Lebesgue measure in such a way that $\boldsymbol{v}(t) \in \boldsymbol{L}^2(\Omega)$ for all $t \in [0,T)$ and satisfies the identity*

$$\int_s^t \left\{ \left(\boldsymbol{v}, \frac{\partial \boldsymbol{\varphi}}{\partial t} \right) - \nu(\nabla \boldsymbol{v}, \nabla \boldsymbol{\varphi}) - (\boldsymbol{v} \cdot \nabla \boldsymbol{v}, \boldsymbol{\varphi}) \right\} d\tau = -\int_s^t (\boldsymbol{f}, \boldsymbol{\varphi}) d\tau + (\boldsymbol{v}(t), \boldsymbol{\varphi}(t)) - (\boldsymbol{v}(s), \boldsymbol{\varphi}(s)), \tag{2.4}$$

for all $s \in [0,t]$, $t < T$, and all $\boldsymbol{\varphi} \in \mathcal{D}_T$.

Proof. It is clear that to show (2.4) for arbitrary $s \in [0,t]$, it is enough to prove it for $s = 0$. We begin to show that (2.4) holds for $s = 0$ and almost every $t \in [0,T)$. Let $\theta \in C^1(\mathbb{R})$ be a monotonic, non-negative function such that

$$\theta(\xi) = \begin{cases} 1 & \text{if } \xi \le 1 \\ 0 & \text{if } \xi \ge 2 \end{cases}$$

For a fixed $t \in [0,T)$ and $h > 0$ with $t + h < T$ we set

$$\theta_h(\tau) = \theta\left(\frac{\tau - t + h}{h} \right).$$

Notice that

$$\left| \frac{d\theta_h}{d\tau} \right| \le C h^{-1}, \quad C > 0, \qquad \int_t^{t+h} \frac{d\theta_h}{d\tau} d\tau = -1. \tag{2.5}$$

Choosing in (2.2) $\boldsymbol{\varphi}(x,t)$ as $\theta_h(t)\boldsymbol{\varphi}(x,t)$, we obtain

$$\int_0^{t+h} \theta_h(\tau) \left\{ \left(\boldsymbol{v}, \frac{\partial \boldsymbol{\varphi}}{\partial t} \right) - \nu(\nabla \boldsymbol{v}, \nabla \boldsymbol{\varphi}) - (\boldsymbol{v} \cdot \nabla \boldsymbol{v}, \boldsymbol{\varphi}) + (\boldsymbol{f}, \boldsymbol{\varphi}) \right\} d\tau = -\int_0^{t+h} \frac{d\theta_h(\tau)}{d\tau} (\boldsymbol{v}, \boldsymbol{\varphi}) d\tau - (\boldsymbol{v}_0, \boldsymbol{\varphi}). \tag{2.6}$$

Letting $h \to 0$ in this relation and recalling Definition 2.1, we easily deduce that the integral on the left-hand side of this relation tends to

$$\int_0^t \left\{ \left(\boldsymbol{v}, \frac{\partial \boldsymbol{\varphi}}{\partial t} \right) - \nu(\nabla \boldsymbol{v}, \nabla \boldsymbol{\varphi}) - (\boldsymbol{v} \cdot \nabla \boldsymbol{v}, \boldsymbol{\varphi}) + (\boldsymbol{f}, \boldsymbol{\varphi}) \right\} d\tau.$$

Let now investigate the behavior of the integral at the right-hand side of (2.6). In view of (2.5) and of a) of Definition 2.1, we have for each fixed t

$$\begin{aligned}
\ell(h,t) &\equiv \left| \int_0^{t+h} \frac{d\theta_h(\tau)}{d\tau} (\boldsymbol{v}(\tau), \boldsymbol{\varphi}(\tau)) d\tau + (\boldsymbol{v}(t), \boldsymbol{\varphi}(t)) \right| \\
&= \left| \int_0^{t+h} \frac{d\theta_h(\tau)}{d\tau} \left\{ (\boldsymbol{v}(\tau) - \boldsymbol{v}(t), \boldsymbol{\varphi}(t)) + (\boldsymbol{v}(\tau), \boldsymbol{\varphi}(\tau) - \boldsymbol{\varphi}(t)) \right\} d\tau \right| \\
&\leq C \|\boldsymbol{\varphi}(t)\|_2 \left(h^{-1} \int_t^{t+h} \|\boldsymbol{v}(\tau) - \boldsymbol{v}(t)\|_2 d\tau \right) \\
&\quad + \max_{\tau \in [t,t+h]} \|\boldsymbol{\varphi}(t) - \boldsymbol{\varphi}(\tau)\|_2 \left(h^{-1} \int_t^{t+h} \|\boldsymbol{v}(\tau)\|_2 d\tau \right) \\
&\leq C \|\boldsymbol{\varphi}(t)\|_2 \left(h^{-1} \int_t^{t+h} \|\boldsymbol{v}(\tau) - \boldsymbol{v}(t)\|_2 d\tau \right) \\
&\quad + M \max_{\tau \in [t,t+h]} \|\boldsymbol{\varphi}(t) - \boldsymbol{\varphi}(\tau)\|_2 .
\end{aligned}$$

Denote by $\mathcal{L}(\boldsymbol{v})$ the set of all those $t \in [0,T)$ for which

$$\lim_{h \to 0} h^{-1} \int_t^{t+h} \|\boldsymbol{v}(\tau) - \boldsymbol{v}(t)\|_2 d\tau = 0.$$

As is well known from the theory of Lebesgue integration (Titchmarsh, 1964, §11.6, Hille and Phillips, 1957, Theorem 38.5), $\mathcal{L}(\boldsymbol{v})$ can differ from $[0,T)$ only by a set of zero Lebesgue measure. Therefore, since

$$\lim_{h \to 0} \max_{\tau \in [t,t+h]} \|\boldsymbol{\varphi}(t) - \boldsymbol{\varphi}(\tau)\|_2 = 0,$$

we obtain

$$\lim_{h \to 0} \ell(h,t) = 0, \quad \text{for all } t \in \mathcal{L}(\boldsymbol{v}),$$

and so identity (2.4) follows for $s = 0$ and all $t \in \mathcal{L}(\boldsymbol{v})$. We set $E_1 = [0,T) - \mathcal{L}(\boldsymbol{v})$. Moreover, by a) of Definition 2.1, there exists a constant $M > 0$ and a set $E_2 \subset [0,T)$ of zero Lebesgue measure such that

$$\|\boldsymbol{v}(t)\|_2 \leq M, \quad \text{for all } t \in [0,T) - E_2. \tag{2.7}$$

Put $E = E_1 \cup E_2$ and pick $\bar{t} \in E$. Then, there exists a sequence $\{t_k\} \subset [0,T) - E$ converging to $\bar{t}$ as $k \to \infty$. By (2.7), $\|\boldsymbol{v}(t_k)\|_2 \leq M$ and so, by the weak compactness of the spaces H we find $\boldsymbol{U}_{\bar{t}} \in H(\Omega)$ such that

$$\lim_{k \to \infty} (\boldsymbol{v}(t_k) - \boldsymbol{U}_{\bar{t}}, \boldsymbol{\psi}) = 0, \quad \text{for all } \boldsymbol{\psi} \in \mathcal{D}(\Omega).$$

Define

$$\boldsymbol{v}^*(x,t) = \begin{cases} \boldsymbol{v}(x,t) & \text{if } t \in [0,T) - E \\ \boldsymbol{U}_t(x) & \text{if } t \in E. \end{cases}$$

(Notice that $\boldsymbol{v}^*(x,0) = \boldsymbol{v}_0(x)$.) Clearly, $\boldsymbol{v}^* \in L^2(\Omega)$, for all $t \in [0,T)$. Furthermore, evaluating (2.4) along the sequence $\{t_k\}$ associated to $\boldsymbol{U}_t$ and letting $k \to \infty$ it is easy to verify the validity of the following statements:

1) $\boldsymbol{v}^*$ satisfies (2.4) for all $t \in [0,T)$;
2) $\boldsymbol{U}_t$ does not depend on the sequence $\{t_k\}$.

The lemma is then completely proved.

As a corollary to this result, we have

Lemma 2.2 *Let* $\boldsymbol{v}$ *be a weak solution in* Ω_T. *Then* $\boldsymbol{v}$ *can be redefined on a set of zero Lebesgue measure in such a way that it satisfies the identity*

$$\begin{aligned}\int_0^t \{-\nu(\nabla \boldsymbol{v}, \nabla \boldsymbol{\psi}) & -(\boldsymbol{v}\cdot\nabla\boldsymbol{v}, \boldsymbol{\psi})\}\, ds \\ & = -\int_0^t (\boldsymbol{f}, \boldsymbol{\psi}) ds + (\boldsymbol{v}(t), \boldsymbol{\psi}) - (\boldsymbol{v}_0, \boldsymbol{\psi}),\end{aligned} \tag{2.8}$$

for all $t \in [0,T)$ *and all* $\boldsymbol{\psi} \in \mathcal{D}(\Omega)$. *Furthermore,* $\boldsymbol{v}$ *is* $\boldsymbol{L}^2$ *weakly continuous, that is,*

$$\lim_{t\to t_0} (\boldsymbol{v}(t) - \boldsymbol{v}(t_0), \boldsymbol{u}) = 0,$$

for all $t_0 \in [0,T)$ *and all* $\boldsymbol{u} \in \boldsymbol{L}^2(\Omega)$.

Proof. We put in (2.4) $s = 0$ and choose $\boldsymbol{\varphi}(x,t) = \theta_h(t)\boldsymbol{\psi}(x)$, where θ_h is the function introduced in the proof of the previous lemma and $\boldsymbol{\psi} \in \mathcal{D}(\Omega)$. Noticing that $\boldsymbol{\varphi} = \boldsymbol{\psi}$ in $t \in [0,t]$, (2.8) follows at once. To show the L^2 weak continuity, we observe that for any fixed $t_0 \in [0,T)$ from (2.8) it follows that

$$\forall \varepsilon > 0,\ \exists\, \delta(\varepsilon) > 0 : |t - t_0| < \delta \Longrightarrow |(\boldsymbol{v}(t) - \boldsymbol{v}(t_0), \boldsymbol{\psi})| < \varepsilon,$$

for all $\boldsymbol{\psi} \in \mathcal{D}(\Omega)$. It is clear that this property continues to hold (by density) for all $\boldsymbol{w} \in H(\Omega)$. Let now $\boldsymbol{u} \in \boldsymbol{L}^2(\Omega)$. By the Helmholtz-Weyl decomposition (2.1) we may write

$$\boldsymbol{u} = \boldsymbol{w} + \nabla q, \quad \boldsymbol{w} \in H(\Omega),\ \nabla q \in G(\Omega)$$

and so, since $\boldsymbol{v} \in H(\Omega)$, we have

$$(\boldsymbol{v}(t) - \boldsymbol{v}(t_0), \boldsymbol{u}) = (\boldsymbol{v}(t) - \boldsymbol{v}(t_0), \boldsymbol{w}),$$

and the lemma follows.

Remark 2.4 Lemma 2.2 tells us, in particular, the way in which a weak solution assumes the initial data, namely, in the sense of the weak $\boldsymbol{L}^2$ convergence.

Throughout the rest of these notes, we shall assume that every weak solution has been modified on a set of zero Lebesgue measure in such a way that it verifies the assertions of Lemma 2.1 *and* Lemma 2.2

Our next concern is to investigate if and in which sense, we can associate a "pressure field" to a weak solution. Let us first assume that $\boldsymbol{v}, p$ is a classical solution to (0.1)–(0.4). Then, multiplying $(0.1)_1$ by $\boldsymbol{\chi} \in \boldsymbol{C}_0^\infty(\Omega)$ and integrating by parts over Ω_t we formally obtain

$$\int_0^t \{-\nu(\nabla \boldsymbol{v}, \nabla \boldsymbol{\chi}) \ -(\boldsymbol{v}\cdot\nabla\boldsymbol{v}, \boldsymbol{\chi}) + (\boldsymbol{f}, \boldsymbol{\chi})\}\, ds = \int_0^t (p, \operatorname{div} \boldsymbol{\chi}) + (\boldsymbol{v}(t), \boldsymbol{\chi}) - (\boldsymbol{v}_0, \boldsymbol{\chi}). \tag{2.9}$$

In the next theorem, we shall show that to any weak solution $\boldsymbol{v}$ in Ω_T we can associate a function $P(t) \in L^2(\omega)$, $t \in [0,T)$, $\omega \subset\subset \Omega$, such that

$$\int_0^t \{-\nu(\nabla \boldsymbol{v}, \nabla \boldsymbol{\chi}) \ -(\boldsymbol{v}\cdot\nabla\boldsymbol{v}, \boldsymbol{\chi}) + (\boldsymbol{f}, \boldsymbol{\chi})\}\, ds = (P(t), \operatorname{div} \boldsymbol{\chi}) + (\boldsymbol{v}(t), \boldsymbol{\chi}) - (\boldsymbol{v}_0, \boldsymbol{\chi}), \tag{2.10}$$

for all $\boldsymbol{\chi} \in \boldsymbol{C}_0^\infty(\Omega)$. We wish to emphasize that, using only (2.8) and the local regularity property of weak solution, in general, we can *not* write

$$P(\cdot, t) = \int_0^t p(\cdot, s) ds, \quad \text{for some } p \in L^1_{loc}([0,T)), \tag{2.11}$$

due to the fact that a weak solution has *a priori* only a mild degree of regularity in time. To see this, let us consider the vector field $\overline{\boldsymbol{v}}$ given in (2.3) and choose $a(t) \in C([0,T))$ but $a' \notin L^1_{loc}([0,T))$. By a straightforward calculation we find that

$$\int_0^t \{-\nu(\nabla \overline{\boldsymbol{v}}, \nabla \boldsymbol{\chi}) \ -(\overline{\boldsymbol{v}}\cdot\nabla\overline{\boldsymbol{v}}, \boldsymbol{\chi})\}\, ds - (\overline{\boldsymbol{v}}(t), \boldsymbol{\chi}) + (\overline{\boldsymbol{v}}_0, \boldsymbol{\chi})$$

$$= \tfrac{1}{2}\int_0^t a^2(s)((\nabla\sigma)^2, \operatorname{div}\boldsymbol{\chi}) ds + (a(t) - a(0))(\sigma, \operatorname{div}\boldsymbol{\chi})$$

and therefore (2.10) is satisfied with $\boldsymbol{f} \equiv 0$ and

$$P = \tfrac{1}{2}\int_0^t a^2(s)(\nabla\sigma)^2 ds + (a(t) - a(0))\sigma.$$

Since $a' \notin L^1_{loc}([0,T))$, P does not verify (2.11); see also Remark 2.6.

Theorem 2.1 *Let $\boldsymbol{v}$ be a weak solution in Ω_T. Then, there exists a scalar field $P : \Omega_T \to \mathbb{R}$ with*

$$P(t) \in L^2(\omega), \quad \textit{for all } t \in [0,T) \textit{ and } \omega \subset\subset \Omega,$$

verifying (2.10) for all $\boldsymbol{\chi} \in \boldsymbol{C}_0^\infty(\Omega)$ and all $t \in [0,T)$. Moreover, if ω satisfies the cone condition, there exist $C = C(t,\omega) \in \mathbb{R}$ and $C_1 = C_1(\omega) > 0$ such that

$$\|P(t) - C\|_{2,\omega} \leq C_1 \left\{ \int_0^t \left(\|\nabla \boldsymbol{v}(s)\|_{2,\omega} + M^\alpha \|\nabla \boldsymbol{v}(s)\|_{2,\omega}^\beta + \|\boldsymbol{f}(s)\|_{2,\omega} \right) ds + M \right\}$$

for all $t \in [0,T)$, where

$$M = \operatorname*{ess\ sup}_{s \in [0,t]} \|\boldsymbol{v}(s)\|_{2,\omega}$$

and

$$\alpha = \begin{cases} 1 & \textit{if } n = 2 \\ 1/2 & \textit{if } n = 3 \end{cases}$$

$$\beta = \begin{cases} 1 & \textit{if } n = 2 \\ 3/2 & \textit{if } n = 3. \end{cases}$$

Proof. Let us consider a sequence of bounded "invading domains" $\{\Omega_k\}$, that is, Ω_k is bounded for each k, and

$$\Omega_{k+1} \supset \Omega_k, \quad \Omega = \bigcup_{k=1}^{\infty} \Omega_k.$$

Without loss, we may assume that Ω_k satisfies the cone condition for each k. For fixed $t \in [0, T)$, consider the functional

$$\mathcal{F}(\boldsymbol{\chi}) = \int_0^t \{-\nu(\nabla \boldsymbol{v}, \nabla \boldsymbol{\chi}) - (\boldsymbol{v} \cdot \nabla \boldsymbol{v}, \boldsymbol{\chi}) + (\boldsymbol{f}, \boldsymbol{\chi})\}\, ds - (\boldsymbol{v}(t), \boldsymbol{\chi}) + (\boldsymbol{v}_0, \boldsymbol{\chi}),$$
$$\boldsymbol{\chi} \in \boldsymbol{W}_0^{1,2}(\Omega_k).$$

It is clear that $\mathcal{F}$ is linear functional on $\boldsymbol{W}_0^{1,2}(\Omega_k)$. Moreover, using the Schwarz inequality and the following ones (see, *e.g.*, Galdi, 1994, Chapter II)

$$\begin{aligned} \|u\|_4 &\le 2^{-1/4} \|u\|_2^{1/2} \|\nabla u\|_2^{1/2}, \quad n = 2, \\ \|u\|_4 &\le \left(\tfrac{4}{3\sqrt{3}}\right)^{3/4} \|u\|_2^{1/4} \|\nabla u\|_2^{3/4}, \quad n = 3, \end{aligned} \tag{2.12}$$

it is easy to see that

$$|\mathcal{F}(\boldsymbol{\chi})| \le c\|\boldsymbol{\chi}\|_2 \left\{ \int_0^t \left(\|\nabla \boldsymbol{v}\|_2 + M_k^{\alpha} \|\nabla \boldsymbol{v}\|^{\beta} \|\boldsymbol{f}\|_2\right) ds + M_k \right\} \tag{2.13}$$

where

$$M_k = \operatorname*{ess\ sup}_{s \in [0,t]} \|\boldsymbol{v}(s)\|_{2,\Omega_k}.$$

As a consequence, $\mathcal{F}$ is a continuous linear functional on $\boldsymbol{W}_0^{1,2}(\Omega_k)$ which, by Lemma 2.2, vanishes on $H^1(\Omega_k)$. Thus, since Ω_k is bounded for all k, by known results (Galdi, 1994, Corollary III.5.1) there exists $P_1 = P_1(t) \in L^2(\Omega_1)$ such that

$$\mathcal{F}(\boldsymbol{\chi}) = (P_1, \operatorname{div} \boldsymbol{\chi}), \quad \text{for all } \boldsymbol{\chi} \in \boldsymbol{W}_0^{1,2}(\Omega_1).$$

Likewise, we show that there exists $P_2 = P_2(t) \in L^2(\Omega_2)$ such that

$$\mathcal{F}(\boldsymbol{\chi}) = (P_2, \operatorname{div} \boldsymbol{\chi}), \quad \text{for all } \boldsymbol{\chi} \in \boldsymbol{W}_0^{1,2}(\Omega_2).$$

Since, for $x \in \Omega_1$, we have $P_2(x,t) = P_1(x,t) + c(\Omega_1, \Omega_2, t)$, $c(\Omega_1, \Omega_2, t) \in \mathbb{R}$, we can modify P_2 by a function of time so that $P_2 \equiv P_1$ in Ω_1. By means of an induction argument, we then prove the existence of a function $P : \Omega_T \to \mathbb{R}$ with $P \in L^2(\Omega_k)$, for all $k \in \mathbb{N}$. [9] Furthermore,

$$\mathcal{F}(\boldsymbol{\chi}) = (P, \operatorname{div} \boldsymbol{\chi}), \quad \text{for all } \boldsymbol{\chi} \in \boldsymbol{W}_0^{1,2}(\Omega_k).$$

and, again by Galdi (1994), Corollary III.5.1, and (2.13), we have

$$\|P(t)\|_{2,\Omega_k} \leq C_1 \left\{ \int_0^t \left(\|\nabla \boldsymbol{v}\|_{2,\Omega_k} + M_k^\alpha \|\nabla \boldsymbol{v}\|_{2,\Omega_k}^\beta \|\boldsymbol{f}\|_{2,\Omega_k} \right) ds + M_k \right\}$$

$$(P(t), 1)_{\Omega_k} = 0$$

which proves the theorem.

Remark 2.5 If Ω has a bounded boundary satisfying the cone condition, one can show that the field P introduced in the previous theorem can be chosen to belong to $L^\infty(0, T; L^2(\Omega))$. In fact, in such a case, assuming some more regularity on Ω one shows that relation (2.11) holds, see Sohr and von Wahl (1986).

Remark 2.6 In a recent paper, J. Simon (1999) has shown that, when Ω is bounded, there exists at least one weak solution satisfying (2.9), with corresponding $p \in W^{-1,\infty}(0, T; L^2_{loc}(\Omega))$, if Ω has no regularity, and with. $p \in W^{-1,\infty}(0, T; L^2(\Omega))$, if Ω is locally lipschitzian. For this result to hold it is sufficient to assume $\boldsymbol{f} \in L^2(0, T; W^{-1,2}(\Omega))$.

We wish now to prove a converse of Lemma 2.2, that is, any function $\boldsymbol{v} \in V_T$ (see Definition 2.1) which satisfies (2.8) for all $t \in [0, T)$ and all $\boldsymbol{\psi} \in \mathcal{D}(\Omega)$ must also satisfy (2.2). This will lead to an equivalent formulation of weak solution involving identity (2.8) instead of (2.2). We begin to show that if $\boldsymbol{v} \in V_T$ solves (2.8) for all $t \in [0, T)$ and all $\boldsymbol{\psi} \in \mathcal{D}(\Omega)$, then it also satisfies (2.2) with

$$\boldsymbol{\varphi}(x, t) \equiv \boldsymbol{\varphi}_N = \sum_{k=1}^N \gamma_k(t) \boldsymbol{\psi}_l(x), \tag{2.14}$$

where $\gamma_l \in C_0^1([0, T))$. By the linearity of (2.8) in $\boldsymbol{\varphi}$, it is enough to show this statement for $N = 1$. Now, (2.2) with $\boldsymbol{\varphi}(x, t) = \gamma(t)\boldsymbol{\psi}(x)$ and (2.8) can be written in the following forms

$$\int_0^T \gamma'(t) g(t) dt = -\int_0^T \gamma(t) G(t) dt - \gamma(0) g(0) \tag{2.2'}$$

and

$$g(t) = \int_0^t G(s) ds + g(0), \quad t \in [0, T) \tag{2.8'}$$

respectively, where

$$\begin{aligned} g(t) &= (\boldsymbol{v}(t), \boldsymbol{\psi}) \\ G(t) &= \{-\nu(\nabla \boldsymbol{v}, \nabla \boldsymbol{\psi}) - (\boldsymbol{v} \cdot \nabla \boldsymbol{v}, \boldsymbol{\psi}) + (\boldsymbol{f}, \boldsymbol{\psi})\} \in L^1(0, T) \end{aligned}$$

[9] $\mathbb{N}$ denotes the set of all positive integers.

From Lemma 2.2 we already know that (2.2′) implies (2.8′). Conversely, from classical results on Lebesgue integration (see, *e.g.*, Titchmarsh, 1964, §11), one shows that (2.8′) implies (2.2′). [10] To complete the equivalence of the two formulations, it remains to show that every $\boldsymbol{\varphi} \in \mathcal{D}_T$ together with their first and second spatial derivatives and first time derivatives, can be approximated in Ω_T by functions of the type (2.14). This is the objective of the following lemma.

Lemma 2.3 *Let Ω be an arbitrary domain in $\mathbb{R}^n$, $n \geq 2$, and let $T > 0$. Then, there exists a sequence of functions $\{\boldsymbol{\psi}_r\} \subset \mathcal{D}(\Omega)$ with the following properties. Given $\boldsymbol{\varphi} \in \mathcal{D}_T$ and $\varepsilon > 0$ there are $N = N(\boldsymbol{\varphi}, \varepsilon) \in \mathbb{N}$ functions $\gamma_k \in C_0^1([0,T))$, $k = 1, \ldots, N$, such that*

$$\max_{t\in[0,T]} \|\boldsymbol{\varphi}_N(t) - \boldsymbol{\varphi}(t)\|_{C^2(\Omega)} + \max_{t\in[0,T]} \left\| \frac{\partial \boldsymbol{\varphi}_N(t)}{\partial t} - \frac{\partial \boldsymbol{\varphi}(t)}{\partial t} \right\|_{C^0(\Omega)} < \varepsilon,$$

with $\boldsymbol{\varphi}_N$ given in (2.14). Moreover, $\{\boldsymbol{\psi}_k\}$ can be chosen to be an orthonormal basis in $H(\Omega)$.

Proof. Let $H^m = H^m(\Omega)$ be the completion of $\mathcal{D}(\Omega)$ in the norm $\|\cdot\|_m$ of the Sobolev space $\boldsymbol{W}^{m,2}(\Omega)$ and let $\{\boldsymbol{\Phi}_r\}$ be a basis of H^m constituted by elements of $\mathcal{D}(\Omega)$. [11] For arbitrary $\eta > 0$, let $0 \equiv t_0 < t_1 < \ldots < t_n \equiv T$ be a partition of $[0,T]$ such that

$$\|\boldsymbol{\varphi}(t') - \boldsymbol{\varphi}(t'')\|_m < \eta, \quad t', t'' \in [t_{k-1}, t_k]. \tag{2.15}$$

Denoting by $(\cdot,\cdot)_m$ the scalar product in H^m and setting

$$\boldsymbol{\varphi}_l(x,t) = \sum_{r=1}^{l} (\boldsymbol{\varphi}, \boldsymbol{\Phi}_r)_m \boldsymbol{\Phi}_r(x),$$

we have

$$\lim_{l\to\infty} \|\boldsymbol{\varphi}_l(t) - \boldsymbol{\varphi}(t)\|_m = 0, \quad \text{for all } \ t \in [0,T], \tag{2.16}$$

and so, by the Schwarz inequality, we find for all $t \in [t_{k-1}, t_k]$

$$\|\boldsymbol{\varphi}_l(t) - \boldsymbol{\varphi}_l(t_k)\|_m \leq \|\boldsymbol{\varphi}(t) - \boldsymbol{\varphi}(t_k)\|_m < \eta. \tag{2.17}$$

Thus, from (2.15)–(2.17), for $t \in [t_{k-1}, t_k]$ and sufficiently large l we find

$$\begin{aligned}\|\boldsymbol{\varphi}_l(t) - \boldsymbol{\varphi}(t)\|_m \leq\ & \|\boldsymbol{\varphi}_l(t) - \boldsymbol{\varphi}_l(t_k)\|_m + \|\boldsymbol{\varphi}_l(t_k) - \boldsymbol{\varphi}(t_k)\|_m \\ & + \|\boldsymbol{\varphi}(t) - \boldsymbol{\varphi}(t_k)\|_m < 3\eta.\end{aligned}$$

Choosing $m > n/2$, by the Sobolev embedding theorem we conclude

$$\max_{t\in[0,T]} \|\boldsymbol{\varphi}_l(t) - \boldsymbol{\varphi}(t)\|_{C^2(\Omega)} < C\eta$$

[10] If $G \in C([0,T))$, equation (2.2′) is obtained from (2.8′) after multiplying this latter by $\gamma'(t)$ and integrating by parts.

[11] This is always possible, owing to the separability of $\boldsymbol{W}^{m,2}$.

with $C = C(\Omega, m, n)$. Moreover, for all $t \in [0,T]$, it is

$$\lim_{l\to\infty} \left\| \frac{\partial \varphi_l(t)}{\partial t} - \frac{\partial \varphi(t)}{\partial t} \right\|_m = 0,$$

and so, by the same kind of argument used before, we show

$$\lim_{l\to\infty} \max_{t\in[0,T]} \left\| \frac{\partial \varphi_l(t)}{\partial t} - \frac{\partial \varphi(t)}{\partial t} \right\|_{C^0(\Omega)} = 0.$$

To the set $\{\boldsymbol{\Phi}_r\}$ we can apply the Schmidt orthogonalization procedure in L^2, thus obtaining another system $\{\boldsymbol{\psi}_r\}$, whose generic element is a linear combination of $\boldsymbol{\Phi}_1, \dots, \boldsymbol{\Phi}_\ell$, $\ell = \ell(r)$. Since H^m is dense in H, it is easy to show that $\{\boldsymbol{\psi}_r\}$ satisfies all requirements stated in the lemma which, consequently, is proved.

From what we have shown, we deduce the following result.

Lemma 2.4 *A measurable function $\boldsymbol{v} : \Omega_T \to \mathbb{R}^n$ is a weak solution of the problem (0.1)–(0.4) in Ω_T if and only if*

a) $\boldsymbol{v} \in V_T$;
b) *$\boldsymbol{v}$ verifies (2.8), for all $t \in [0,T)$ and all $\boldsymbol{\psi} \in \mathcal{D}(\Omega)$.*

Remark 2.7 Differentiating (2.8) with respect to t and recalling that $\boldsymbol{v} \in V_T$, we find

$$\frac{d}{dt}(\boldsymbol{v}(t), \boldsymbol{\psi}) = -\nu(\nabla \boldsymbol{v}(t), \nabla \boldsymbol{\psi}) - (\boldsymbol{v}(t) \cdot \nabla \boldsymbol{v}(t), \boldsymbol{\psi}) + (\boldsymbol{f}, \boldsymbol{\psi}) \tag{2.18}$$

for a.a. $t \in [0,T)$ and all $\boldsymbol{\psi} \in \mathcal{D}(\Omega)$. It is easily seen that the right-hand side of (2.18) defines a linear, bounded functional in $\boldsymbol{\psi} \in H^1(\Omega)$. In fact, denoting by $\mathcal{F}$ such a functional, by the Schwarz inequality and by (2.12) we have,

$$\begin{aligned} |\mathcal{F}(\boldsymbol{\psi})| &\le (\nu\|\nabla \boldsymbol{v}\|_2 + \|\boldsymbol{v}\|_4^2 + \|\boldsymbol{f}\|_2)\, \|\boldsymbol{\psi}\|_{1,2} \\ &\le c\left(\|\nabla \boldsymbol{v}\|_2 + \|\boldsymbol{v}\|_2^\alpha \|\nabla \boldsymbol{v}\|_2^\beta\right) \|\boldsymbol{\psi}\|_{1,2} \end{aligned} \tag{2.19}$$

where

$$\alpha = \beta = 1, \quad \text{if } n = 2$$

$$\alpha = 1/2, \;\; \beta = 3/2, \quad \text{if } n = 3.$$

Thus, denoting by $H^{-1}(\Omega)$ the dual space of $H^1(\Omega)$, for almost all $t \in [0,T)$, there exists $\boldsymbol{v}_t \in H^{-1}(\Omega)$ such that

$$\frac{d}{dt}(\boldsymbol{v}(t), \boldsymbol{\psi}) = \langle \boldsymbol{v}_t, \boldsymbol{\psi} \rangle, \quad \boldsymbol{\psi} \in H^1(\Omega),$$

where $\langle \cdot, \cdot \rangle$ denotes the duality pairing between H^{-1} and H^1. Notice that $\boldsymbol{v}_t$ is in H^{-1} but *not* necessarily in $W^{-1,2}$. Moreover, by (2.18), (2.19) and the assumption $\boldsymbol{v} \in V_T$, we find

$$\langle \boldsymbol{v}_t, \boldsymbol{\psi} \rangle = -\nu(\nabla \boldsymbol{v}(t), \nabla \boldsymbol{\psi}) - (\boldsymbol{v}(t) \cdot \nabla \boldsymbol{v}(t), \boldsymbol{\psi}) + (\boldsymbol{f}, \boldsymbol{\psi}), \quad \boldsymbol{v}_t \in L^\sigma(0,T; H^{-1}(\Omega)),$$

where $\sigma = 2$ if $n = 2$ and $\sigma = 4/3$ if $n = 3$.

Remark 2.8 The method of proof used for Lemma 2.3 enables us to give a density result which will be used several times in the next sections. To this end, we recall standard facts concerning the theory of approximation of functions. Let $w \in L^q(0,T;X)$, $1 \leq q < \infty$. For $T > h > 0$, the mollifier w_h (in the sense of Friederichs) of w is defined by

$$w_h(t) = \int_0^T j_h(t-s)w(s)ds \tag{2.20}$$

where $j_h(s)$ is an even, positive, infinitely differentiable function with support in $(-h,h)$, and $\int_{-\infty}^{\infty} j_h(s)ds = 1$. We have (see, *e.g.*, Hille and Phillips, 1957)

Lemma 2.5 Let $w \in L^q(0,T;X)$, $1 \leq q < \infty$. Then $w_h \in C^k([0,T];X)$, for all $k \geq 0$. Moreover

$$\lim_{h\to 0} \|w_h - w\|_{L^q(0,T;X)} = 0.$$

Finally, if $\{w_k\} \subset L^q(0,T;X)$ converges to w in the norm of $L^q(0,T;X)$ then

$$\lim_{k\to\infty} \|(w_k)_h - w_h\|_{L^q(0,T;X)} = 0.$$

We also have.

Lemma 2.6 *$\mathcal{D}_T$ is dense in $L^2(0,T;H^1(\Omega))$.*

Proof. Let $\{\mathbf{\Phi}_r\} \subset \mathcal{D}(\Omega)$ be a basis of H^1 and let $\boldsymbol{w} \in L^2(0,T;X)$. Denoting by $(\cdot,\cdot)_1$ the scalar product in H^1, and setting

$$\boldsymbol{w}_{l,h}(x,t) = \sum_{r=1}^{l} (\boldsymbol{w}_h, \mathbf{\Phi}_r)_1 \mathbf{\Phi}_r(x),$$

we have

$$\lim_{l\to\infty} \|\boldsymbol{w}_{l,h}(t) - \boldsymbol{w}_h(t)\|_1 = 0, \quad \text{for all } \; t \in [0,T] \text{ and } h < T. \tag{2.21}$$

Clearly, $\boldsymbol{w}_{l,h} \in \mathcal{D}_T$. By Lemma 2.5, for a given $\varepsilon > 0$, there is $h > 0$

$$\int_0^T \|\boldsymbol{w}_h(t) - \boldsymbol{w}(t)\|_1^2 < \varepsilon.$$

On the other hand, from (2.21) and the Lebesgue dominated convergence theorem, we have for all fixed h

$$\lim_{l\to\infty} \int_0^T \|\boldsymbol{w}_{l,h}(t) - \boldsymbol{w}_h(t)\|_1^2 = 0.$$

Thus the result follows from the last two displayed relations and the triangle inequality.

3. Existence of weak solutions

The aim of this section is to prove the following existence theorem of weak solutions.

Theorem 3.1 *Let Ω be any domain in $\mathbb{R}^n$ and let $T > 0$. Then, for any given*

$$\boldsymbol{v}_0 \in H(\Omega), \quad \boldsymbol{f} \in \boldsymbol{L}^2(\Omega_T),$$

there exists at least one weak solution to (0.1)–(0.3) in Ω_T. This solution verifies, in addition, the following properties

i) *The energy inequality:*

$$\|\boldsymbol{v}(t)\|_2^2 + 2\nu \int_0^t \|\nabla \boldsymbol{v}(\tau)\|_2^2 d\tau \le 2\int_0^t (\boldsymbol{v}(\tau), \boldsymbol{f}(\tau))d\tau + \|\boldsymbol{v}_0\|_2^2, \quad t \in [0,T]. \tag{EI}$$

ii) $\lim_{t\to 0}\|\boldsymbol{v}(t) - \boldsymbol{v}_0\|_2 = 0.$

Proof. We shall use the so called "Faedo-Galerkin" method. Let $\{\boldsymbol{\psi}_r\} \subset \mathcal{D}(\Omega)$ be the basis of $H(\Omega)$ given in Lemma 2.3. We shall look for approximating solutions $\boldsymbol{v}_k$ of the form

$$\boldsymbol{v}_k(x,t) = \sum_{r=1}^{k} c_{kr}(t)\boldsymbol{\psi}_r(x), \quad k \in \mathbb{N}, \tag{3.1}$$

where the coefficients c_{kr} are required to satisfy the following system of ordinary differential equations

$$\frac{dc_{kr}}{dt} + \sum_{i=1}^{k} a_{ir}c_{ki} + \sum_{i,s=1}^{k} a_{isr}c_{ki}c_{ks} = f_r, \quad r = 1,\dots,k, \tag{3.2}$$

with the initial condition

$$c_{kr}(0) = C_{0r} \quad r = 1,\dots,k, \tag{3.3}$$

where

$$a_{ir} = \nu(\nabla\boldsymbol{\psi}_i, \nabla\boldsymbol{\psi}_r), \quad a_{isr} = (\boldsymbol{\psi}_i \cdot \nabla\boldsymbol{\psi}_s, \nabla\boldsymbol{\psi}_r),$$
$$f_r = (\boldsymbol{f}, \boldsymbol{\psi}_r), \quad C_{0r} = (\boldsymbol{v}_0, \boldsymbol{\psi}_r).$$

Since $f_r \in L^2(0,T)$ for all $r = 1,\dots,k$, from the elementary theory of ordinary differential equations, we know that problem (3.1)–(3.3) admits a unique solution $c_{kr} \in W^{1,2}(0,T_k)$, $r = 1,\dots,k$, where $T_k \le T$. Multiplying (3.2) by c_{kr}, summing over r and employing the orthonormality conditions on $\{\boldsymbol{\psi}_r\}$ along with the identity

$$(\boldsymbol{\psi}\cdot\nabla\boldsymbol{\psi}, \boldsymbol{\psi}) = 0, \quad \text{for all } \boldsymbol{\psi} \in \mathcal{D}(\Omega),$$

we obtain for all $t \in [0,T)$

$$\|\boldsymbol{v}_k(t)\|_2^2 + 2\nu\int_0^t \|\nabla\boldsymbol{v}_k(\tau)\|_2^2 d\tau = 2\int_0^t (\boldsymbol{v}_k(\tau), \boldsymbol{f}(\tau))d\tau + \|\boldsymbol{v}_{0k}\|_2^2 \tag{3.4}$$

with $\boldsymbol{v}_{0k} = \boldsymbol{v}_k(0)$. Since $\|\boldsymbol{v}_{0k}\|_2 \leq \|\boldsymbol{v}_0\|_2$, Using in (3.4) the Schwarz inequality along with Gronwall's lemma, we easily deduce the following bound

$$\|\boldsymbol{v}_k(t)\|_2^2 + \int_0^t \|\nabla \boldsymbol{v}_k(\tau)\|_2^2 d\tau \leq M, \quad \text{for all } t \in [0,T) \tag{3.5}$$

with M independent of t and k. From this inequality it follows, in particular, that $|c_{kr}(t)| \leq M^{1/2}$ for all $r = 1, \ldots, k$ which in turn, by standard results on ordinary differential equations, implies $T_k = T$, for all $k \in \mathbb{N}$. We shall now investigate the properties of convergence of the sequence $\{\boldsymbol{v}_k\}$ when $k \to \infty$. To this end, we begin to show that, for any fixed $r \in \mathbb{N}$, the sequence of functions

$$G_k^{(r)}(t) \equiv (\boldsymbol{v}_k(t), \boldsymbol{\psi}_r)$$

is uniformly bounded and uniformly continuous in $t \in [0,T]$. The uniform boundedness follows at once from (3.5). To show the uniform continuity, we observe that from (3.2), (3.5), with the help of the Schwarz inequality it easily follows that

$$\begin{aligned} |G_k^{(r)}(t) - G_k^{(r)}(s)| \leq S_1 \int_s^t (\|\nabla \boldsymbol{v}_k(\tau)\|_2 + \|\boldsymbol{f}(\tau)\|_2)\, d\tau \\ + S_2 M^{1/2} \int_s^t \|\nabla \boldsymbol{v}_k(\tau)\|_2 d\tau, \end{aligned} \tag{3.6}$$

where

$$S_1 \equiv \|\boldsymbol{\psi}_r\|_2, \quad S_2 = \max_{x \in \Omega} |\boldsymbol{\psi}_r(x)|.$$

Thus, using the Schwarz inequality into (3.6) and recalling (3.5), we readily show the equicontinuity of $G_k^{(r)}(t)$. By the Ascoli-Arzelà theorem, from the sequence $\{G_k^{(r)}(t)\}_{k\in\mathbb{N}}$ we may then select a subsequence – which we continue to denote by $\{G_k^{(r)}(t)\}_{k\in\mathbb{N}}$ – uniformly converging to a continuous function $G^{(r)}(t)$. The selected sequence $\{G_k^{(r)}(t)\}_{k\in\mathbb{N}}$ may depend on r. However, using the classical Cantor diagonalization method, we end up with a sequence – again denoted by $\{G_k^{(r)}(t)\}_{k\in\mathbb{N}}$ – converging to $G^{(r)}$, for all $r \in \mathbb{N}$, uniformly in t$\in [0,T)$. This information, together with (3.5) and the weak compactness of the space H, allows us to infer the existence of $\boldsymbol{v}(t) \in H(\Omega)$ such that

$$\lim_{k\to\infty} (\boldsymbol{v}_k(t) - \boldsymbol{v}(t), \boldsymbol{\psi}_r) = 0 \ \text{ uniformly in } t \in [0,T] \text{ and for all } r \in \mathbb{N}. \tag{3.7}$$

Let us now prove that $\boldsymbol{v}_k(t)$ converges weakly in $\boldsymbol{L}^2$ to $\boldsymbol{v}(t)$, uniformly in $t \in [0,T]$, that is,

$$\lim_{k\to\infty} (\boldsymbol{v}_k(t) - \boldsymbol{v}(t), \boldsymbol{u}) = 0, \ \text{ uniformly in } t \in [0,T] \text{ and for all } \boldsymbol{u} \in \boldsymbol{L}^2(\Omega). \tag{3.8}$$

By the Helmholtz-Weyl orthogonal decomposition (2.1), it is enough to show (3.8) for $\boldsymbol{u} \in H(\Omega)$. To this end, writing

$$\boldsymbol{u} = \sum_{r=1}^{\infty} u_r \boldsymbol{\psi}_r \equiv \sum_{r=1}^{N} u_r \boldsymbol{\psi}_r + \boldsymbol{u}^{(N)}$$

and using the Schwarz inequality together with (3.5), we find

$$|(\boldsymbol{v}_k(t) - \boldsymbol{v}(t), \boldsymbol{u})| \le \sum_{r=1}^{N} |(\boldsymbol{v}_k(t) - \boldsymbol{v}(t), u_r \boldsymbol{\psi}_r)| + |(\boldsymbol{v}_k(t) - \boldsymbol{v}(t), \boldsymbol{u}^{(N)})|$$

$$\le \sum_{r=1}^{N} \|\boldsymbol{u}\|_2 |(\boldsymbol{v}_k(t) - \boldsymbol{v}(t), \boldsymbol{\psi}_r)| + 2M^{1/2} \|\boldsymbol{u}^{(N)}\|_2.$$

For $\varepsilon > 0$, we choose N so large that

$$\|\boldsymbol{u}^{(N)}\|_2 < \varepsilon.$$

Further, by (3.6) we can pick $k = k(\boldsymbol{u}, \varepsilon)$ so that

$$\sum_{r=1}^{N} \|\boldsymbol{u}\|_2 |(\boldsymbol{v}_k(t) - \boldsymbol{v}(t), \boldsymbol{\psi}_r)| < \varepsilon,$$

and (3.8) follows from (3.7) and the last two displayed inequalities. In view of (3.5) we clearly have $\boldsymbol{v} \in L^\infty(0,T; H(\Omega))$. Again from (3.5), by the weak compactness of the space $L^2(\Omega_T)$ it follows the existence of $\widetilde{\boldsymbol{v}} \in L^2(0,T; H^1(\Omega))$ such that for $m = 1, \dots, n$ (with $\partial_m = \partial/\partial x_m$)

$$\lim_{k\to\infty} \int_0^T (\boldsymbol{v}_k - \widetilde{\boldsymbol{v}}, \boldsymbol{w}) ds = \lim_{k\to\infty} \int_0^T (\partial_m(\boldsymbol{v}_k - \widetilde{\boldsymbol{v}}), \boldsymbol{w}) ds = 0, \text{ for all } \boldsymbol{w} \in \boldsymbol{L}^2(\Omega_T).$$

Choosing in this inequality $\boldsymbol{w} \in \mathcal{D}_T$ and using (3.8), it is easy to show that $\boldsymbol{v} = \widetilde{\boldsymbol{v}}$. Thus, in particular, we find

$$\lim_{k\to\infty} \int_0^T (\partial_m(\boldsymbol{v}_k - \boldsymbol{v}), \boldsymbol{w}) ds = 0, \quad \text{for all } \boldsymbol{w} \in \boldsymbol{L}^2(\Omega_T), \quad m = 1, \dots, n. \tag{3.9}$$

We wish now to prove that (3.8) and (3.9) imply the *strong* convergence of $\{\boldsymbol{v}_k\}$ to $\boldsymbol{v}$ in $\boldsymbol{L}^2(\omega \times [0,T])$, for all $\omega \subset\subset \Omega$. To show this, we need the following *Friederichs inequality*, see, *e.g.*, Galdi (1994, Lemma II.4.2): *Let C be a cube in $\mathbb{R}^n$, then for any $\eta > 0$, there exist $K(\eta, C) \in \mathbb{N}$ functions $\boldsymbol{\omega}_i \in L^\infty(C)$, $i = 1, \dots, K$ such that*

$$\int_0^T \|\boldsymbol{w}(t)\|_{2,C}^2 dt \le \sum_{i=1}^{K} \int_0^T (\boldsymbol{w}(t), \boldsymbol{\omega}_i)_C^2 dt + \eta \int_0^T \|\nabla \boldsymbol{w}(t)\|_{2,C}^2 dt.$$

If we apply this inequality with $\boldsymbol{w} \equiv \boldsymbol{v}_k - \boldsymbol{v}$ and use (3.5), (3.7) we find

$$\lim_{k\to\infty} \int_0^T \|\boldsymbol{v}_k(t) - \boldsymbol{v}(t)\|_{2,C}^2 dt = 0 \tag{3.10}$$

With the help of (3.8)–(3.10), we shall now show that $\boldsymbol{v}$ is a weak solution to (0.1)–(0.3). Since we already proved that $\boldsymbol{v} \in V_T$, by Lemma 2.4, it remains to show that $\boldsymbol{v}$ satisfies (2.8). Integrating (3.1) from 0 to $t \le T$ we find

$$\begin{aligned}\int_0^t \{-\nu(\nabla \boldsymbol{v}_k, \nabla \boldsymbol{\psi}_r) \ -(\boldsymbol{v}_k \cdot \nabla \boldsymbol{v}_k, \boldsymbol{\psi}_r)\}\, ds \\ = -\int_0^t (\boldsymbol{f}, \boldsymbol{\psi}_r) ds + (\boldsymbol{v}_k(t), \boldsymbol{\psi}_r) - (\boldsymbol{v}_0, \boldsymbol{\psi}_r).\end{aligned} \tag{3.11}$$

From (3.8), (3.9) we at once obtain

$$\lim_{k\to\infty} (\boldsymbol{v}_k(t) - \boldsymbol{v}(t), \boldsymbol{\psi}_r) = 0, \quad \lim_{k\to\infty} \int_0^t (\nabla \boldsymbol{v}_k(s) - \nabla \boldsymbol{v}(s), \boldsymbol{\psi}_r) ds = 0. \tag{3.12}$$

Furthermore, denoting by C a cube containing the support of $\boldsymbol{\psi}_r$, we have

$$\begin{aligned}\left| \int_0^t \ (\boldsymbol{v}_k \cdot \nabla \boldsymbol{v}_k, \boldsymbol{\psi}_r) - (\boldsymbol{v} \cdot \nabla \boldsymbol{v}, \boldsymbol{\psi}_r) ds \right| \\ \le \left| \int_0^t ((\boldsymbol{v}_k - \boldsymbol{v}) \cdot \nabla \boldsymbol{v}_k, \boldsymbol{\psi}_r)_C ds \right| + \left| \int_0^t (\boldsymbol{v} \cdot \nabla (\boldsymbol{v}_k - \boldsymbol{v}), \boldsymbol{\psi}_r)_C ds \right| .\end{aligned} \tag{3.13}$$

Setting $S = \max_{x\in C} |\boldsymbol{\psi}(x)|$, by (3.5) we also have

$$\begin{aligned}\left| \int_0^t ((\boldsymbol{v}_k - \boldsymbol{v}) \cdot \nabla \boldsymbol{v}_k, \boldsymbol{\psi}_r)_C \right| &\le S \left(\int_0^t \|\boldsymbol{v}_k - \boldsymbol{v}\|_{2,C}^2 dt \right)^{1/2} \left(\int_0^t \|\nabla \boldsymbol{v}_k\|_2^2 dt \right)^{1/2} \\ &\le S M^{1/2} \left(\int_0^t \|\boldsymbol{v}_k - \boldsymbol{v}\|_{2,C}^2 dt \right)^{1/2}\end{aligned}$$

and so, using (3.10), we find

$$\lim_{k\to\infty} \left| \int_0^t ((\boldsymbol{v}_k - \boldsymbol{v}) \cdot \nabla \boldsymbol{v}_k, \boldsymbol{\psi}_r) ds \right| = 0. \tag{3.14}$$

Furthermore, we have

$$\left| \int_0^t (\boldsymbol{v} \cdot \nabla (\boldsymbol{v}_k - \boldsymbol{v}), \boldsymbol{\psi}_r)_C ds \right| \le \sum_{m=1}^n \left| \int_0^t (\partial_m (\boldsymbol{v}_k - \boldsymbol{v}), v_m \boldsymbol{\psi}_r)_C ds \right|$$

and since $v_m \boldsymbol{\psi} \in \boldsymbol{L}^2(\Omega_T)$, from (3.9) we deduce

$$\lim_{k\to\infty} \left| \int_0^t (\boldsymbol{v} \cdot \nabla (\boldsymbol{v}_k - \boldsymbol{v}), \boldsymbol{\psi}_r)_C ds \right| = 0. \tag{3.15}$$

Therefore, passing into the limit $k \to \infty$ in (3.11), from (3.12)–(3.15) we conclude

$$\begin{aligned}\int_0^t \{-\nu(\nabla \boldsymbol{v}, \nabla \boldsymbol{\psi}_r) \ -(\boldsymbol{v} \cdot \nabla \boldsymbol{v}, \boldsymbol{\psi}_r)\}\, ds \\ = -\int_0^t (\boldsymbol{f}, \boldsymbol{\psi}_r) ds + (\boldsymbol{v}(t), \boldsymbol{\psi}_r) - (\boldsymbol{v}_0, \boldsymbol{\psi}_r).\end{aligned} \tag{3.16}$$

However, from Lemma 2.3 we know that every function $\psi \in \mathcal{D}(\Omega)$ can be uniformly approximated in $C^2(\overline{\Omega})$ by functions of the form

$$\psi_N(x) = \sum_{r=1}^{N} \gamma_r \psi_r(x), \quad N \in \mathbb{N},\ \gamma_r \in \mathbb{R}.$$

So, writing (3.16) with ψ_N in place of ψ, we may pass to the limit $N \to \infty$ in this new relation and use the fact that $\boldsymbol{v} \in V_T$ to show the validity of (2.8) for all $\psi \in \mathcal{D}(\Omega)$. We shall now prove the energy inequality (EI). To this end, we shall take the lim inf as $k \to \infty$ of both sides of (3.4). By the definition of $\boldsymbol{v}_{0k}$, the properties of $\boldsymbol{f}$, and (3.8) we deduce

$$\lim_{k\to\infty} \left\{ \int_0^t (\boldsymbol{v}_k(\tau), \boldsymbol{f}(\tau))d\tau + \|\boldsymbol{v}_{0k}\|_2^2 \right\} = \int_0^t (\boldsymbol{v}(\tau), \boldsymbol{f}(\tau))d\tau + \|\boldsymbol{v}_0\|_2^2.$$

Moreover, by (3.8), (3.9) and a classical property of weak limits, we find that

$$\liminf_{k\to\infty} \left\{ \|\boldsymbol{v}_k(t)\|_2^2 + 2\nu \int_0^t \|\nabla \boldsymbol{v}_k(\tau)\|_2^2 d\tau \right\} \geq \|\boldsymbol{v}(t)\|_2^2 + 2\nu \int_0^t \|\nabla \boldsymbol{v}(\tau)\|_2^2 d\tau, \tag{3.17}$$

and (EI) follows from (3.4) and the last two displayed relations. From (EI) we deduce at once

$$\limsup_{t\to 0} \|\boldsymbol{v}(t)\|_2^2 \leq \|\boldsymbol{v}_0\|_2^2.$$

On the other hand, $\boldsymbol{v}(t)$ is weakly continuous in $\boldsymbol{L}^2$ (see Lemma 2.2), and so we have

$$\liminf_{t\to 0} \|\boldsymbol{v}(t)\|_2^2 \geq \|\boldsymbol{v}_0\|_2^2,$$

which implies

$$\lim_{t\to 0} \|\boldsymbol{v}(t)\|_2^2 = \|\boldsymbol{v}_0\|_2^2.$$

This relation together with the $\boldsymbol{L}^2$ weak continuity of $\boldsymbol{v}$ allows us to conclude

$$\lim_{t\to 0} \|\boldsymbol{v}(t) - \boldsymbol{v}_0\|_2 = 0,$$

and the theorem is thus proved.

Remark 3.1 In the literature, one may find many different definitions of weak solution (see, *e.g.*, Lions, 1969; Masuda, 1984; von Wahl (1985)). The one chosen here is due to Leray and Hopf. Likewise, there are many different constructive procedures of weak solutions (see Leray 1934a, 1934b; Kiselev and Ladyzhenskaya, 1957; Shinbrot, 1973). Since, as we shall see in the next section, there is no uniqueness guaranteed for weak solutions in dimension 3 (or higher), these procedures may conceivably lead to different solutions.

4. The energy equality and uniqueness of weak solutions

An interesting feature of weak solutions that should be emphasized, is that they obey only an energy *inequality* rather than the energy *equality* (that is (EI) with the equality sign), as should be expected from the physical point of view. To analyze this fact in more detail, let us take, for simplicity, $\boldsymbol{f} \equiv 0$. Then, any "physically reasonable" solution should be such that the associated kinetic energy $E(t)$ at a certain time t $(= \frac{1}{2}\|\boldsymbol{v}(t)\|_2^2)$ is equal to $E(\sigma)$ $(\sigma < t)$ minus the amount of energy dissipated by viscosity in the time interval $[\sigma, t]$ $(= \nu \int_\sigma^t \|\nabla \boldsymbol{v}(\tau)\|^2 d\tau)$. According to (EI), however, a weak solution not only does not satisfy *a priori* this property but, in fact, its kinetic energy could even *increase* in certain time intervals. Therefore, a first question to ask is if it is possible to construct weak solutions for which the corresponding kinetic energy is a decreasing function of time. To this end, it would be enough that weak solutions would satisfy the following relation

$$\|\boldsymbol{v}(t)\|_2^2 + 2\nu \int_\sigma^t \|\nabla \boldsymbol{v}(\tau)\|_2^2 d\tau \leq \|\boldsymbol{v}(\sigma)\|_2^2, \qquad \text{for almost all } \sigma \geq 0, \text{ and all } t \in [\sigma, T). \tag{4.1}$$

Inequality (4.1) is usually called the *strong energy inequality* (SEI).

It is easy to see that if Ω is bounded, then the solutions constructed in Theorem 3.1 obey the (SEI). In fact, from (3.10), by taking $C \supset \Omega$, it follows that

$$\lim_{k\to\infty} \|\boldsymbol{v}_k(\sigma) - \boldsymbol{v}(\sigma)\|_2 = 0, \quad \text{for almost all } \sigma \in [0, T). \tag{4.2}$$

On the other hand, from (3.4) (with $\boldsymbol{f} \equiv 0$) we have

$$\|\boldsymbol{v}_k(t)\|_2^2 + 2\nu \int_\sigma^t \|\nabla \boldsymbol{v}_k(\tau)\|_2^2 d\tau = \|\boldsymbol{v}_k(\sigma)\|_2^2, \qquad \text{for all } \sigma \in [0, T) \text{ and } t \in [\sigma, T).$$

and so, passing to the limit $k \to \infty$ in this relation and using (3.17) (with 0 replaced by σ) and (4.2), we recover (SEI). With much more effort, one can show existence of weak solutions obeying (SEI) when Ω is either the whole $\mathbb{R}^n$ (Leray 1934b), or an exterior domain (Galdi and Maremonti, 1986; Sohr, von Wahl and Wiegner, 1986; Miyakawa and Sohr, 1988), or a half space (Borchers and Miyakawa, 1988). It is interesting to observe that all proofs given by these authors rely on a certain estimate for the pressure field, which implies, in particular, the following property for p:

$$p \in L^r(0, T; L^q(\Omega)), \quad \text{for suitable exponents } r, q.$$

This is much more than the regularity property proved in Theorem 2.1. On the other hand, one knows how to prove this estimate only for a certain type of domains and, therefore, *it is not known if* (SEI) *holds for an arbitrary* Ω *(no matter how smooth).*

The strong energy inequality, even though more reasonable than the energy inequality, still presents an undesired feature, in all time intervals I (*if any*) where it holds as a *strict* inequality. Actually, in any of such intervals, the kinetic energy *is* decreased by a certain amount, say M_I, which is *not* due to the dissipation. It seems therefore interesting to furnish sufficient conditions on a weak solution in order that it verifies an *energy equality* and to compare them with those ensuring its *uniqueness*. As we shall see, the former (see Theorem 4.1) are weaker than the latter (see Theorem 4.2), and they are both verified by a weak solution in dimension 2, but not *a priori* in dimension 3. Thus, the question of the existence of a three dimensional weak solution which 1) satisfies the energy equality and 2) is unique, remains open.

In this section we provide conditions on a weak solution under which 1) and 2) above are met. The following theorem holds.

Theorem 4.1 *Let $\boldsymbol{v}$ be a weak solution in Ω_T. Assume*

$$\boldsymbol{v} \in L^4(0,T;\boldsymbol{L}^4(\Omega)). \tag{4.3}$$

Then $\boldsymbol{v}$ verifies the energy equality

$$\|\boldsymbol{v}(t)\|_2^2 + 2\nu \int_0^t \|\nabla \boldsymbol{v}(\tau)\|_2^2 d\tau = 2\int_0^t (\boldsymbol{v}(\tau), \boldsymbol{f}(\tau))d\tau + \|\boldsymbol{v}_0\|_2^2, \tag{4.4}$$

for all $t \in [0,T)$.

Proof. Let $\{\boldsymbol{v}_k\} \subset \mathcal{D}_T$ be a sequence converging to $\boldsymbol{v}$ in $L^2(0,T;H^1(\Omega))$, see Lemma 2.6. We choose in (2.4) (with $s=0$) $\boldsymbol{\varphi} = (\boldsymbol{v}_k)_h \equiv \boldsymbol{v}_{h,k}$, where $(\,\cdot\,)_h$ is the mollification operator defined in (2.20), see Lemma 2.5. Observing that

$$\begin{aligned} &\left| \int_0^T \{(\boldsymbol{v}\cdot \nabla \boldsymbol{v}_{k,h}, \boldsymbol{v}) - (\boldsymbol{v}\cdot\nabla \boldsymbol{v}_h, \boldsymbol{v})\}\, dt \right| \\ &\qquad \le \int_0^T \|\boldsymbol{v}\|_4^2 \|\nabla(\boldsymbol{v}_{k,h} - \boldsymbol{v}_h)\|_2 \\ &\qquad \le \|\boldsymbol{v}_{k,h} - \boldsymbol{v}_h\|_{L^2(0,T;H^1)} \left(\int_0^T \|\boldsymbol{v}(t)\|_4^4 dt \right)^{1/2}, \end{aligned} \tag{4.5}$$

by a standard procedure which makes use of Lemma 2.5, we find in the limit $k \to \infty$

$$\begin{aligned} &\int_0^t \left\{ \left(\boldsymbol{v}, \frac{\partial \boldsymbol{v}_h}{\partial t}\right) - \nu(\nabla \boldsymbol{v}, \nabla \boldsymbol{v}_h) - (\boldsymbol{v}\cdot\nabla \boldsymbol{v}_h, \boldsymbol{v}) \right\} ds \\ &\qquad\qquad = -\int_0^t (\boldsymbol{f}, \boldsymbol{v}_h) ds + (\boldsymbol{v}(t), \boldsymbol{v}_h(t)) - (\boldsymbol{v}_0, (\boldsymbol{v}_0)_h). \end{aligned} \tag{4.6}$$

Since the kernel $j_h(s)$ in (2.20) is even in $(-h,h)$, we obtain

$$\int_0^t \left(\boldsymbol{v}, \frac{\partial \boldsymbol{v}_h}{\partial t}\right) = \int_0^t\int_0^t \frac{dj_h(t-t')}{dt}\,(\boldsymbol{v}(t),\boldsymbol{v}(t'))\,dtdt' = 0.$$

Moreover, by Lemma 2.5 and (4.5) with $\boldsymbol{v}_h$ in place of $\boldsymbol{v}_{k,h}$ and $\boldsymbol{v}$ in place of $\boldsymbol{v}_h$, respectively, we obtain

$$\lim_{h\to 0}\int_0^t (\nabla\boldsymbol{v},\nabla\boldsymbol{v}_h)ds = \int_0^t (\nabla\boldsymbol{v},\nabla\boldsymbol{v})ds$$

$$\lim_{h\to 0}\int_0^t (\boldsymbol{f},\boldsymbol{v}_h)ds = \int_0^t (\boldsymbol{f},\boldsymbol{v})ds$$

$$\lim_{h\to 0}\int_0^T (\boldsymbol{v}\cdot\nabla\boldsymbol{v}_h,\boldsymbol{v})ds = \int_0^t (\boldsymbol{v}\cdot\nabla\boldsymbol{v},\boldsymbol{v})ds.$$

Now, $\boldsymbol{v}(t)\in H^1(\Omega)$, for a.a. $t\in[0,T)$ and so, for any such fixed t, denoting by $\{\boldsymbol{\psi}_k\}$ a sequence from $\mathcal{D}(\Omega)$ converging to $\boldsymbol{v}$ in H^1 we have

$$\begin{aligned}|(\boldsymbol{v}\cdot\nabla\boldsymbol{v},\boldsymbol{v})-(\boldsymbol{v}\cdot\nabla\boldsymbol{\psi}_k,\boldsymbol{\psi}_k)| &\le |(\boldsymbol{v}\cdot\nabla\boldsymbol{v},(\boldsymbol{v}-\boldsymbol{\psi}_k))|+|(\boldsymbol{v}\cdot\nabla(\boldsymbol{v}-\boldsymbol{\psi}_k),\boldsymbol{v})|\\ &\le +\|\boldsymbol{v}\|_4\|\nabla\boldsymbol{v}\|_2\|\boldsymbol{v}-\boldsymbol{\psi}_k\|_4+\|\boldsymbol{v}\|_4^2\|\nabla(\boldsymbol{v}-\boldsymbol{\psi}_k)\|_2.\end{aligned}$$

By the Sobolev embedding theorem it follows that [12]

$$\|u\|_4\le c\left(\|u\|_2+\|\nabla u\|_2\right),\quad u\in W^{1,2}(\Omega)$$

and so we deduce

$$\lim_{k\to\infty}(\boldsymbol{v}\cdot\nabla\boldsymbol{\psi}_k,\boldsymbol{\psi}_k)=(\boldsymbol{v}\cdot\nabla\boldsymbol{v},\boldsymbol{v}).$$

However, since $\boldsymbol{v}(t)\in H(\Omega)$ for a.a. t, we get

$$(\boldsymbol{v}\cdot\nabla\boldsymbol{\psi}_k,\boldsymbol{\psi}_k)=\tfrac{1}{2}(\boldsymbol{v},\nabla(\boldsymbol{\psi}_k)^2)=0,\quad \text{for all } k\in\mathbb{N},$$

which furnishes

$$\int_0^t(\boldsymbol{v}\cdot\nabla\boldsymbol{v},\boldsymbol{v})ds=0.$$

Finally, by the weak L^2 continuity, and recalling that $\int_0^h j_h(s)ds=1/2$, we have

$$\begin{aligned}(\boldsymbol{v}(t),\boldsymbol{v}_h(t)) &= \int_0^h j_h(s)\,(\boldsymbol{v}(t),\boldsymbol{v}(t+s))\,ds\\ &= \int_0^h j_h(s)\left(\|\boldsymbol{v}(t)\|_2^2+(\boldsymbol{v}(t),(\boldsymbol{v}(t+s)-\boldsymbol{v}(t))\right)ds\\ &= \tfrac{1}{2}\|\boldsymbol{v}(t)\|_2^2+O(h).\end{aligned}$$

Likewise,

$$(\boldsymbol{v}_0,(\boldsymbol{v}_0)_h)=\tfrac{1}{2}\|\boldsymbol{v}_0\|_2^2+O(h).$$

Therefore, the theorem follows by letting $h\to 0$ in (4.6).

[12] Recall that the space dimension is 2 or 3.

Remark 4.1 From $(2.12)_1$, for a weak solution $\boldsymbol{v}$ we have

$$\int_0^T \|\boldsymbol{v}(t)\|_4^4 dt \le c\int_0^T \|\boldsymbol{v}(t)\|_2^2\|\nabla\boldsymbol{v}(t)\|_2^2 dt < \infty, \quad n=2,$$

and so every weak solution, in dimension 2, obeys the energy equality. However, by $(2.12)_2$, we have only

$$\boldsymbol{v} \in L^{8/3}(0,T;\boldsymbol{L}^4(\Omega)), \quad n=3$$

and the question of whether a weak solution obeys the energy equality remains open.

Remark 4.2 Recalling that every weak solution is L^2 weakly continuous in time, all weak solutions satisfying (4.4) – and so, all weak solutions in dimension 2 – belong to $C^0([0,T);\boldsymbol{L}^2(\Omega))$.

Remark 4.3 Serrin (1963, Theorem 5) proves (4.4) for $n=3$ under the assumption

$$\boldsymbol{v} \in L^r(0,T;\boldsymbol{L}^s(\Omega)), \quad \frac{3}{s}+\frac{2}{r}=1, \ s\in[3,\infty]. \tag{*}$$

This condition, however, is stronger than (4.3) for any choice of r and s in their ranges. Actually, for $s=4$, it furnishes $\boldsymbol{v} \in L^8(0,T;\boldsymbol{L}^4(\Omega))$ which implies (4.3). For $s>4$, by the convexity inequality we find

$$\|\boldsymbol{v}\|_4^{2(r+4)/3} \le \|\boldsymbol{v}\|_s^r\|\boldsymbol{v}\|_2^{(8-r)/3},$$

and so $(*)$ implies (4.3), since $r\ge 2$ and $\boldsymbol{v}\in V_T$. If $s<4$, by the Sobolev embedding theorem we have [13]

$$\|\boldsymbol{v}\|_4 \le c\|\boldsymbol{v}\|_s^{s/2(6-s)}\|\nabla\boldsymbol{v}\|_2^{3(4-s)/2(6-s)},$$

which, by the Hölder inequality, gives

$$\int_0^T \|\boldsymbol{v}\|_4^{4(6-s)/(9-2s)} dt \le c\left(\int_0^T \|\boldsymbol{v}\|_s^r dt\right)^{(s-3)/(9-2s)}\left(\int_0^T \|\nabla\boldsymbol{v}\|_2^2 dt\right)^{3(4-s)/(9-2s)}$$

Since $s\ge 3$, also in this case $(*)$ implies (4.3).

Remark 4.4 The result proved in Theorem 4.1 is due to Lions (1960) and is a particular case of that stated in Shinbrot (1974, Theorem 4.4), where assumption (4.3) is replaced by the following one:

$$\boldsymbol{v} \in L^r(0,T;\boldsymbol{L}^q(\Omega)), \quad \frac{1}{r}+\frac{1}{q}=\frac{1}{2}, \quad q\ge 4.$$

However, unlike Theorem 4.1, the proof given by Shinbrot requires certain restrictions on the domain Ω (such as boundedness of its boundary) which are not explicitly formulated by the author. For related questions, we also refer to Taniuchi (1997).

---

[13] Recall that the space dimension is 2 or 3.

Our next objective is to give sufficient conditions under which a weak solution is unique in the class of weak solutions. The basic idea is due to Leray (1934b, pp. 242–244), who gave this result for the Cauchy problem ($\Omega \equiv \mathbb{R}^n$). The generalization to an arbitrary domain is due to Serrin (1963, Theorem 6). The procedure to prove uniqueness is essentially the same as that we have just used for proving the energy equality and, here as there, one approximates the solutions by a suitable sequence of functions from $\mathcal{D}_T$. The main difficulty is to show the convergence of the nonlinear terms along these sequences. Apparently, the condition $\boldsymbol{v} \in V_T$ satisfied by a weak solution does not guarantee this convergence in dimension 3, while it does in dimension 2. The following lemmas play a fundamental role in estimating the nonlinear term. The first one is a simple consequence of the Hölder and Sobolev inequalities (see Serrin 1963, Lemma 1; Masuda 1984, Lemma 2.4); the second one is a clever application of Dini's theorem on the uniform convergence of sequences of monotonically decreasing functions (Masuda, 1984, Lemma 2.7).

Lemma 4.1 *Let r, s satisfy*

$$\frac{n}{s} + \frac{2}{r} = 1, \quad s \in [n, \infty].$$

and let $\boldsymbol{v}, \boldsymbol{w} \in V_T$, $\boldsymbol{u} \in L^r(0,T; \boldsymbol{L}^s(\Omega))$. Then,

$$\left|\int_0^T (\boldsymbol{v} \cdot \nabla \boldsymbol{w}, \boldsymbol{u}) dt\right| \le c \left(\int_0^T \|\nabla \boldsymbol{w}\|_2^2 dt\right)^{1/2} \left(\int_0^T \|\nabla \boldsymbol{v}\|_2^2 dt\right)^{n/2s} \left(\int_0^T \|\boldsymbol{u}\|_s^r \|\boldsymbol{v}\|_2^2 dt\right)^{1/r}.$$

with the exception of the single case $s = n = 2$.

Lemma 4.2 *Let $\boldsymbol{w} \in L^2(\tau, T; H^1(\Omega))$, $\boldsymbol{v} \in L^\infty(\tau, T; \boldsymbol{L}^n(\Omega))$. Assume that*

$$\int_\tau^t \|\nabla \boldsymbol{w}\|_2^2 ds > 0, \quad \textit{for all } t \in (\tau, T)$$

and that $\boldsymbol{v}$ is right continuous at $t = \tau$ in the L^n-norm. Then, for any $\varepsilon > 0$ there exists $M = M(\boldsymbol{w}, \boldsymbol{v}, \varepsilon) > 0$ such that

$$\left|\int_\tau^t (\boldsymbol{w} \cdot \nabla \boldsymbol{w}, \boldsymbol{v}) ds\right| \le \varepsilon \int_\tau^t \|\nabla \boldsymbol{w}\|_2^2 ds + M \int_\tau^t \|\boldsymbol{w}\|_2^2 ds, \quad \textit{for all } t \in (\tau, T).$$

We also have

Lemma 4.3 *Let $\boldsymbol{v} \in V_T$. Then, there exists a sequence of functions $\{\boldsymbol{v}_k\} \subset L^2(0,T; H^1(\Omega))$ such that*

(i) *$\boldsymbol{v}_k$ tends to $\boldsymbol{v}$ in $L^2(0,T; H^1(\Omega))$*
(ii) *$\boldsymbol{v}_k(t) \in \mathcal{D}(\Omega)$ for a.a. $t \in [0,T)$*

Moreover, their mollifiers $(\boldsymbol{v}_k)_h \equiv \boldsymbol{v}_{h,k}$ ($\in \mathcal{D}_T$), see (2.20), satisfy the following properties

$$\lim_{k\to\infty} \int_0^t (\boldsymbol{u} \cdot \nabla \boldsymbol{u}, \boldsymbol{v}_{h,k}) ds = \int_0^t (\boldsymbol{u} \cdot \nabla \boldsymbol{u}, \boldsymbol{v}_h) ds,$$

for all $\boldsymbol{u} \in V_T$.

Proof. Let $(\,\cdot,\cdot\,)_1$ denote the scalar product in H^1. Let $\{\boldsymbol{\Phi}_r\}$ be an orthonormal basis in $H^1(\Omega)$ constituted by elements of $\mathcal{D}(\Omega)$, and set

$$\boldsymbol{v}_k(t) = \sum_{r=1}^{k} (\boldsymbol{v}(t), \boldsymbol{\Phi}_r)_1 \boldsymbol{\Phi}_r.$$

Clearly, $\boldsymbol{v}_k$ satisfies (i), by the Lebesgue dominated convergence theorem, and (ii). Now, we have

$$\boldsymbol{v}_{h,k}(t) = \sum_{r=1}^{k} (\boldsymbol{v}_h(t), \boldsymbol{\Phi}_r)_1 \boldsymbol{\Phi}_r \quad \text{and} \quad \lim_{k\to\infty} \|\boldsymbol{v}_{h,k}(t) - \boldsymbol{v}_h(t)\|_{1,2} = 0, \text{ for all } t \in [0,T).$$

By the Sobolev embedding theorem and by the property of mollifiers, we also have

$$\begin{aligned} \|\boldsymbol{v}_{h,k}(t) - \boldsymbol{v}_h(t)\|_4 &\le c\|\boldsymbol{v}_{h,k}(t) - \boldsymbol{v}_h(t)\|_{1,2} \le c \max_{t\in[0,T]} \|\boldsymbol{v}_h(t)\|_{1,2} \quad n=2, \\ \|\boldsymbol{v}_{h,k}(t) - \boldsymbol{v}_h(t)\|_3 &\le c\|\boldsymbol{v}_{h,k}(t) - \boldsymbol{v}_h(t)\|_{1,2} \le c \max_{t\in[0,T]} \|\boldsymbol{v}_h(t)\|_{1,2} \quad n=3, \end{aligned} \tag{4.7}$$

from which we deduce, in particular, for all $t \in [0,T)$

$$\begin{aligned} \lim_{k\to\infty} \|\boldsymbol{v}_{h,k}(t) - \boldsymbol{v}_h(t)\|_4 &= 0 \quad \text{for } n=2 \\ \lim_{k\to\infty} \|\boldsymbol{v}_{h,k}(t) - \boldsymbol{v}_h(t)\|_3 &= 0 \quad \text{for } n=3. \end{aligned} \tag{4.8}$$

Let us first consider the case $n=2$. We know from (2.12) and Remark 4.2 that $\|\boldsymbol{u}\|_4^2 \le C\|\nabla \boldsymbol{u}\|_2$ and so, by the Hölder inequality,

$$\int_0^t |(\boldsymbol{u}\cdot\nabla\boldsymbol{u}, \boldsymbol{v}_{h,k} - \boldsymbol{v})|ds \le C \int_0^t \|\nabla\boldsymbol{u}\|_2^2 \|\boldsymbol{v}_{h,k}(t) - \boldsymbol{v}_h(t)\|_4^2. \tag{4.9}$$

The result then follows from (4.9), (4.7), $(4.8)_1$ and the Lebesgue dominated convergence theorem. In the case $n=3$, by the Sobolev theorem, we have $\|\boldsymbol{u}\|_6 \le c\|\nabla\boldsymbol{u}\|_2$ and, in place of (4.9), we find

$$\int_0^t |(\boldsymbol{u}\cdot\nabla\boldsymbol{u}, \boldsymbol{v}_{h,k} - \boldsymbol{v})|ds \le C \int_0^t \|\nabla\boldsymbol{u}\|_2^2 \|\boldsymbol{v}_{h,k}(t) - \boldsymbol{v}_h(t)\|_3^2,$$

and the result follows as in the case $n=2$.

We are now in a position to show the following uniqueness theorem.

Theorem 4.2 *Let $\boldsymbol{v}$, $\boldsymbol{u}$ be two weak solutions in Ω_T corresponding to the same data $\boldsymbol{v}_0$ and $\boldsymbol{f}$. Assume that $\boldsymbol{u}$ satisfies the energy inequality* (EI) *and that $\boldsymbol{v}$ satisfies at least one of the next two conditions:*

(i) $\boldsymbol{v} \in L^r(0,T;\boldsymbol{L}^s(\Omega))$, *for some r, s such that* $\dfrac{n}{s} + \dfrac{2}{r} = 1$, $s \in (n,\infty]$;

(ii) $\boldsymbol{v} \in L^\infty(0,T;\boldsymbol{L}^n(\Omega))$, *and $\boldsymbol{v}(t)$ is right continuous for $t \in [0,T)$ in the L^n-norm.*

Then $\boldsymbol{v} = \boldsymbol{u}$ a.e. *in Ω_T.*

Proof. Let $\{\boldsymbol{u}_{h,k}\}$ be a sequence of functions of the type introduced in the previous theorem, and let $\{\boldsymbol{v}_{h,k}\}$ be the sequence of Lemma 4.3 . We choose $\boldsymbol{\varphi} = \boldsymbol{u}_{h,k}$ in (2.4), with $s = 0$, and $\boldsymbol{\varphi} = \boldsymbol{v}_{h,k}$ in (2.4), with $s = 0$ and with $\boldsymbol{u}$ in place of $\boldsymbol{v}$. We thus obtain

$$\begin{aligned}\int_0^t \left\{ \left(\boldsymbol{v}, \frac{\partial \boldsymbol{u}_{h,k}}{\partial \tau} \right) - \nu(\nabla \boldsymbol{v}, \nabla \boldsymbol{u}_{h,k}) - (\boldsymbol{v} \cdot \nabla \boldsymbol{v}, \boldsymbol{u}_{h,k}) \right\} d\tau \\ = -\int_0^t (\boldsymbol{f}, \boldsymbol{u}_{h,k}) d\tau + (\boldsymbol{v}(t), \boldsymbol{u}_{h,k}(t)) - (\boldsymbol{v}_0, (\boldsymbol{v}_0)_{h,k}),\end{aligned} \tag{4.10}$$

$$\begin{aligned}\int_0^t \left\{ \left(\boldsymbol{u}, \frac{\partial \boldsymbol{v}_{h,k}}{\partial \tau} \right) - \nu(\nabla \boldsymbol{u}, \nabla \boldsymbol{v}_{h,k}) - (\boldsymbol{u} \cdot \nabla \boldsymbol{u}, \boldsymbol{v}_{h,k}) \right\} d\tau \\ = -\int_0^t (\boldsymbol{f}, \boldsymbol{v}_{h,k}) d\tau + (\boldsymbol{u}(t), \boldsymbol{v}_{h,k}(t)) - (\boldsymbol{v}_0, (\boldsymbol{v}_0)_{h,k}).\end{aligned} \tag{4.11}$$

We wish to let $k \to \infty$ in these relations. The only terms which need a little care are the nonlinear ones. From Lemma 4.1 and the assumptions made on $\boldsymbol{v}, \boldsymbol{u}$ it follows that

$$\begin{aligned}\left| \int_0^t (\boldsymbol{v} \cdot \nabla \boldsymbol{v}, \boldsymbol{u}_{h,k} - \boldsymbol{u}_h) d\tau \right| &= \left| \int_0^t (\boldsymbol{v} \cdot \nabla (\boldsymbol{u}_{h,k} - \boldsymbol{u}_h), \boldsymbol{v}) d\tau \right| \\ &\le C \left(\int_0^t \|\nabla(\boldsymbol{u}_{h,k} - \boldsymbol{u}_h)\|_2^2 d\tau \right)^{\frac{1}{2}}\end{aligned}$$

where C depends on $\boldsymbol{v}$. Therefore, from Lemma 2.5, we find

$$\lim_{k \to \infty} \int_0^t (\boldsymbol{v} \cdot \nabla \boldsymbol{v}, \boldsymbol{u}_{h,k}) d\tau = -\int_0^t (\boldsymbol{v} \cdot \nabla \boldsymbol{u}_h, \boldsymbol{v}) d\tau. \tag{4.12}$$

Moreover, from Lemma 4.3, we have

$$\lim_{k \to \infty} \int_0^t (\boldsymbol{u} \cdot \nabla \boldsymbol{u}, \boldsymbol{v}_{h,k}) d\tau = \int_0^t (\boldsymbol{u} \cdot \nabla \boldsymbol{u}, \boldsymbol{v}_h) d\tau. \tag{4.13}$$

Thus, letting $k \to \infty$ in (4.10), (4.11) and using (4.12), (4.13) and Lemma 2.5, we find

$$\begin{aligned}\int_0^t \left\{ \left(\boldsymbol{v}, \frac{\partial \boldsymbol{u}_h}{\partial \tau} \right) - \nu(\nabla \boldsymbol{v}, \nabla \boldsymbol{u}_h) - (\boldsymbol{v} \cdot \nabla \boldsymbol{v}, \boldsymbol{u}_h) \right\} d\tau \\ = -\int_0^t (\boldsymbol{f}, \boldsymbol{u}_h) d\tau + (\boldsymbol{v}(t), \boldsymbol{u}_h(t)) - (\boldsymbol{v}_0, (\boldsymbol{v}_0)_h),\end{aligned} \tag{4.14}$$

$$\int_0^t \left\{ \left(\boldsymbol{u}, \frac{\partial \boldsymbol{v}_h}{\partial \tau} \right) - \nu(\nabla \boldsymbol{u}, \nabla \boldsymbol{v}_h) \; - (\boldsymbol{u} \cdot \nabla \boldsymbol{u}, \boldsymbol{v}_h) \right\} d\tau$$
$$= - \int_0^t (\boldsymbol{f}, \boldsymbol{v}_h) d\tau + (\boldsymbol{u}(t), \boldsymbol{v}_h(t)) - (\boldsymbol{v}_0, (\boldsymbol{v}_0)_h). \tag{4.15}$$

By Fubini's theorem and the properties of the mollifier, we show

$$\int_0^t \left(\boldsymbol{v}, \frac{\partial \boldsymbol{u}_h}{\partial \tau} \right) d\tau = - \int_0^t \left(\boldsymbol{u}, \frac{\partial \boldsymbol{v}_h}{\partial \tau} \right) d\tau,$$

and so, adding (4.14) and (4.15) furnishes

$$\begin{aligned} -\int_0^t \{ \; \nu(\nabla \boldsymbol{v}, \nabla \boldsymbol{u}_h) + \nu(\nabla \boldsymbol{u}, \nabla \boldsymbol{v}_h) - (\boldsymbol{v} \cdot \nabla \boldsymbol{u}_h, \boldsymbol{v}) + (\boldsymbol{u} \cdot \nabla \boldsymbol{u}, \boldsymbol{v}_h) \} \, d\tau \\ = - \int_0^t (\boldsymbol{f}, \boldsymbol{u}_h + \boldsymbol{v}_h) d\tau + (\boldsymbol{v}(t), \boldsymbol{u}_h(t)) + (\boldsymbol{u}(t), \boldsymbol{v}_h(t)) \\ -(\boldsymbol{v}_0, (\boldsymbol{v}_0)_h) - (\boldsymbol{v}_0, (\boldsymbol{v}_0)_h). \end{aligned} \tag{4.16}$$

We now want to let $h \to 0$ in this relation. Again, the main difficulty is given by the nonlinear terms, the other terms being easily treated by means of Lemma 2.5. By the same reasoning leading to (4.12) we find

$$\lim_{h \to 0} \int_0^t (\boldsymbol{v} \cdot \nabla \boldsymbol{u}_h, \boldsymbol{v}) d\tau = \int_0^t (\boldsymbol{v} \cdot \nabla \boldsymbol{u}, \boldsymbol{v}) d\tau. \tag{4.17}$$

Concerning the other nonlinear term, we shall distinguish the three cases:

a) $s > n$;
b) $s = n$;
c) $s = \infty$.

In case a), since $\boldsymbol{u} \in V_T$, from Lemma 4.1 we obtain

$$\int_0^t |(\boldsymbol{u} \cdot \nabla \boldsymbol{u}, \boldsymbol{v}_h - \boldsymbol{v})| d\tau \le C \|\boldsymbol{v}_h - \boldsymbol{v}\|_{L^r(0,T;L^s(\Omega))}$$

with $C = C(\boldsymbol{u})$, and so, by Lemma 2.5 we find

$$\lim_{h \to 0} \int_0^t (\boldsymbol{u} \cdot \nabla \boldsymbol{u}, \boldsymbol{v}_h) d\tau = \int_0^t (\boldsymbol{u} \cdot \nabla \boldsymbol{u}, \boldsymbol{v}) d\tau. \tag{4.18}$$

In case b), we shall consider only the case $n = 3$, the case $n = 2$ being treated in a similar way. We thus observe that by the Hölder and Sobolev inequalities, and recalling that $\boldsymbol{u} \in V_T$, it follows that

$$\int_0^t \|\boldsymbol{u} \cdot \nabla \boldsymbol{u}\|_{3/2} d\tau \le \int_0^t \|\boldsymbol{u}\|_6 \|\nabla \boldsymbol{u}\|_2 d\tau \le c \int_0^t \|\nabla \boldsymbol{u}\|_2^2 d\tau \le C. \tag{4.19}$$

Therefore, setting $\boldsymbol{w} \equiv \boldsymbol{u} \cdot \nabla \boldsymbol{u}$, by the property of the mollifier, we obtain

$$\begin{aligned}\left|\int_0^t (\boldsymbol{u}\cdot\nabla\boldsymbol{u}, \boldsymbol{v}_h - \boldsymbol{v})d\tau\right| &\equiv \left|\int_0^t (\boldsymbol{w}, \boldsymbol{v}_h - \boldsymbol{v})d\tau\right| = \left|\int_0^t (\boldsymbol{w} - \boldsymbol{w}_h, \boldsymbol{v})d\tau\right| \\ &\leq \operatorname*{ess\,sup}_{t\in[0,T]} \|\boldsymbol{v}(t)\|_3 \int_0^t \|\boldsymbol{w} - \boldsymbol{w}_h\|_{3/2} d\tau.\end{aligned}$$

By (4.19), we have $\boldsymbol{w} \in L^1(0,T;\boldsymbol{L}^{3/2}(\Omega))$ and so, by Lemma 2.5, we conclude the validity of (4.18). Finally, in case c), from the Schwarz inequality and the fact that $\boldsymbol{u} \in V_T$, we easily establish that $\boldsymbol{w} \in L^2(0,T;\boldsymbol{L}^1(\Omega))$ and so, using the following relation

$$\begin{aligned}\left|\int_0^t (\boldsymbol{w}, \boldsymbol{v}_h - \boldsymbol{v})d\tau\right| &= \left|\int_0^t (\boldsymbol{w} - \boldsymbol{w}_h, \boldsymbol{v})d\tau\right| \\ &\leq \left(\int_0^t \|\boldsymbol{w} - \boldsymbol{w}_h\|_1^2\right)^{1/2} \left(\int_0^t \|\boldsymbol{v}\|_\infty^2\right)^{1/2},\end{aligned}$$

we again arrive at (4.18). Letting $h \to 0$ in (4.16), and using (4.17), (4.18), we obtain

$$\begin{aligned}&-\int_0^t \{\ 2\nu(\nabla\boldsymbol{v}, \nabla\boldsymbol{u}) + (\boldsymbol{w}\cdot\nabla\boldsymbol{u}, \boldsymbol{v})\}\, d\tau \\ &\qquad = -\int_0^t (\boldsymbol{f}, \boldsymbol{u} + \boldsymbol{v})d\tau + 2[(\boldsymbol{v}(t), \boldsymbol{u}(t)) - (\boldsymbol{v}_0, \boldsymbol{v}_0)],\end{aligned} \tag{4.20}$$

with $\boldsymbol{w} = \boldsymbol{u} - \boldsymbol{v}$. By Remark 4.3, $\boldsymbol{v}$ obeys the energy equality

$$\|\boldsymbol{v}(t)\|_2^2 + 2\nu\int_0^t \|\nabla\boldsymbol{v}\|_2^2 d\tau = 2\int_0^t (\boldsymbol{v}, \boldsymbol{f})d\tau + \|\boldsymbol{v}_0\|_2^2, \tag{4.21}$$

while, by assumption, $\boldsymbol{u}$ obeys the energy inequality

$$\|\boldsymbol{u}(t)\|_2^2 + 2\nu\int_0^t \|\nabla\boldsymbol{u}\|_2^2 d\tau \leq 2\int_0^t (\boldsymbol{u}, \boldsymbol{f})d\tau + \|\boldsymbol{v}_0\|_2^2. \tag{4.22}$$

Adding 2 × (4.20), (4.21) and (4.22), and observing that

$$\int_0^t (\boldsymbol{w}\cdot\nabla\boldsymbol{v}, \boldsymbol{v})d\tau = 0$$

we obtain

$$\|\boldsymbol{w}(t)\|_2^2 + 2\nu\int_0^t \|\nabla\boldsymbol{w}\|_2^2 d\tau \leq 2\int_0^t (\boldsymbol{w}\cdot\nabla\boldsymbol{w}, \boldsymbol{v})d\tau. \tag{4.23}$$

If $s > n$, we employ Lemma 4.1 on the term on the right-hand side of (4.23) together with the Young inequality to deduce

$$\begin{aligned}\int_0^t (\boldsymbol{w}\cdot\nabla\boldsymbol{w},\boldsymbol{v})d\tau &\le c\left(\int_0^t \|\nabla\boldsymbol{w}\|^2 d\tau\right)^{1-1/r}\left(\int_0^t \|\boldsymbol{v}\|_s^r\|\boldsymbol{w}\|_2^2 d\tau\right)^{1/r}\\ &\le \nu\int_0^t \|\nabla\boldsymbol{w}\|^2 d\tau + c_1\int_0^t \|\boldsymbol{v}\|_s^r\|\boldsymbol{w}\|_2^2 d\tau.\end{aligned}$$

Replacing this inequality into (4.23), we find

$$\|\boldsymbol{w}(t)\|_2^2 \le c_1\int_0^t \|\boldsymbol{v}\|_s^r\|\boldsymbol{w}\|_2^2 d\tau,$$

which, with the help of Gronwall's lemma, allows us to conclude $\boldsymbol{v}=\boldsymbol{u}$ a.e. in Ω_T. If $s=n$, we set

$$\mathcal{T}=\{\tau\in[0,T]:\ \|\boldsymbol{w}(s)\|_2=0,\quad \text{for all } s\in[0,\tau]\}.$$

Clearly, $\mathcal{T}$ is not empty and, in virtue of the L^2 weak continuity of $\boldsymbol{w}$, it is also closed. Let us denote by τ_0 its maximum. If $\tau_0=T$, there is nothing to prove. Therefore, assuming $\tau_0<T$, we have

$$\int_{\tau_0}^t \|\nabla\boldsymbol{w}\|_2^2 ds>0,\quad \text{for all } t\in[\tau_0,T).$$

By Lemma 4.2, it then follows

$$\left|\int_{\tau_0}^t (\boldsymbol{w}\cdot\nabla\boldsymbol{w},\boldsymbol{v})d\tau\right| \le \varepsilon\int_{\tau_0}^t \|\nabla\boldsymbol{w}\|_2^2 ds + M\int_{\tau_0}^t \|\boldsymbol{w}\|_2^2 ds,\quad \text{for all } t\in(\tau_0,T).$$

Replacing this inequality into (4.23), and recalling that $\boldsymbol{w}(s)=0$ for all $s\le\tau_0$, we find

$$\|\boldsymbol{w}(t)\|_2^2 \le M\int_{\tau_0}^t \|\boldsymbol{w}\|_2^2 ds,$$

which, with the help of Gronwall's lemma, again implies $\boldsymbol{v}=\boldsymbol{u}$ a.e. in Ω_T. The theorem is thus proved.

Remark 4.5 If Ω is a bounded or an exterior domain with a sufficiently smooth boundary, or a half space, one can furnish an important generalization of the uniqueness result given in the previous theorem. Such a generalization, instead of hypothesis (ii), requires only

$$\boldsymbol{v}\in L^\infty(0,T;\boldsymbol{L}^n(\Omega)). \tag{4.24}$$

This result, due to Kozono and Sohr (1996a) (see also Sohr and von Wahl (1984), under more restrictive assumptions on $\boldsymbol{v}$, and the review article of Kozono (1998)) will be proved in Section 7, Theorem 7.2, in the case $\Omega=\mathbb{R}^n$.

Remark 4.6 Since in dimension 2 every weak solution belongs to the class $C^0([0,T);\boldsymbol{L}^2(\Omega))$, see Remark 4.2, by Theorem 4.2 it follows that every such weak solution is unique in the class of weak solutions assuming the same data, a fact discovered for the first time by Lions and Prodi (1959). In dimension 3, by the Sobolev inequality, we have

$$\|\boldsymbol{v}\|_s \le c\|\boldsymbol{v}\|_2^{(6-s)/2s}\|\nabla\boldsymbol{v}\|_2^{3(s-2)/2s}, \quad s\in[2,6]$$

and so, for $\boldsymbol{v}\in V_T$, we find

$$\boldsymbol{v}\in L^r(0,T;\boldsymbol{L}^s(\Omega)),\quad \frac{3}{s}+\frac{2}{r}=\frac{3}{2},$$

and the condition in Theorem 4.2 is *not* satisfied. The problem of whether a three dimensional weak solution obeying the energy inequality is unique in its class is an outstanding open question. In this respect, we wish to mention the contribution of Ladyzhenskaya (1969), in her effort to *disprove* uniqueness. Specifically, using a method introduced by Golovkin (1964) in a different context, she constructs two distinct three dimensional solutions $\boldsymbol{v}_i$, $i=1,2$, with rotational symmetry, corresponding to the same data, in a *non-cylindrical* domain Q_T of the space-time. This latter is defined as

$$Q_T=\{(r,z,t):\ t\in[0,T],\ r\in[\eta\sqrt{t},\ell\sqrt{t}],\ z\in[-\ell\sqrt{t},\ell\sqrt{t}],\ \eta<<\ell\},$$

where (r,z) denote cylindrical coordinates. Both solutions belong to the Leray-Hopf class in the sense that

$$\max_{t\in[0,T]}\int_{\Omega(t)}|\boldsymbol{v}_i|^2+\int_{Q_T}|\nabla\boldsymbol{v}_i|^2<\infty$$

where

$$\Omega(t)\equiv\{r\in[\eta\sqrt{t},\ell\sqrt{t}],\ z\in[-\ell\sqrt{t},\ell\sqrt{t}]\}.$$

Moreover, they match the (vanishing) initial data in the following sense

$$\lim_{t\to0}\int_{\Omega(t)}|\boldsymbol{v}_i|^2=0$$

and obey "stress-free" boundary conditions. [14] Finally, they satisfy the following condition

$$\int_0^T\left(\int_{\Omega(t)}|\boldsymbol{v}_i|^s\right)^{r/q}dt<M(\varepsilon)<\infty,\quad i=1,2,$$

with exponents s,r such that

$$\frac{3}{s}+\frac{2}{r}=1+\varepsilon,\quad \varepsilon>0,$$

($M(\varepsilon)\to\infty$ as $\varepsilon\to0$). However, this result can not be considered completely satisfactory, in that the space-time domain Q_T where the solutions $\boldsymbol{v}_i$ exist is *not* cylindrical (that is, of the type $\Omega\times I$ with Ω a fixed spatial domain and I a time interval). Rather, it expands when time increases and reduces to a single point when time goes to zero. In the same paper, Ladyzhenskaya furnishes another counter example to uniqueness in a class of solutions slightly weaker than the

[14] That is, the normal component of $\boldsymbol{v}$ is prescribed, together with the tangential component of the vorticity field.

Leray-Hopf one, in that the spatial derivatives are summable with an exponent *strictly* less than 2. This time the boundary conditions are the usual adherence conditions, but the space-time domain is still non-cylindrical.

Remark 4.7 For later purposes, we wish to notice that the condition

$$\boldsymbol{v} \in L^{r_1}(0,T;\boldsymbol{L}^{s_1}(\Omega)), \quad \text{for some } s_1 \in (n,\infty],\ r_1 \in [2,\infty), \ \text{with } \frac{n}{s_1}+\frac{2}{r_1}=1 \tag{A}$$

is weaker than

$$\boldsymbol{v} \in L^{r_2}(0,T;\boldsymbol{L}^{s_2}(\Omega)), \quad \text{for some } s_2 \in (n,\infty],\ r_2 \in (2,\infty), \text{ with } \frac{n}{s_2}+\frac{2}{r_2}<1 \tag{B}$$

in the sense that if $\boldsymbol{v}$ satisfies (B), then, by the Hölder inequality, $\boldsymbol{v}$ satisfies (A) with $s_1 = s_2$ and $r_1 = 2s_2/(s_2-n)$ $(< r_2)$.

Remark 4.8 In the literature, one may find many other uniqueness theorems for weak solutions, see, among others, Prodi (1959), Lions and Prodi (1959), Ladyzhenskaya (1967). However, in all these papers one compares two weak solutions *each of which* possesses more regularity than that established in the existence Theorem 3.1. It is therefore worth emphasizing that Theorem 4.2 compares two weak solutions of which *only one* possesses extra regularity. For uniqueness results related to Theorem 4.2, in a class of "very weak" solutions, see Foiaş (1961), Fabes, Jones and Riviére (1972), H. Kato (1993), Chemin (1999), Monniaux (1999), Amann (1999).

5. Regularity of weak solutions

The regularity theory for weak solutions to the Navier-Stokes equations presents different features, according to whether one looks for *interior regularity* or *regularity for the initial-boundary value problem.* In the first case, denoting by $R = \omega \times (t_1,t_2)$ a bounded domain strictly contained in Ω_T, one considers a field $\boldsymbol{v}$ that satisfies the identity (2.2) for all solenoidal test functions $\boldsymbol{\varphi} \in \boldsymbol{C}_0^\infty(R)$, (hereafter denoted by $(2.2)_0$), which is divergence free in R and, further, verifies the following condition

$$\boldsymbol{v} \in L^2(t_1,t_2;W^{1,2}(\omega)) \cap L^\infty(t_1,t_2;L^2(\omega)). \tag{5.1}$$

In the second case, one requires that $\boldsymbol{v}$ is a weak solution to the initial-boundary value problem, in the sense of Definition 2.1. Now, let us consider the field $\overline{\boldsymbol{v}}$ defined in (2.3). As already observed in Section 2, $\overline{\boldsymbol{v}}$ satisfies $(2.2)_0$ with $\boldsymbol{f} \equiv 0$. However, this field – though infinitely differentiable in the space variables – need have no time derivative at all and, in fact, it may even have (integrable) singularities in the time interval $[0,T)$. This example, due to Serrin (1962), leads us to the following considerations. First, for interior regularity, one can *not* expect to prove a result where the amount of regularity in time is more than that assumed at the outset.

Second, the existence of such "bad" solutions is due to the fact that the possible singularities are absorbed by the pressure term. For instance, the field (2.3) with a "bad" behavior in time could also be a solution to the quasi-linear (vector) heat equation (0.1″), on condition that, however, the force $\boldsymbol{f}$ is chosen to have an equally "bad" behavior. On the other hand, $\overline{\boldsymbol{v}}$ does not meet the boundary conditions (0.3) hidden in requirement a) of Definition 2.1, unless it is identically zero, and so there is hope that one can "gain" regularity in time by dealing with solutions of the initial-boundary value problem.

The aim of this section is to furnish sufficient conditions for regularity of weak solutions. As we shall see, these conditions do not overlap completely with those ensuring uniqueness, and there is an interesting question which is still left open. Moreover, as in the case of uniqueness, one shows that every weak solution in dimension 2 is regular, provided the data are regular enough. In dimension 3, the regularity of weak solutions is an outstanding open problem. We shall report, without proof, the interior regularity results, due essentially to Ohyama (1960), Serrin (1962) and Struwe (1988), see Theorem 5.1. Successively, in Theorem 5.2, we shall give a result concerning the regularity of weak solutions of the initial-boundary value problem (in the sense of Definition 2.1). In doing this, we shall follow the method of Galdi and Maremonti (1988). For further regularity results, see H. Kato (1977/78,1986,1989,1993), Tanaka (1987).

Theorem 5.1 *Let $\boldsymbol{v}$ be a solenoidal field in $\omega \times (t_1, t_2)$, satisfying $(2.2)_0$ with $\boldsymbol{f} = 0$,* [15] *and (5.1). Assume, in addition, that $\boldsymbol{v}$ verifies at least one of the following two conditions:*

(i) *$\boldsymbol{v} \in L^r(t_1, t_2; \boldsymbol{L}^s(\omega))$, for some r, s such that $\dfrac{n}{s} + \dfrac{2}{r} = 1$, $s \in (n, \infty)$;*

(ii) *$\boldsymbol{v} \in L^\infty(t_1, t_2; \boldsymbol{L}^n(\omega))$, and, given $\epsilon > 0$ there is $\rho > 0$ such that*

$$\int_{B_\rho \cap \omega} |\boldsymbol{v}(x,t)|^n dx < \epsilon, \quad \text{for all } t \in (t_1, t_2)$$

where B_ρ is a ball of radius ρ.

Then, $\boldsymbol{v}$ is of class $C^\infty(\omega)$, and each space derivative is bounded in compact subregions of $\omega \times (t_1, t_2)$. If, in addition,

$$\frac{\partial \boldsymbol{v}}{\partial t} \in L^2(t_1, t_2; \boldsymbol{L}^q(\omega)), \quad \text{for some } q \geq 1,$$

then, the space derivatives of $\boldsymbol{v}$ are absolutely continuous functions of time.

Remark 5.1 For $n = 2$, a possible choice of exponents is $s = r = 4$. Therefore, from Remark 4.1, we conclude that every two dimensional weak solution is regular in the sense specified in Theorem 5.1. On the other hand, three dimensional weak solutions do not satisfy either of assumption (i), (ii), see Remark 4.6, and nothing can be said about their regularity. An interesting variant of Theorem 5.1(i) has been given by Takahashi (1990, 1992), who replaces the Lebesgue space L^r with

[15] For the general case $\boldsymbol{f} \not\equiv 0$, we refer the reader to the papers of Serrin and Struwe.

the Lorentz space $L^{(r)}$ ("L^r-*weak*"), requiring, however, that the corresponding "norm" be sufficiently small. In particular, denoting by $B_R(x_0)$ a ball of radius R centered at x_0, he shows that a sufficient condition for a weak solution $\boldsymbol{v}$ to be of class L^∞ in $B_R(x_0) \times (-R^2 + t_1, t_1)$ [16] is that it satisfies an estimate of the type

$$\|\boldsymbol{v}(t)\|_{\sigma, B_R(x_0)} \leq \frac{\varepsilon}{(t_1 - t)^{(\sigma-n)/2\sigma}}, \quad t \in (-R^2 + t_1, t_1), \ \sigma \in (n, \infty]$$

with a "small" ε. As we shall see in Theorem 7.3, a necessary condition for $\boldsymbol{v}$ to become irregular at a time t_1 [17] is that

$$\|\boldsymbol{v}(t)\|_\sigma \geq \frac{C}{(t_1 - t)^{(\sigma-n)/2\sigma}}, \quad t < t_1,$$

with $C = C(n, \sigma, \nu) > 0$; Takahashi also extends Theorem 5.1(i) to the case $s = \infty$.

We shall now be concerned with the regularity of weak solutions to the initial-boundary value problem, in the sense of Definition 2.1. For simplicity, we shall assume that $\boldsymbol{f} \equiv 0$. Before going into details, we wish to outline the main idea underlying the proof. To this end, let $\boldsymbol{v}$ be a weak solution in Ω_T and let $\boldsymbol{u}$ be a weak solution in Ω_T to the following initial-boundary value problem

$$\begin{aligned} &\frac{\partial \boldsymbol{u}}{\partial t} + \boldsymbol{v} \cdot \nabla \boldsymbol{u} = \nu \Delta \boldsymbol{u} + \nabla \pi \\ &\operatorname{div} \boldsymbol{u} = 0 \\ &\boldsymbol{u}(x, 0) = \boldsymbol{v}_0, \quad x \in \Omega \\ &\boldsymbol{u}(y, t) = 0, \quad y \in \partial\Omega, \ t > 0. \end{aligned} \tag{5.2}$$

By this we mean that $\boldsymbol{u} \in V_T$ and that it satisfies the following relation

$$\int_0^\infty \left\{ \left(\boldsymbol{u}, \frac{\partial \varphi}{\partial t} \right) - \nu(\nabla \boldsymbol{u}, \nabla \varphi) - (\boldsymbol{v} \cdot \nabla \boldsymbol{u}, \varphi) \right\} dt = -(\boldsymbol{v}_0, \varphi(0)), \quad \text{for all } \varphi \in \mathcal{D}_T. \tag{5.3}$$

Thus, $\boldsymbol{v}$ becomes the coefficient of a "linearized" Navier-Stokes equation. Notice, also, that $\boldsymbol{v}$ and $\boldsymbol{u}$ are both weak solutions to the *same* Navier-Stokes problem with the *same* data $\boldsymbol{v}_0$. The question is now to determine the weakest conditions on $\boldsymbol{v}$ in order that:

a) $\boldsymbol{v} = \boldsymbol{u}$, a.a. in Ω_T.

b) $\boldsymbol{u}$ has more regularity than that originally assumed for $\boldsymbol{v}$.

If b) is met, then, by a), $\boldsymbol{v}$ becomes more regular and then $\boldsymbol{u}$ becomes more regular too and so, by a boot-strap argument, we can conclude that $\boldsymbol{v}$ becomes as much regular as allowed by the data. In this latter respect, we wish to emphasize that this method only requires $\boldsymbol{v}_0 \in H(\Omega)$, since regularity is established in the semi-open interval $(0, T]$. On the other hand, we shall prove regularity up to the boundary of Ω which, therefore, will be assumed suitably smooth.

---

[16] And hence regular, in the sense of Theorem 5.1.

[17] See Definition 6.1.

Remark 5.2 Instead of the linearized problem (5.2), we could consider the following one:

$$
\begin{aligned}
&\frac{\partial \boldsymbol{u}}{\partial t} + \boldsymbol{u}\cdot\nabla\boldsymbol{v} = \nu\Delta\boldsymbol{u} + \nabla\pi \\
&\operatorname{div}\boldsymbol{u} = 0 \\
&\boldsymbol{u}(x,0) = \boldsymbol{v}_0, \quad x\in\Omega \\
&\boldsymbol{u}(y,t) = 0, \quad y\in\partial\Omega,\ t>0.
\end{aligned}
$$

With such a choice, one could find conditions on $\nabla\boldsymbol{v}$ (instead of $\boldsymbol{v}$) under which the weak solution $\boldsymbol{v}$ becomes regular. This can be done exactly along the same lines we shall follow hereafter for problem (5.2). We shall limit ourselves to state the corresponding results, without proof, in Remarks 5.3 and 5.6.

Let us first consider condition a). Since the system (5.2) is *linear* in $\boldsymbol{u}$, we expect that the conditions on $\boldsymbol{v}$ which ensure a), should be weaker than those ensuring the uniqueness of a weak solution to the full nonlinear Navier-Stokes problem. Actually, we have

Lemma 5.1 *Let $\boldsymbol{v}\in V_T$ and let $\boldsymbol{u}$ be a weak solution to (5.2) in Ω_T. Then, if*

$$
\boldsymbol{v}\in L^4(0,T;\boldsymbol{L}^4(\Omega)), \tag{5.4}
$$

we have $\boldsymbol{v}=\boldsymbol{u}$, a.a. in Ω_T.

Proof. Reasoning exactly as in the proof of Lemma 2.1, we show that $\boldsymbol{u}$ satisfies the following relation

$$
\int_0^t \left\{ \left(\boldsymbol{u},\frac{\partial\boldsymbol{\varphi}}{\partial t}\right) - \nu(\nabla\boldsymbol{u},\nabla\boldsymbol{\varphi}) - (\boldsymbol{v}\cdot\nabla\boldsymbol{u},\boldsymbol{\varphi})\right\} ds = (\boldsymbol{u}(t),\boldsymbol{\varphi}(t)) - (\boldsymbol{v}_0,\boldsymbol{\varphi}(0)),
$$

$$
\text{for all } t\in[0,T) \text{ and all } \boldsymbol{\varphi}\in\mathcal{D}_T.
$$

Subtracting the integral equation in the previous relation from that in (2.4) with $\boldsymbol{f}\equiv 0$, and setting $\boldsymbol{w}=\boldsymbol{u}-\boldsymbol{v}$ we find

$$
\int_0^t \left\{ \left(\boldsymbol{w},\frac{\partial\boldsymbol{\varphi}}{\partial t}\right) - \nu(\nabla\boldsymbol{w},\nabla\boldsymbol{\varphi}) - (\boldsymbol{v}\cdot\nabla\boldsymbol{w},\boldsymbol{\varphi})\right\} ds = (\boldsymbol{w}(t),\boldsymbol{\varphi}(t)). \tag{5.5}
$$

From now on, the proof is the same as that of Theorem 4.1. Specifically, we denote by $\{\boldsymbol{w}_k\}\subset\mathcal{D}_T$ a sequence converging to $\boldsymbol{v}$ in $L^2(0,T;H^1(\Omega))$. We then choose in (5.5) $\boldsymbol{\varphi}=\boldsymbol{w}_{h,k}$, and pass to the limits $k\to\infty$ and $h\to 0$. Reasoning as in Theorem 4.1, we show

$$
\tfrac{1}{2}\|\boldsymbol{w}(t)\|_2^2 = \int_0^t \{(\boldsymbol{v}\cdot\nabla\boldsymbol{w},\boldsymbol{w}) - \nu\|\nabla\boldsymbol{w}\|_2^2\}\, ds.
$$

However, since $\boldsymbol{v}(t)\in H(\Omega)$ for a.a. t, we get

$$
\int_0^t (\boldsymbol{v}\cdot\nabla\boldsymbol{w},\boldsymbol{w})ds = 0,
$$

and the lemma follows.

The major assumption on the weak solution $\boldsymbol{v}$ comes into the proof of point b). To show this, however, we need some preliminary considerations. The first one concerns well known results for the steady *Stokes system*, obtained as a suitable linearization of the full steady-state Navier-Stokes system (0.4). Specifically, we have (see, *e.g.*, Galdi, 1994, Theorem IV.6.1).

Lemma 5.2 *Let Ω be a bounded domain of $\mathbb{R}^n$, of class C^{m+2}, $m \geq 0$. For any $\boldsymbol{F} \in \boldsymbol{W}^{m,q}(\Omega)$, $1 < q < \infty$, there exists one and only one solution $\boldsymbol{u}, \phi$* [18] *to the following Stokes problem*

$$-\nu\Delta \boldsymbol{u} = \nabla\phi + \boldsymbol{F}$$

$$\operatorname{div} \boldsymbol{u} = 0$$

$$\boldsymbol{u}(y) = 0, \quad y \in \partial\Omega,$$

such that

$$\boldsymbol{u} \in \boldsymbol{W}^{m+2,q}(\Omega), \quad \phi \in \boldsymbol{W}^{m+1,q}(\Omega).$$

This solution satisfies the estimate:

$$\|\boldsymbol{u}\|_{m+2,q} + \|\phi\|_{m+1,q} \leq c\|\boldsymbol{F}\|_{m,q}.$$

Moreover, the problem

$$-\nu\Delta \boldsymbol{a} = \nabla\phi + \lambda \boldsymbol{a}$$

$$\operatorname{div} \boldsymbol{a} = 0$$

$$\boldsymbol{a}(y) = 0, \quad y \in \partial\Omega,$$

admits a denumerable number of positive eigenvalues $\{\lambda_r\}$ clustering at infinity, and the corresponding eigenfunctions $\{\boldsymbol{a}_r\}$ form an orthonormal basis in H.

Our second preliminary result concerns an estimate for the nonlinear term, which strengthens that given in Lemma 4.2.

Lemma 5.3 *Let*

$$\boldsymbol{v} \in C^0([0,T]; \boldsymbol{L}^n(\Omega)), \quad \boldsymbol{u} \in \boldsymbol{W}^{2,2}(\Omega), \quad \boldsymbol{a} \in L^2(\Omega).$$

Then, given $\eta > 0$ there exists $M = M(\boldsymbol{v}, \eta) > 0$ such that

$$|(\boldsymbol{v} \cdot \nabla \boldsymbol{u}, \boldsymbol{a})| \leq \varepsilon \left(\|P\Delta \boldsymbol{u}\|_2^2 + \|\boldsymbol{a}\|_2^2\right) + M\|\nabla \boldsymbol{u}\|_2^2,$$

where P is the orthogonal projection operator from $\boldsymbol{L}^2$ to H (see Section 2).

Proof. We extend $\boldsymbol{v}$ to zero outside Ω, and let $\boldsymbol{v}_\eta$ be the spatial mollifier of $\boldsymbol{v}$, that is,

$$\boldsymbol{v}_\eta(x,t) = \int_{\mathbb{R}^3} J_\eta(x-\xi)\boldsymbol{v}(\xi,t)d\xi,$$

with $J_\eta(\sigma)$ an infinitely differentiable function vanishing for $|\sigma| > \eta$ and normalized to 1. It is well known that

$$\sup_{x\in\Omega} |\boldsymbol{v}_\eta(x,t)| \leq c(\eta)\|\boldsymbol{v}(t)\|_3$$

---

[18] With the normalization condition $\int_\Omega \phi = 0$.

and that

$$\lim_{\eta\to 0} \|\boldsymbol{v}_\eta(t) - \boldsymbol{v}(t)\|_3 = 0, \quad \text{for all } t \in [0,T].$$

Using the continuity assumption on $\boldsymbol{v}$, by an argument completely analogous to that adopted in the proof of Lemma 2.3, we show that this limit is taken uniformly in $t \in [0,T]$. In view of this, by Sobolev's theorem and Lemma 5.2, we thus have

$$\begin{aligned} |(\boldsymbol{v}\cdot\nabla\boldsymbol{u},\boldsymbol{a})| &\le |((\boldsymbol{v}-\boldsymbol{v}_\eta)\cdot\nabla\boldsymbol{u},\boldsymbol{a})| + |(\boldsymbol{v}_\eta\cdot\nabla\boldsymbol{u},\boldsymbol{a})| \\ &\le \|\boldsymbol{v}-\boldsymbol{v}_\eta\|_3\|\nabla\boldsymbol{u}\|_6\|\boldsymbol{a}\|_2 + \sup\nolimits_{x\in\Omega}|\boldsymbol{v}_\eta(x,t)|\|\nabla\boldsymbol{u}\|_2\|\boldsymbol{a}\|_2 \\ &\le \varepsilon\|P\Delta\boldsymbol{u}\|_2\|\boldsymbol{a}\|_2 + M\|\nabla\boldsymbol{u}\|_2\|\boldsymbol{a}\|_2, \end{aligned}$$

and the result follows after using Cauchy's inequality on the last line of this inequality.

Using these lemmas we can now show the first regularity result for $\boldsymbol{v}$.

Lemma 5.4 *Let Ω be an arbitrary domain in $\mathbb{R}^n$, uniformly of class C^2.* [19] *Assume that $\boldsymbol{v}\in V_T$ and that it satisfies at least one of the following two conditions:*

(i) $\boldsymbol{v}\in L^r(0,T;\boldsymbol{L}^s(\Omega))$, *for some r,s such that* $\dfrac{n}{s}+\dfrac{2}{r}=1, \quad s\in(n,\infty]$;

(ii) $\boldsymbol{v}\in C^0([0,T];\boldsymbol{L}^n(\Omega))$.

Then, for any $\boldsymbol{v}_0\in H(\Omega)$, there exists one weak solution $\boldsymbol{u}$ to (5.2) in Ω_T such that

$$\boldsymbol{u}\in C^0((\varepsilon,T];H^1(\Omega))\cap L^2(\varepsilon,T;\boldsymbol{W}^{2,2}(\Omega))$$

$$\frac{\partial\boldsymbol{u}}{\partial t}\in L^2(\varepsilon,T;H(\Omega)),$$

where ε is an arbitrary positive number. Moreover, by Lemma 5.1 *and* Remarks 4.1, 4.3, $\boldsymbol{v}=\boldsymbol{u}$ a.e. *in Ω_T.* [20]

Proof. To avoid unessential technical difficulties, we limit ourselves to give the proof in the case Ω bounded and $n=3$, referring the reader to Galdi and Maremonti (1988) for the proof in the general case. We shall use the Faedo-Galerkin method of Theorem 3.1, with the basis $\{\boldsymbol{a}_r\}$ of H constituted by the eigenvectors of the Stokes problem (Lemma 5.3). Thus, we shall look for approximating solutions $\boldsymbol{u}_k$ of the form

$$\boldsymbol{u}_k(x,t)=\sum_{r=1}^{k}c_{kr}(t)\boldsymbol{a}_r(x), \quad k\in\mathbb{N},$$

[19] Ω is said *uniformly of class C^m*, $m\ge 0$, if Ω lies on one part of its boundary $\partial\Omega$ and, for each $x_0\in\partial\Omega$, there exists a ball B centered at x_0 and of radius independent of x_0, such that $\partial\Omega\cap B$ admits a Cartesian representation of the form $x_n=\gamma(x_1,\dots,x_{n-1})$, where γ is a function of class C^m in its domain, with its derivatives up to order m inclusive uniformly bounded by the same constant, independently of x_0. If Ω is uniformly of class C^m, for all $m\ge 0$, we shall say that Ω is uniformly of class C^∞.

[20] Of course, in case (ii), we have $\boldsymbol{v}_0\in H(\Omega)\cap\boldsymbol{L}^n(\Omega)$.

where the coefficients c_{kr} are required to satisfy the following system of ordinary differential equations

$$\frac{d}{dt}(\boldsymbol{u}_k, \boldsymbol{a}_r) + \nu(\nabla \boldsymbol{u}_k, \nabla \boldsymbol{a}_r) + (\boldsymbol{v} \cdot \nabla \boldsymbol{u}_k, \boldsymbol{a}_r) = 0 \quad r = 1, \dots, k, \tag{5.6}$$

with the initial condition

$$c_{kr}(0) = (\boldsymbol{v}_0, \boldsymbol{a}_r) \quad r = 1, \dots, k.$$

As in Theorem 3.1, we show that this system of ordinary differential equations admits a (unique) solution in the time interval $[0, T]$, as a consequence of the following relation, which is obtained by multiplying (5.6) by c_{kr} and summing over the index r:

$$\|\boldsymbol{u}(t)\|_2^2 + 2\nu \int_0^t \|\nabla \boldsymbol{u}(\tau)\|_2^2 d\tau = \|\boldsymbol{v}_0\|_2^2. \tag{5.7}$$

For simplicity, here as in the following relations, we shall omit the subscript "k". We next multiply (5.6) by $\lambda_k c_{kr}$ and by dc_{kr}/dt, respectively, sum over r, and employ the properties of the eigenfunctions $\boldsymbol{a}_k$ to obtain

$$\frac{1}{2}\frac{d}{dt}\|\nabla \boldsymbol{u}(t)\|_2^2 + \nu\|P\Delta \boldsymbol{u}\|_2^2 = (\boldsymbol{v} \cdot \nabla \boldsymbol{u}, P\Delta \boldsymbol{u}), \tag{5.8}$$

and

$$\frac{1}{2}\frac{d}{dt}\|\nabla \boldsymbol{u}(t)\|_2^2 + \nu\|D_t \boldsymbol{u}\|_2^2 = -(\boldsymbol{v} \cdot \nabla \boldsymbol{u}, D_t \boldsymbol{u}), \tag{5.9}$$

where P is the orthogonal projection operator from $\boldsymbol{L}^2$ to H (see Section 2) and D_t denotes differentiation with respect to t. We wish now to increase the trilinear form $(\boldsymbol{v} \cdot \nabla \boldsymbol{u}, \boldsymbol{a})$. Let us first consider the case (i), *i.e.*, $\infty \geq s > 3$ $(=n)$. By the Hölder inequality we have

$$|(\boldsymbol{v} \cdot \nabla \boldsymbol{u}, \boldsymbol{a})| \leq \|\boldsymbol{v}\|_s \|\nabla \boldsymbol{u}\|_{2s/(s-2)} \|\boldsymbol{a}\|_2;$$

Furthermore, since $2s/(s-2) \in [2, 6)$, by the Sobolev theorem and Lemma 5.3 we obtain

$$\|\nabla \boldsymbol{u}\|_{2s/(s-2)} \leq c\|\boldsymbol{u}\|_{2,2}^{3/s} \|\nabla \boldsymbol{u}\|_2^{(s-3)/s} \leq c_1 \|P\Delta \boldsymbol{u}\|_2^{3/s} \|\nabla \boldsymbol{u}\|_2^{(s-3)/s}$$

and so, it follows that

$$|(\boldsymbol{v} \cdot \nabla \boldsymbol{u}, \boldsymbol{a})| \leq c\|\boldsymbol{v}\|_s \|\nabla \boldsymbol{u}\|_2^{(s-3)/s} \|P\Delta \boldsymbol{u}\|_2^{3/s} \|\boldsymbol{a}\|_2.$$

Employing Young's inequality, with exponents $2s/(s-3)$, $2s/3$ and $1/2$ we thus conclude

$$|(\boldsymbol{v} \cdot \nabla \boldsymbol{u}, \boldsymbol{a})| \leq c\|\boldsymbol{v}\|_s^{2(s-3)/s} \|\nabla \boldsymbol{u}\|_2^2 + \eta\|P\Delta \boldsymbol{u}\|_2^2 + \eta\|\boldsymbol{a}\|_2^2 \tag{5.10}$$

with arbitrary $\eta > 0$ and $c = c(\Omega, s, \eta)$. Summing (5.8) and (5.9), and using (5.10) with $\boldsymbol{a} = P\Delta \boldsymbol{u}$ and $\boldsymbol{a} = D_t \boldsymbol{u}$, respectively, for sufficiently small η we find

$$\frac{d}{dt}\|\nabla \boldsymbol{u}\|_2^2 + c_1(\|P\Delta \boldsymbol{u}\|_2^2 + \|D_t \boldsymbol{u}\|_2^2) \leq c_2 \|\boldsymbol{v}\|_s^{2(s-3)/s} \|\nabla \boldsymbol{u}\|_2^2.$$

Integrating this relation furnishes

$$\|\nabla \boldsymbol{u}(t)\|_2^2 + c_3 \int_\tau^t (\|P\Delta \boldsymbol{u}\|_2^2 + \|D_\tau \boldsymbol{u}\|_2^2) d\tau \le \|\nabla \boldsymbol{u}(\tau)\|_2^2 \exp\left[c_2 \int_0^T \|\boldsymbol{v}(\tau)\|_s^r d\tau\right]$$

for all $t \in [s, T]$, $s \ge \varepsilon$.

If we integrate this inequality on $\tau \in [\varepsilon, t]$ and use (5.7), we obtain the following limitations

$$\|\boldsymbol{u}(t)\|_{1,2} + \int_\varepsilon^t (\|P\Delta \boldsymbol{u}\|_2^2 + \|\boldsymbol{u}_\tau\|_2^2) \le M, \quad \text{for all } t \in [\varepsilon, T], \tag{5.11}$$

where M depends on $\|\boldsymbol{v}_0\|_2$, ε, Ω, and s. Using these *a priori* estimates on the approximating solutions $\{\boldsymbol{u}_k\}$ and proceeding as in the proof of Theorem 3.1, we easily show that from the sequence $\{\boldsymbol{u}_k\}$ we can select a subsequence which converges to a weak solution $\boldsymbol{u}$ of the problem (5.2) and which, in addition, satisfies

$$\boldsymbol{u} \in L^\infty(\varepsilon, T; H^1(\Omega)) \cap L^2(\varepsilon, T; \boldsymbol{W}^{2,2}(\Omega))$$

$$\frac{\partial \boldsymbol{u}}{\partial t} \in L^2(\varepsilon, T; H(\Omega)).$$

From these properties and the identity:

$$\|\boldsymbol{u}(t+h) - \boldsymbol{u}(t)\|_{1,2}^2 = \int_0^h \left(\frac{d}{ds}\|\boldsymbol{u}(s+h) - \boldsymbol{u}(t)\|_2^2 \right.$$
$$\left. -(\Delta(\boldsymbol{u}(s+h) - \boldsymbol{u}(t)), \frac{\partial}{\partial s}(\boldsymbol{u}(s+h) - \boldsymbol{u}(t)))\right) ds,$$

we deduce

$$\boldsymbol{u} \in C^0((\varepsilon, T]; H^1(\Omega)),$$

and so the result follows under the assumption (i). In case (ii), instead of (5.10), we use the estimate showed in Lemma 5.3 and proceed exactly as in case (i). The lemma is thus proved.

Remark 5.3 The same conclusion of Lemma 5.4 can be obtained under the following alternative assumptions, see Remark 5.2,

(i)$'$ $\nabla \boldsymbol{v} \in L^{r'}(0, T; \boldsymbol{L}^{s'}(\Omega))$, $\frac{2}{r'} + \frac{n}{s'} = 2$, $s' \in (n, \infty]$,

(i)$''$ $\nabla \boldsymbol{v} \in C^0([0, T]; \boldsymbol{L}^{n/2}(\Omega))$.

A similar result, for the case $\Omega \equiv \mathbb{R}^n$, was first obtained by Beirão da Veiga (1995a, 1995b).

Remark 5.4 Once we have established that $\boldsymbol{v}$ has the "minimum" regularity ensured by Lemma 5.4, we shall prove, in the next two lemmas that, in fact, $\boldsymbol{v}$ must be of class C^∞ in $\overline{\Omega} \times (\varepsilon, T]$, if Ω is uniformly of class C^∞. Now, while the assumption (i) coincides with that made for uniqueness when $s > n$, the assumption (ii) for $s = n$ is stronger. Actually, if we compare it with the analogous assumption for uniqueness in a domain with a compact boundary (see Remark 4.5), we see that regularity requires *continuity* in time, while uniqueness only requires *essential* *boundedness*. Though it may be very likely that this latter weaker condition also ensures regularity, no proof is so far available. To add more weight to this conjecture, there is the recent remarkable contribution of Nečas, Růžička and Šverák (1996) who rule out a possible counter example to regularity proposed by Leray (1934b, pp. 225, 245) (see Remark 7.4). This weak solution (whose existence has been disproved by the previous authors) would satisfy *neither* conditions (i), (ii) of Lemma 5.4 but *only* the weaker assumption of being in the class $L^\infty(0,T;\boldsymbol{L}^n(\Omega))$. We shall return on the importance of this condition in Section 7.

In the next lemma we show that a weak solution satisfying either (i) or (ii) of Lemma 5.4, possesses time derivative of arbitrary order. The method of proof is borrowed from Heywood (1980).

Lemma 5.5 *Let Ω and $\boldsymbol{v}$ satisfy the assumption of* Lemma 5.4. *Then,*

$$D_t^\ell \boldsymbol{v} \in L^2(\varepsilon,T;\boldsymbol{W}^{2,2}(\Omega)), \quad \textit{for all } \ell \geq 0. \tag{5.12}$$

Proof. By Lemmas 5.1 and 5.4, it is enough to prove (5.12) for the solution $\boldsymbol{u}$ to (5.2). For $\ell = 0$ the result has already proved in Lemma 5.4. We then construct a solution $\boldsymbol{u}$ to (5.2) satisfying (5.12) for $\ell = 1$. By uniqueness, it will coincide with $\boldsymbol{v}$ which will then verify (5.12) wit $\ell = 1$. With this information on the coefficient, we shall then construct a solution $\boldsymbol{u}$ to (5.2) which satisfies (5.12) with $\ell = 2$. By uniqueness, it will coincide with $\boldsymbol{v}$ and so, by induction, we can prove (5.12) for arbitrary $k \in \mathbb{N}$. Here, for simplicity, we shall prove (5.12) for $\ell = 1$, referring the reader to the paper of Galdi and Maremonti (1988) for a proof in the general case. To construct the solution $\boldsymbol{u}$ we shall use the Faedo-Galerkin method. So, in addition to the estimates on the approximating solution that we have already obtained in the proof of Lemma 5.4, we obtain the following ones. We differentiate (5.6) with respect to time, multiply by dc_{kr}/dt, and sum over r from 1 to k to obtain

$$\tfrac{1}{2}\frac{d}{dt}\|D_t\boldsymbol{u}\|_2^2 + \nu\|\nabla D_t\boldsymbol{u}\|_2^2 = -(D_t\boldsymbol{v}\cdot\nabla\boldsymbol{u}, D_t\boldsymbol{u}), \tag{5.13}$$

where, as before, we have omitted the subscript "k". From the Hölder inequality, the Sobolev theorem, and Lemma 5.2 we find

$$\begin{aligned} |(D_t\boldsymbol{v}\cdot\nabla\boldsymbol{u}, D_t\boldsymbol{u})| \leq & \ \|D_t\boldsymbol{v}\|_2\|\nabla\boldsymbol{u}\|_3\|D_t\boldsymbol{u}\|_6 \leq c\|D_t\boldsymbol{v}\|_2\|\boldsymbol{u}\|_{2,2}\|\nabla D_t\boldsymbol{u}\|_2 \\ \leq & \ \|c\|D_t\boldsymbol{v}\|_2^2\|P\Delta\boldsymbol{u}\|_2^2 + \eta\|\nabla D_t\boldsymbol{u}\|_2^2, \end{aligned} \tag{5.14}$$

where η is a small positive number. Multiplying (5.6) by $\lambda_r c_{kr}$, summing over r and recalling the second part of Lemma 5.2, we find

$$(P\Delta\boldsymbol{u}, D_t\boldsymbol{u}) = \|P\Delta\boldsymbol{u}\|_2^2 + (\boldsymbol{v}\cdot\nabla\boldsymbol{u}, P\Delta\boldsymbol{u}).$$

From Lemma 5.5 and by the Sobolev theorem, we know that $\boldsymbol{v} \in C^0(\varepsilon,T;\boldsymbol{L}^3(\Omega))$ and so, we may use Lemma 5.3 in the preceding relation to obtain

$$\|P\Delta\boldsymbol{u}\|_2 \leq c(\|D_t\boldsymbol{u}\|_2 + \|\nabla\boldsymbol{u}\|_2), \tag{5.15}$$

with a constant c independent of $t \in [\varepsilon, T]$. Replacing this inequality into (5.14), and recalling that $\|\nabla \boldsymbol{u}\|_2 \le C$ with C independent of t, we deduce

$$|(D_t\boldsymbol{v} \cdot \nabla \boldsymbol{u}, D_t\boldsymbol{u})| \le c\|D_t\boldsymbol{v}\|_2^2\|D_t\boldsymbol{u}\|_2^2 + \eta\|\nabla D_t\boldsymbol{u}\|_2^2.$$

With this estimate, equation (5.13) furnishes

$$\frac{d}{dt}\|D_t\boldsymbol{u}\|_2^2 + c_1\|\nabla D_t\boldsymbol{u}\|_2^2 \le c_2\|D_t\boldsymbol{v}\|_2^2\|D_t\boldsymbol{u}\|_2^2.$$

Integrating this inequality from τ to t and then on τ from ε to t, and recalling Lemma 5.4, we obtain that the weak solution $\boldsymbol{u}$ satisfies

$$D_t\boldsymbol{u} \in L^\infty(\varepsilon, T; \boldsymbol{L}^2(\Omega)) \cap L^2(\varepsilon, T; H^1(\Omega)). \tag{5.16}$$

By uniqueness, the same properties hold for $\boldsymbol{v}$. Notice that, by virtue of Lemmas 5.2 and 5.4, and (5.16) it also follows that

$$\boldsymbol{v}, \boldsymbol{u} \in L^\infty(\varepsilon, T; \boldsymbol{W}^{2,2}(\Omega)). \tag{5.17}$$

We now differentiate (5.6) with respect to t, multiply by $\lambda_r dc_{kr}/dt$ and sum over r. We get

$$\tfrac{1}{2}\frac{d}{dt}\|\nabla D_t\boldsymbol{u}\|_2^2 + \nu\|P\Delta D_t\boldsymbol{u}\|_2^2 = (D_t\boldsymbol{v} \cdot \nabla \boldsymbol{u}, P\Delta D_t\boldsymbol{u}) + (\boldsymbol{v} \cdot \nabla D_t\boldsymbol{u}, P\Delta D_t\boldsymbol{u}). \tag{5.18}$$

By the Hölder inequality, the Sobolev theorem, (5.17), and Lemma 5.2 it easily follows that

$$\begin{aligned}
|(D_t\boldsymbol{v} \cdot \nabla \boldsymbol{u}, P\Delta D_t\boldsymbol{u})| &\le \|D_t\boldsymbol{v}\|_3\|\nabla \boldsymbol{u}\|_6\|P\Delta D_t\boldsymbol{u}\|_2 \le c|D_t\boldsymbol{v}\|_{1,2}\|\boldsymbol{u}\|_{2,2}\|P\Delta D_t\boldsymbol{u}\|_2 \\
&\le c_1\|D_t\boldsymbol{v}\|_{1,2}\|P\Delta D_t\boldsymbol{u}\|_2 \le c_2\|D_t\boldsymbol{v}\|_{1,2}^2 + \eta\|P\Delta D_t\boldsymbol{u}\|_2^2 \\
|(\boldsymbol{v} \cdot \nabla D_t\boldsymbol{u}, P\Delta D_t\boldsymbol{u})| &\le \|\boldsymbol{v}\|_\infty\|\nabla D_t\boldsymbol{u}\|_2\|P\Delta D_t\boldsymbol{u}\|_2 \\
&\le c\|D_t\boldsymbol{u}\|_{1,2}\|P\Delta D_t\boldsymbol{u}\|_2 \le c_2\|D_t\boldsymbol{u}\|_{1,2}^2 + \eta\|P\Delta D_t\boldsymbol{u}\|_2^2.
\end{aligned}$$

We now replace these inequalities into (5.18) and integrate with respect to time twice, as we already did many times previously. If we then use (5.16), we arrive at (5.12) with $\ell = 1$. As we noticed, the general case is treated by an elementary induction procedure.

The next lemma provides regularity in space and time for a weak solution, for sufficiently smooth Ω.

Lemma 5.6 *Let $\boldsymbol{v}$ be a weak solution satisfying the assumption of Lemma 5.4. Assume, further, Ω uniformly of class C^m, $m \ge 2$. Then*

$$D_t^\ell \boldsymbol{v} \in L^2(\varepsilon, T; \boldsymbol{W}^{k,2}(\Omega)), \quad \text{for all } \ell \ge 0 \text{ and all } k = 2, \ldots, m. \tag{5.19}$$

Proof. The main idea is to write (5.2) as a Stokes system of the following type

$$\begin{aligned} &-\nu\Delta\boldsymbol{u} = -\frac{\partial \boldsymbol{u}}{\partial t} - \boldsymbol{v}\cdot\nabla\boldsymbol{u} - \nabla\pi \equiv \boldsymbol{F} - \nabla\pi \\ &\operatorname{div}\boldsymbol{u} = 0 \\ &\boldsymbol{u}(y,t) = 0, \quad y\in\partial\Omega,\ t>0. \end{aligned} \tag{5.20}$$

Then, as in the previous lemma, the proof is again based on an inductive argument and the "interplay" between $\boldsymbol{v}$ and $\boldsymbol{u}$. Specifically, knowing that

$$D_t^\ell \boldsymbol{F} \in L^2(\varepsilon, T; \boldsymbol{W}^{k,2}(\Omega)),$$

by Lemma 5.2 we deduce

$$D_t^\ell \boldsymbol{u} \in L^2(\varepsilon, T; \boldsymbol{W}^{k+2,2}(\Omega)),$$

and so, by uniqueness,

$$D_t^\ell \boldsymbol{v} \in L^2(\varepsilon, T; \boldsymbol{W}^{k+2,2}(\Omega)).$$

If we plug this information back into (5.20), we obtain that $\boldsymbol{F}$ has more spatial regularity than that assumed at the outset and, by induction, we obtain the proof. Referring the reader to the paper of Galdi and Maremonti (1988) for full details, we wish here to give a proof of the lemma for the case $m = 3$, $\ell = 1$. By what we said, it is enough to show that

$$(D_t^2\boldsymbol{u} + D_t\boldsymbol{v}\cdot\nabla\boldsymbol{u} + \boldsymbol{v}\cdot\nabla D_t\boldsymbol{u}) \in L^2(\varepsilon, T; \boldsymbol{W}^{1,2}(\Omega)).$$

From Lemma 5.5 we know already [21]

$$D_t^2 \boldsymbol{u} \in L^2(\varepsilon, T; \boldsymbol{W}^{1,2}(\Omega)).$$

By the Hölder inequality and the Sobolev theorem, we have ($\partial_i = \partial/\partial x_i$)

$$\begin{aligned} \|D_t\boldsymbol{v}\cdot\nabla\boldsymbol{u}\|_2 &\le \|D_t\boldsymbol{v}\|_4\|\nabla\boldsymbol{u}\|_4 \le \|D_t\boldsymbol{v}\|_{1,2}\|\boldsymbol{u}\|_{2,2} \\ \|D_t\partial_i\boldsymbol{v}\cdot\nabla\boldsymbol{u}\|_2 &\le \|D_t\boldsymbol{v}\|_{2,2}\|\boldsymbol{u}\|_{2,2} \\ \|D_t\boldsymbol{v}\cdot\partial_i\nabla\boldsymbol{u}\|_2 &\le \|D_t\boldsymbol{v}\|_{1,2}\|\boldsymbol{u}\|_{2,2} \\ \|\boldsymbol{v}\cdot\nabla D_t\boldsymbol{u}\|_2 &\le \|\boldsymbol{v}\|_\infty\|\nabla D_t\boldsymbol{u}\|_2 \le \|\boldsymbol{v}\|_{2,2}\|D_t\boldsymbol{u}\|_{1,2} \\ \|\partial_i\boldsymbol{v}\cdot\nabla D_t\boldsymbol{u}\|_2 &\le \|\partial_i\boldsymbol{v}\|_4\|\nabla D_t\boldsymbol{u}\|_4 \le \|\boldsymbol{v}\|_{2,2}\|D_t\boldsymbol{u}\|_{2,2} \\ \|\boldsymbol{v}\cdot\partial_i\nabla D_t\boldsymbol{u}\|_2 &\le \|\boldsymbol{v}\|_\infty\|\partial_i\nabla D_t\boldsymbol{u}\|_2 \le \|\boldsymbol{v}\|_{2,2}\|D_t\boldsymbol{u}\|_{2,2}, \end{aligned}$$

and the result follows from these inequalities and Lemma 5.5.

From the preceding lemma and the Sobolev theorem we at once deduce the following result

Theorem 5.2 *Let $\boldsymbol{v}$ be a weak solution in Ω_T, corresponding to $\boldsymbol{f} \equiv 0$ and to $\boldsymbol{v}_0 \in H(\Omega)$. Assume that $\boldsymbol{v}$ satisfies at least one of the following two conditions:*

---

[21] Recall that, by uniqueness, $\boldsymbol{v} = \boldsymbol{u}$.

(i) $\boldsymbol{v} \in L^r(0,T;\boldsymbol{L}^s(\Omega))$, *for some* r, s *such that* $\dfrac{n}{s} + \dfrac{2}{r} = 1$, $s \in (n, \infty]$;
(ii) $\boldsymbol{v} \in C^0([0,T];\boldsymbol{L}^n(\Omega))$.

Then, if Ω *is uniformly of class* C^∞, *we have*

$$\boldsymbol{v} \in \boldsymbol{C}^\infty(\overline{\Omega} \times (0,T]).$$

Remark 5.5 Intermediate regularity results, with Ω only of class C^m, $m \geq 2$, can be directly obtained from Lemma 5.6, and the Sobolev theorem. We leave it to the reader as an exercise.

Remark 5.6 The same conclusion of Theorem 5.2 can be obtained under the following alternative assumptions, see Remarks 5.2, 5.3

(i)′ $\nabla \boldsymbol{v} \in L^{r'}(0,T;\boldsymbol{L}^{s'}(\Omega))$, $\frac{2}{r'} + \frac{n}{s'} = 2$, $s' \in (n/2, \infty]$,
(i)″ $\nabla \boldsymbol{v} \in C^0([0,T);\boldsymbol{L}^{n/2}(\Omega))$.

Remark 5.7 Every weak solution in dimension 2 is L^2 strongly continuous, and, thus, by Theorem 5.2, it is regular in space and time. Regularity of weak solutions in dimension two was first obtained by Leray (1934a), Ladyzhenskaya (1958)

Remark 5.8 Theorem 5.2(i), for $\Omega = \mathbb{R}^3$ was proved for the first time by Leray (1934b, pp. 224–227), while for $\Omega = \mathbb{R}^n$, $n \geq 2$, and $s < \infty$ it is due to Fabes, Jones and Riviere (1972); see also Fabes, Lewis and Riviere (1977a, 1977b). Sohr (1983) proved Theorem 5.2(i) with $s < \infty$, for domains with a bounded boundary. An attempt to prove Sohr's result was already made by Kaniel and Shinbrot (1967). However, their proof contains an oversight which leads to the Corollary at p. 323 of their paper, where it is stated that, if Ω is of class C^∞, then any weak solution corresponding to initial data in $C^\infty(\overline{\Omega})$ and satisfying condition (i) is in $C^\infty(\overline{\Omega} \times [0,T])$. This result can not hold as stated, due to the fact that if a solution is regular *up to the time* $t = 0$ *included*, then certain compatibility conditions have to be met, see Solonnikov (1964, p. 97 of the english translation), Heywood (1980, Remark at p. 677). The same oversight is contained in the book of Temam (1977, pp. 303, 307). The question of "how much smooth" a solution can be up to the time $t = 0$, *without* compatibility conditions is studied by Rautmann (1983), von Wahl (1983) and Temam (1980). That condition (ii) implies regularity was first discovered by von Wahl (1986), in the case of a bounded domain. This latter result was extended to domains with a bounded boundary by Giga (1986). The case $n = 3$, $s' = 2$, $r' = 4$ of Remark 5.6 for $\Omega = \mathbb{R}^3$, is due to Leray (1934b, p. 227); see also Section 6.

Remark 5.9 Regularity results involving assumptions on the pressure, rather than the velocity, have been given by Kaniel (1969) and, more recently, by Berselli (1999).

6. More regular solutions and the "théorème de structure"

The aim of this section is two-fold. On one hand, we would like to show that regular solutions do exist in three dimension if either we restrict ourselves to a "short" time interval, or if we choose initial data "small" compared to viscosity. [22] On the other hand, we wish to give more information about the possible formation of singularities for a weak solution, along the lines of the so-called "théorème de structure", Leray (1934b, pp. 244–245).

We have the following result due to Heywood (1980) [23].

Theorem 6.1 *Let $\Omega \subset \mathbb{R}^3$ be uniformly of class C^2. Then, for any $\boldsymbol{v}_0 \in H^1(\Omega)$, there exists $T > 0$ and at least one weak solution in Ω_T such that*

$$\boldsymbol{v} \in L^\infty(0,T; H^1(\Omega)) \cap L^2(0,T; \boldsymbol{W}^{2,2}(\Omega)).$$

The number T is bounded from below by a constant depending only on $\|\nabla \boldsymbol{v}_0\|_2$, ν and the C^2-regularity of Ω. In the case when Ω is bounded or $\Omega = \mathbb{R}^n$ we have

$$T \geq \nu^3 C / \|\nabla \boldsymbol{v}_0\|_2^4,$$

where C depends only on Ω. Moreover, there is a decreasing function $G = G(\lambda)$, $\lambda > 0$, such that if

$$\|\boldsymbol{v}_0\|_2 \leq G(\|\nabla \boldsymbol{v}_0\|_2),$$

T can be taken as an arbitrary positive number. In the case when Ω is bounded or $\Omega = \mathbb{R}^n$ we have $G = C\nu^2/\|\nabla \boldsymbol{v}_0\|_2$, with C depending only on Ω.

Proof. We shall show the result for the case Ω bounded, referring the reader to the paper of J. Heywood for the general case. To show the existence of such a solution, we then use the Faedo-Galerkin method of Theorem 3.1, with the basis of the eigenfunctions of the Stokes problem, see Lemma 5.3. In addition to the estimate (3.4) with $\boldsymbol{f} \equiv 0$, we obtain the following one. We multiply (3.2) (with $\boldsymbol{f} \equiv 0$) by $\lambda_r c_{kr}$ and sum over r, to get (as usual, we omit the subscript "k")

$$\frac{1}{2}\frac{d}{dt}\|\nabla \boldsymbol{v}(t)\|_2^2 + \nu \|P\Delta \boldsymbol{v}\|_2^2 = (\boldsymbol{v} \cdot \nabla \boldsymbol{v}, P\Delta \boldsymbol{v}). \tag{6.1}$$

Using the Hölder inequality, the Sobolev theorem, and Lemma 5.2 we have the following two different ways of increasing the term N (say) on the right-hand side of this equation, namely,

a) $N \leq \|\boldsymbol{v}\|_6 \|\nabla \boldsymbol{v}\|_3 \|P\Delta \boldsymbol{v}\|_2 \leq c\|\nabla \boldsymbol{v}\|_2^{3/2} \|P\Delta \boldsymbol{v}\|_2^{3/2} \leq c\nu^{-3}\|\nabla \boldsymbol{v}\|_2^6 + \frac{1}{2}\nu \|P\Delta \boldsymbol{v}\|_2^2$

b) $N \leq \|\boldsymbol{v}\|_3 \|\nabla \boldsymbol{v}\|_6 \|P\Delta \boldsymbol{v}\|_2 \leq c\|\boldsymbol{v}\|_2^{1/2} \|\nabla \boldsymbol{v}\|_2^{1/2} \|P\Delta \boldsymbol{v}\|_2^2$.

Replacing a) in (6.1) and setting $y(t) = \|\nabla \boldsymbol{v}(t)\|_2^2$ we find

$$\frac{dy}{dt} \leq \nu^{-3} c y^3, \tag{6.2}$$

[22] We assume hereafter, for simplicity, that $\boldsymbol{f} \equiv 0$. We also notice that existence of regular and global solutions in dimension 2 has been established in Theorem 5.2, see Remark 5.7.

[23] Actually, Heywood requires more regularity on the boundary than that requested in Theorem 6.1.

which, by Gronwall's lemma, and (6.1) and a), in turns gives

$$\|\nabla \boldsymbol{v}(t)\|_2 + \int_0^t \|P\Delta \boldsymbol{v}(s)\|_2^2 ds \le M, \quad \text{for all } t \in [0, T) \tag{6.3}$$

where $[0, T)$ is the maximal interval of existence of the differential inequality (6.2). By classical comparison theorems for differential inequalities, we have $T \ge \nu^3/2c\|\nabla \boldsymbol{v}_0\|_2^4$. In case b) we find

$$\frac{d}{dt}\|\nabla \boldsymbol{v}(t)\|_2^2 + (\nu - c\|\boldsymbol{v}\|_2^{1/2}\|\nabla \boldsymbol{v}\|_2^{1/2})\|P\Delta \boldsymbol{v}\|_2^2 \le 0,$$

which, once integrated, furnishes (6.3) for arbitrary $T > 0$, provided

$$\nu > c\|\boldsymbol{v}_0\|_2^{1/2}\|\nabla \boldsymbol{v}_0\|_2^{1/2}.$$

Using the estimate (6.3) along the approximating solutions, together with the procedure employed in Theorem 3.1, we then show the result.

From this theorem, Theorem 5.2, and $(2.12)_2$ we then obtain the following result.

Theorem 6.2 *Let $\Omega \subset \mathbb{R}^3$ be uniformly of class C^∞.* [24] *Then, for any $\boldsymbol{v}_0 \in H^1(\Omega)$ there exist $T > 0$ and a unique solution to (0.1)–(0.3) with $\boldsymbol{f} \equiv 0$, which assumes the data $\boldsymbol{v}_0$ and which is of class $\boldsymbol{C}^\infty(\overline{\Omega} \times (0, T))$. Moreover, there exists a positive constant $C(\Omega)$ such that, if*

$$\|\boldsymbol{v}_0\|_2 \le G(\|\nabla \boldsymbol{v}_0\|_2),$$

with G defined in Theorem 6.1, *we can take T arbitrarily large.*

We shall now derive some other consequences of Theorem 6.1. Following Leray, we are able to specify better the set of times where a weak solution can be irregular. This can be done for all those Ω for which a *strong energy inequality* holds (see (4.1)). Specifically, we have the following result of "partial regularity".

Theorem 6.3 (Théorème de Structure) *Let Ω satisfy the assumption of* Theorem 6.2. [25] *Assume $\boldsymbol{v}$ is a weak solution in Ω_T, for all $T > 0$, corresponding to $\boldsymbol{f} \equiv 0$ and satisfying the strong energy inequality* (4.1)*. Then, there exists a union $\mathcal{T}$ of disjoint open time intervals such that:*

(i) *The Lebesgue measure of $(0, \infty) - \mathcal{T}$ is zero;*
(ii) *$\boldsymbol{v}$ is of class C^∞ in $\overline{\Omega} \times \mathcal{T}$,*
(iii) *There exists $T^* \in (0, \infty)$* [26] *such that $\mathcal{T} \supset (T^*, \infty)$;*
(iv) *If $\boldsymbol{v}_0 \in H^1(\Omega)$ then $\mathcal{T} \supset (0, T_1)$ for some $T_1 > 0$.*

---

[24] See Remark 5.5.

[25] See Remark 5.5.

[26] T^* can be estimated from above by a quantity depending only on $\|\boldsymbol{v}_0\|_2$ and Ω, see Heywood (1980, Theorem 8 (ii)). See also Remark 6.3.

Proof. Since

$$\|\boldsymbol{v}(t)\|_2^2 + \int_0^\infty \|\nabla \boldsymbol{v}(\tau)\|_2^2 d\tau \le \|\boldsymbol{v}_0\|_2^2 \text{ for all } T > 0,$$

and since $\boldsymbol{v}$ verifies (4.1) for almost all $s > 0$, we can find T^* with the following properties:

a) $\|\boldsymbol{v}(T^*)\|_2 \le G(\|\nabla \boldsymbol{v}(T^*)\|_2)$,
b) The strong energy inequality (4.1) holds with $s = T^*$,

where G is the function introduced in Theorem 6.1. Let us denote by $\widetilde{\boldsymbol{v}}$ the solution of Theorem 6.1 corresponding to the data $\boldsymbol{v}(T^*)$. By a), $\widetilde{\boldsymbol{v}}$ exists for all times $t \ge T^*$ and, by Theorem 6.2, it is of class C^∞ in $\Omega \times (T^*, \infty)$. By the uniqueness Theorem 4.2 we must have $\boldsymbol{v} = \widetilde{\boldsymbol{v}}$ in $\Omega \times (T^*, \infty)$, and part (iii) is proved. Next, denote by I the subset of $(0, T^*)$ where the following conditions are met:

a) $\|\boldsymbol{v}(t)\|_{1,2} < \infty$, for $t \in I$,
b) The strong energy inequality (4.1) holds with $s \in I$.

Clearly, $(0, T^*) - I$ is of zero Lebesgue measure. Moreover, for every $t_0 \in I$ we can construct in the time interval $(t_0, t_0 + T(t_0))$ a solution $\widetilde{\boldsymbol{v}}$ assuming at t_0 the initial data $\boldsymbol{v}(t_0)$ $(\in H^1(\Omega))$. From Theorems 6.1 and 4.1, we know that $\widetilde{\boldsymbol{v}}$ is of class C^∞ in $\Omega \times (t_0, t_0 + T(t_0))$ and that it coincides with $\boldsymbol{v}$, since this latter satisfies the energy inequality with $s = t_0$. It is obvious that the set $\bigcup_{t_0 \in I} (t_0, t_0 + T(t_0)) - I$ has zero Lebesgue measure. Finally, if $\boldsymbol{v}_0 \in H^1(\Omega)$, by Theorems 6.1 and 6.2, there exists $T_1 > 0$ such that $\boldsymbol{v}$ is of class C^∞ in $\Omega \times (0, T_1)$. The theorem thus follows with $\mathcal{T} \equiv \bigcup_{t_0 \in I} (t_0, t_0 + T(t_0)) \cup (T^*, \infty)$.

Remark 6.1 It is likely that Theorem 6.3 holds for any (sufficiently smooth) domain. However, no proof is so far available, since one can prove the strong energy inequality only for certain domains (see Section 4). On the other hand, Heywood (1988) has shown that for any Ω, uniformly of class C^2, and any $\boldsymbol{v}_0 \in H(\Omega)$ there exists at least one corresponding weak solution $\boldsymbol{v}$ satisfying the following condition: There exists an open set $R \subseteq [0, \infty)$ such that

a) $[0, \infty) - R$ has zero Lebesgue measure;
b) For every compact interval $[\alpha, \beta] \subset R$ there holds

$$\sup_{t \in [\alpha, \beta]} \|\boldsymbol{v}(t)\|_{1,2}^2 + \int_\alpha^\beta \left(\|\boldsymbol{v}(\tau)\|_{2,2}^2 + \|D_\tau \boldsymbol{v}(\tau)\|_2^2\right) ds < \infty.$$

Since it is not known if weak solutions in dimension 3 are unique in their class, we can not conclude from this result that *any* weak solution satisfies a) and b). Notice that, by Theorem 5.2, every weak solution satisfying b) is of class $\boldsymbol{C}^\infty(\overline{\Omega} \times (\alpha, \beta])$, if Ω is uniformly of class C^∞.

Our next objective is to investigate when and in which way a weak solution $\boldsymbol{v}$ can become irregular, and to give a more precise estimate of the set of the possible irregular times. From Theorem 6.3, we know that this set is the complement to $(0,\infty)$ of a union $\mathcal{T}$ of intervals, and that, under suitable assumptions on the smoothness of Ω, $\boldsymbol{v} \in C^\infty(\overline{\Omega} \times \mathcal{T})$.

For simplicity, *in the remaining part of this section, we shall assume that the domain $\Omega \subset \mathbb{R}^3$ is either bounded and uniformly of class C^∞, or $\Omega = \mathbb{R}^3$, and that* $\boldsymbol{f} \equiv 0$.

Following Leray (1934b, p. 224) we give the following

Definition 6.1 We shall say that a solution $\boldsymbol{v}$, *becomes irregular at the time* t_1 if and only if

a) t_1 is finite;
b) $\boldsymbol{v} \in C^\infty(\overline{\Omega} \times (t_0, t_1))$, for some $t_0 < t_1$;
c) It is not possible to extend $\boldsymbol{v}$ to a regular solution in (t_0, t') with $t' > t_1$.

The number t_1 will be called *epoch of irregularity* ("époque de irrégularité", Leray, *loc. cit.*).

We shall denote by $\mathcal{I} = \mathcal{I}(\boldsymbol{v})$ the set of all possible epochs of irregularity. As we know from Theorem 6.3, the one-dimensional Lebesgue measure of $\mathcal{I}$ is zero.

We have the following result which is essentially due to Leray (1934b, pp. 245–246) and Scheffer (1976a).

Theorem 6.4 *Let $\boldsymbol{v}$ be a weak solution in Ω_T, for all $T > 0$, corresponding to the initial data $\boldsymbol{v}_0 \in H(\Omega)$, and satisfying the strong energy inequality (4.1). Let t_1 be an epoch of irregularity for $\boldsymbol{v}$. Then, the following properties hold:*

(i) $\|\nabla \boldsymbol{v}(t)\|_2$ *diverges as $t \to t_1^-$ in such a way that*

$$\|\nabla \boldsymbol{v}(t)\|_2 \geq \frac{C\nu^{3/4}}{(t_1 - t)^{1/4}}, \quad t < t_1,$$

with $C = C(\Omega) > 0$;

(ii) *There exists a constant $C > 0$, depending only on Ω, such that*

$$t_1 \leq C\nu^{-5}\|\boldsymbol{v}_0\|_2^4.$$

(iii) *The one-half dimensional Hausdorff measure of $\mathcal{I}(\boldsymbol{v})$ is equal to zero.* [27]

---

[27] Let S be a subset of $\mathbb{R}^n$. The m-dimensional (spherical) Hausdorff measure $\mathcal{H}^m$ of S is given by

$$\mathcal{H}^m(S) = \lim_{\delta \to 0} \mathcal{H}^m_\delta(S),$$

where

$$\mathcal{H}^m_\delta(S) = \inf \sum_i (2^{-1} \operatorname{diam} B_i)^m,$$

the infimum being taken over all at most countable coverings $\{B_i\}$ of S constituted by closed balls B_i with

$$\operatorname{diam} B_i < \delta, \quad \text{for all } i,$$

see, *e.g.*, Simon (1983).

Proof. Let t_1 be an epoch of irregularity. Then,

$$\lim_{t \to t_1^-} \|\nabla \boldsymbol{v}(t)\|_2 = \infty. \tag{6.4}$$

Actually, assuming that (6.4) does not hold, there would exist a sequence $\{\tau_k\}$ tending to t_1, $\tau_k < t_1$ for all $k \in \mathbb{N}$, and a number $M > 0$ such that

$$\|\nabla \boldsymbol{v}(\tau_k)\|_2 \le M.$$

Since $\boldsymbol{v}(\tau_k) \in H^1(\Omega)$, by Theorem 6.1 we may construct a solution $\overline{\boldsymbol{v}}$ with initial data $\boldsymbol{v}(\tau_k)$, in a time interval $(\tau_k, \tau_k + T_1)$ where

$$T_1 \ge A/\|\nabla \boldsymbol{v}(\tau_k)\|_2^4 \ge AM \equiv T_0,$$

and A depends only on Ω and ν. The solution $\overline{\boldsymbol{v}}$ belongs to $L^\infty(\tau_k, \tau_k + T_0; H^1(\Omega))$ and so, by the Sobolev theorem, it satisfies Theorem 5.2(i) with $s = 6$ and $r = 4$ (for instance). Therefore, $\overline{\boldsymbol{v}} \in C^\infty(\overline{\Omega} \times (\tau_k, \tau_k + T_0])$. Moreover, by the uniqueness Theorem 4.2, $\boldsymbol{v} = \overline{\boldsymbol{v}}$ in $[\tau_k, \tau_k + T_0]$. We may now select τ_k such that $\tau_k + T_0 > t_1$, contradicting the assumption that t_1 is an epoch of irregularity, and (6.4) follows. We next operate as in the proof of Theorem 6.1, to show that $y(t) \equiv \|\nabla \boldsymbol{v}(t)\|_2^2$ satisfies (6.2) in the time interval (t_0, t_1). Integrating (6.2) we then find

$$\frac{1}{\|\nabla \boldsymbol{v}(t)\|_2^4} - \frac{1}{\|\nabla \boldsymbol{v}(\tau)\|_2^4} \le \nu^{-3} c(\tau - t), \quad t_0 < t < \tau < t_1.$$

Letting $\tau \to t_1$ and recalling (6.4), we prove (i). Property (ii) is simply obtained, by integrating the inequality in (i) from 0 to t_1, and then using the energy inequality (EI) in Theorem 3.1. To show (iii) we observe that the set $\mathcal{T}$ introduced in Theorem 6.3, can be decomposed as follows

$$\mathcal{T} = \left(\bigcup_{i \in I} (\tau_i, s_i) \right) \cup (T^*, \infty), \quad \tau_i < s_i,$$

where $T^* < \infty$, each s_i is an epoch of irregularity, and

$$\begin{aligned} &(\tau_i, s_i) \subset [0, T^*], \quad \text{for all } i \in I; \\ &(\tau_i, s_i) \cap (\tau_j, s_j) = \emptyset, \quad i \neq j. \end{aligned} \tag{6.5}$$

From (i) and the energy inequality (EI) we at once deduce that

$$\sum_{i \in I} (\tau_i - s_i)^{1/2} \le C \sum_{i \in I} \int_{\tau_i}^{s_i} \|\nabla \boldsymbol{v}(\tau)\|_2^2 dt \le C_1 \|\boldsymbol{v}_0\|_2^2.$$

Thus, for every $\delta > 0$ we can find a finite part I_δ of I such that

$$\sum_{i \notin I_\delta} (\tau_i - s_i) < \delta, \qquad \sum_{i \notin I_\delta} (\tau_i - s_i)^{1/2} < \delta. \tag{6.6}$$

By $(6.5)_1$, $\cup_{i \in I} (\tau_i, s_i) \subset [0, T^*]$ and so the set

$$[0, T^*] - \cup_{i \in I_\delta} (\tau_i, s_i)$$

consists of a finite number of disjoint closed intervals B_j, $j = 1, \dots, N$. Clearly,

$$\bigcup_{j=1}^{N} B_j \supset \mathcal{I}(\boldsymbol{v}). \tag{6.7}$$

By $(6.5)_2$, we have that each interval (τ_i, s_i), $i \notin I_\delta$, is included in one and only one B_j. Denote by I_j the set of all indeces i satisfying $B_j \supset (\tau_i, s_i)$. We thus have

$$\begin{aligned} I &= I_\delta \cup \left(\bigcup_{j=1}^{N} I_j \right) \\ B_j &= \left(\bigcup_{i \in I_j} (\tau_i, s_i) \right) \cup (B_j \cap \mathcal{I}(\boldsymbol{v})) . \end{aligned} \tag{6.8}$$

By Theorem 6.3, the set $\mathcal{I}$ has zero Lebesgue measure and so, from $(6.8)_2$ we have

$$\operatorname{diam} B_j = \sum_{i \in I_j} (\tau_i - s_i).$$

Thus, by (6.6),

$$\operatorname{diam} B_j \leq \sum_{i \notin I_\delta} (\tau_i - s_i) < \delta \tag{6.9}$$

and, again by (6.6) and $(6.8)_1$,

$$\sum_{j=1}^{N} (\operatorname{diam} B_j)^{1/2} \leq \sum_{j=1}^{N} \left(\sum_{i \in I_j} (\tau_i - s_i) \right)^{1/2} \leq \sum_{i \notin I_\delta} (\tau_i - s_i)^{1/2} < \delta. \tag{6.10}$$

Therefore, property (iii) follows from (6.7), (6.9) and (6.10).

Remark 6.2 From Theorem 6.4(i) it follows that a sufficient condition for the absence of epochs of irregularity is that

$$\nabla \boldsymbol{v} \in L^4(0, T; \boldsymbol{L}^2(\Omega)),$$

a fact discovered for the first time by Leray (1934b, p. 227) when $\Omega = \mathbb{R}^3$. As we already noticed, this is a particular case of the more general conditions furnished in Remark 5.6.

Remark 6.3 From Theorem 6.4(ii) it follows that the number T^* introduced in Theorem 6.3 is bounded above by $\nu^{-5} C \|\boldsymbol{v}_0\|_2^4$, with $C = C(\Omega)$. Moreover, assume $\boldsymbol{v}_0 \in H^1(\Omega)$. By Theorem 6.1 we then know that any epoch of irregularity t_1 satisfies the following estimate

$$t_1 \geq \nu^3 C / \|\nabla \boldsymbol{v}_0\|_2^4,$$

with $C = C(\Omega)$. Thus, from this inequality and Theorem 6.4(ii), it follows that there exists $A = A(\Omega) > 0$ such that if

$$\|\boldsymbol{v}_0\|_2 \|\nabla \boldsymbol{v}_0\|_2 \le A\nu^2,$$

the set $\mathcal{I}(\boldsymbol{v})$ is empty, and we reobtain the second part of Theorem 6.2.

Remark 6.4 There is a wide range of results concerning "partial regularity" of "suitable" weak solutions, that we will not treat here. In this regard, we refer the reader to the work of Scheffer (1976a, 1976b, 1977, 1978, 1980, 1982, 1985), Foiaş and Temam (1979), Caffarelli, Kohn and Nirenberg (1982), Maremonti (1987), Wu (1991), Lin (1998), and Ladyzhenskaya and Seregin (1999).

7. Existence in the class $L^r(0,T;\boldsymbol{L}^s(\Omega))$, $2/r + n/s = 1$, and further regularity properties

Theorem 5.1 has revealed that the functional class

$$L^{s,r}(\Omega_T) \equiv L^r(0,T;\boldsymbol{L}^s(\Omega)), \quad 2/r + n/s = 1, \tag{7.1}$$

plays a crucial role in the study of regularity of weak solutions. However, as we have seen in Remark 4.6, unless $n = 2$, it is not known whether a *generic* weak solution belongs to this class, for a suitable choice of r and s. It seems, therefore, of the utmost importance to investigate under which assumptions on the initial data $\boldsymbol{v}_0$ [28] one can *construct* a weak solution which, in addition, belongs to such a class. For example, in Theorem 6.1 we have shown that this happens if $\boldsymbol{v}_0 \in H^1(\Omega)$. Our main objective in this section is to show existence of weak solutions in the class (7.1), [29] under mild assumption on $\boldsymbol{v}_0$, namely, that it belongs to Lebesgue spaces $\boldsymbol{L}^\sigma(\Omega)$. Though obvious, it is worth noticing that, in order to show regularity of weak solutions, it would *not* really matter if existence in the class (7.1) is proved for a *short* time T (say) only, on condition that one could take σ and T suitably. For instance, regularity would trivially follow if we could take $\sigma = 2$ and T a decreasing function of $\|\boldsymbol{v}_0\|_2$. However, the existence theory known so far, with data in L^q, requires $\sigma \ge n$. [30, 31]

In order to avoid technical difficulties, in what follows we shall assume that $\Omega = \mathbb{R}^n$, referring to Giga (1986, Theorem 4) for the more general case when Ω has a (non-empty) compact boundary. [32] The results we shall prove will be then an immediate consequence of suitable estimates for solutions to the heat equation and of the classical successive approximation method applied to the *linearized Stokes*

[28] Throughout this section, for the sake of simplicity, we shall assume $\mathbf{f} \equiv 0$.

[29] See Remark 4.7.

[30] This is another way of obtaining regularity of weak solution for $n = 2$.

[31] Weak solutions with data in L^σ, $2 < \sigma < n$, have been constructed by Calderon (1990a). For existence of strong solutions with data in suitable Besov spaces, larger than L^n, see Cannone (1997), Kozono (1998), Kozono and Yamazaki (1998), Amann (1999), and the extensive literature cited therein.

[32] For the Cauchy problem, see also T. Kato (1984).

problem (see (7.3) below). In fact, using a decomposition lemma of the Helmholtz-Weyl type, we shall see that the assumption $\Omega = \mathbb{R}^n$ allows us to treat this latter problem as a (vector) heat equation.

We have the following.

Lemma 7.1 *Let $\boldsymbol{F} = \{F_{ij}\}$ be a second order tensor field with*

$$F_{ij} \in L^r(\mathbb{R}^n),\ \partial_i F_{ij} \in L^s(\mathbb{R}^n),\ \ j = 1, \dots, n,\quad 1 < r, s < \infty.$$

Then, there exists a second order tensor field $\boldsymbol{G} = \{G_{ij}\}$ with $\partial_j \partial_i G_{ij} = 0$, [33] *and a scalar field p such that, for all $i, j = 1, \dots, n$,*

$$G_{ij} \in L^r(\mathbb{R}^n),\quad \partial_i G_{ij} \in L^s(\mathbb{R}^n),\quad p \in L^r(\mathbb{R}^n),\quad \partial_i p \in L^s(\mathbb{R}^n)$$

$$\partial_i F_{ij} = \partial_i G_{ij} + \partial_j p$$

$$\|G_{ij}\|_r \leq c(n,r)\|F_{ij}\|_r$$

$$\|\partial_i G_{ij}\|_s \leq c(n,s)\|\partial_i F_{ij}\|_s.$$

Proof. Without loss of generality, we may assume that F_{ij} are smooth functions with compact support in $\mathbb{R}^n$, see Galdi (1994, Lemma VII.4.3). We set

$$\begin{aligned} p(x) &= \int_{\mathbb{R}^n} \mathcal{E}(x-y)\partial_i \partial_j F_{ij}(y)dy \\ G_{ij} &= \delta_{ij} p - F_{ij}, \end{aligned}$$

where $\mathcal{E}(\xi)$ is the fundamental solution of Laplace's equation. It is clear that

$$\begin{aligned} \partial_i F_{ij} &= \partial_i G_{ij} + \partial_j p \\ \partial_j \partial_i G_{ij} &= 0. \end{aligned}$$

Moreover, from the Calderon-Zygmund theorem on singular integrals we find that

$$\begin{aligned} \|\nabla p\|_s &\leq c(n,s)\|\partial_i F_{ij}\|_s \\ \|p\|_r &\leq c(n,r)\|F_{ij}\|_r, \end{aligned} \tag{7.2}$$

and the lemma is proved.

Our next objective is to prove some existence results for weak solutions $\boldsymbol{u}$ to the following Cauchy problem for the linearized *Stokes system*:

$$\left.\begin{aligned} &\frac{\partial \boldsymbol{u}}{\partial t} = \nu \Delta \boldsymbol{u} + \nabla p + \operatorname{div} \boldsymbol{F} \\ &\operatorname{div} \boldsymbol{u} = 0 \end{aligned}\right\} \text{ in } \mathbb{R}^n_T \\ \boldsymbol{u}(x,0) = \boldsymbol{u}_0(x) \tag{7.3}$$

[33] We shall use the Einstein summation convention over repeated indeces. This condition on G_{ij} has to be understood in the distributional sense.

where $\boldsymbol{F}$ is a given second-order tensor field, and $\{\operatorname{div} \boldsymbol{F}\}_j = \partial_i F_{ij}$. As usual, we shall say that $\boldsymbol{u}$ is a weak solution to (7.3) if $\boldsymbol{u} \in V_T$ (see Definition 2.1), and it satisfies the following relation

$$\int_0^\infty \left\{ \left(\boldsymbol{u}, \frac{\partial \boldsymbol{\varphi}}{\partial t} \right) - \nu(\nabla \boldsymbol{u}, \nabla \boldsymbol{\varphi}) \right\} dt = \int_0^\infty (\boldsymbol{F}, \nabla \boldsymbol{\varphi}) dt - (\boldsymbol{u}_0, \boldsymbol{\varphi}(0)), \quad \text{for all } \boldsymbol{\varphi} \in \mathcal{D}_T. \tag{7.4}$$

Before proving our results, however, we wish to recall some well-known properties concerning the heat equation and classical inequalities. Denote by $W(x,t)$, $(x,t) \in \mathbb{R}^n_T$, the *Weierstrass function*, that is,

$$W(x,t) = \frac{1}{(4\pi\nu t)^{n/2}} \exp\left\{ \frac{x^2}{4\nu t} \right\}.$$

By a direct computation, we show that

$$\begin{aligned} |W(x,t)| &\leq \frac{c}{(x^2+t)^{n/2}} \\ |\partial_k W(x,t)| &\leq \frac{c}{(x^2+t)^{(n+1)/2}}, \end{aligned} \tag{7.5}$$

where $c = c(n, \nu)$. For $u_0 \in L^\sigma(\mathbb{R}^n)$ and $f \in L^{s,r}(\mathbb{R}^n_T)$ the convolutions

$$U(x,t) = \int_{\mathbb{R}^n} W(x-y,t) u_0(y) dy \equiv W * u_0$$

and

$$U_1(x,t) = \int_0^t \left(\int_{\mathbb{R}^n} W(x-y, t-\tau) f(y,\tau) dy \right) d\tau$$

are called the *volume potential* and *volume heat potential*, respectively. It is well known, see, *e.g.* Ladyzhenskaya, Ural´ceva and Solonnikov (1968, Chapter IV, §1), that the volume potentials solve the following Cauchy problems for the *heat equation*

$$\begin{aligned} \frac{\partial U}{\partial t} &= \nu \Delta U \quad \text{in } \mathbb{R}^n \times \{t > 0\}, \\ &\lim_{t \to 0} \|U(t) - u_0\|_2 = 0 \end{aligned}$$

and

$$\begin{aligned} \frac{\partial U_1}{\partial t} &= \nu \Delta U_1 \quad \text{in } \mathbb{R}^n \times \{t > 0\}, \\ &\lim_{t \to 0} U_1(x,t) = 0, \quad x \in \mathbb{R}^n. \end{aligned}$$

For $f \in L^{q_1}(\mathbb{R}^n)$, $1 < q_1 < \infty$, $n \geq 1$, we set

$$Tf(x) = \int_{\mathbb{R}^n} \frac{f(y)}{|x-y|^{n-\lambda}} dy, \quad 0 < \lambda < n.$$

Then, the following *Hardy-Littlewood-Sobolev inequality* holds (see, *e.g.* Stein (1970))

$$\|Tf\|_p \leq c\|f\|_{q_1}, \quad \frac{1}{q_1} = \frac{1}{p} + \frac{\lambda}{n}. \tag{7.6}$$

Finally, if $t^\alpha u \in BC([0,T); L^q(\mathbb{R}^n))$, and $w \in L^{r,s}(\mathbb{R}^n_T)$, we set

$$\langle\langle u \rangle\rangle_{q,\alpha,T} = \sup_{t\in[0,T)} t^\alpha \|u(t)\|_q$$

$$\|w\|_{r,s,T} \equiv \|w\|_{L^{r,s}(\mathbb{R}^n_T)}.$$

We are now in a position to prove the following result.

Lemma 7.2 *Let $n < s < \infty$, $1 < q_1 < \infty$, $1/s_1 = 1/s + 1/2$, $1/q_2 = 1 - n/2s$, $\alpha = (1 - n/s)/2$. Assume*

$$\begin{aligned} &\boldsymbol{F} \in L^{s/2,q_1}(\mathbb{R}^n_T) \\ &\operatorname{div} \boldsymbol{F} \in L^{s_1,q_2}(\mathbb{R}^n_T) \\ &t^{2\alpha}\boldsymbol{F} \in BC([0,T); \boldsymbol{L}^{s/2}(\mathbb{R}^n)) \\ &t^{\alpha}\boldsymbol{F} \in BC([0,T); \boldsymbol{L}^{s_1}(\mathbb{R}^n)). \end{aligned}$$

Assume also that $\boldsymbol{u}_0 \in H(\mathbb{R}^n)$ and that

$$\begin{aligned} &W * \boldsymbol{u}_0 \in L^{s,r}(\mathbb{R}^n_T) \\ &t^\alpha W * \boldsymbol{u}_0 \in BC([0,T); \boldsymbol{L}^s(\mathbb{R}^n)) \end{aligned}$$

where W is the Weierstrass function and

$$\frac{1}{r} = \frac{1}{q_1} - \frac{1}{2}\left(1 - \frac{n}{s}\right).$$

Then, there exists a unique weak solution $\boldsymbol{u}$ to (7.3) such that

$$\begin{aligned} &\boldsymbol{u} \in L^{s,r}(\mathbb{R}^n_T) \\ &t^\alpha \boldsymbol{u} \in BC([0,T); \boldsymbol{L}^s(\mathbb{R}^n)). \end{aligned}$$

This solution satisfies the following estimates

$$\begin{aligned} \|\boldsymbol{u}\|_{s,r,T} &\leq \|W * \boldsymbol{u}_0\|_{s,r,T} + C\|\boldsymbol{F}\|_{s/2,q_1,T} \\ \langle\langle \boldsymbol{u} \rangle\rangle_{s,\alpha,T} &\leq \langle\langle W * \boldsymbol{u}_0 \rangle\rangle_{s,\alpha,T} + C\langle\langle \boldsymbol{F} \rangle\rangle_{s/2,2\alpha,T} \\ \|\boldsymbol{u}\|_{2,\infty,T} &\leq \|\boldsymbol{u}_0\|_2 + C\langle\langle \boldsymbol{F} \rangle\rangle_{s_1,\alpha,T} \\ \|\nabla \boldsymbol{u}\|_{2,2,T} &\leq C(\|\boldsymbol{u}_0\|_2 + \|\operatorname{div} \boldsymbol{F}\|_{s_1,q_2}), \end{aligned} \tag{7.7}$$

with $C = C(\nu, n, s, q_1)$.

Proof. Uniqueness of the solution in the class V_T is easy to show along the same lines of the proof of Theorem 4.2, and we leave it to the reader. In view of the

Helmholtz-Weyl decomposition result given in Lemma 7.1, it is enough to give the proof of existence for the following non-homogeneous heat equation problem

$$\frac{\partial u}{\partial t} = \nu \Delta u + \operatorname{div} \boldsymbol{F} \quad \text{in } \mathbb{R}^n \times (0,T)$$
$$u(x,0) = u_0(x),$$

where u is the j-th component of the velocity field, and $\boldsymbol{F} = (G_{j1}, \dots, G_{jn})$, with $\boldsymbol{G}$ given in Lemma 7.1. A solution to this problem may be written as the sum of the volume potential corresponding to u_0 and to the heat volume potential corresponding to $\operatorname{div} \boldsymbol{F}$, namely,

$$u(t) = W(t) * u_0 + \int_0^t W(t-\tau) * \operatorname{div} \boldsymbol{F}(\tau) d\tau. \tag{7.8}$$

Integrating by parts in the space variables in the last integral, we have

$$u(t) = W(t) * u_0 - \int_0^t \partial_k W(t-\tau) * F_k(\tau) d\tau. \tag{7.9}$$

From (7.9), we find

$$\|u(t)\|_\sigma \le \|W(t) * u_0\|_\sigma + \int_0^t \|\partial_k W(t-\tau) * F_k(\tau)\|_\sigma \, d\tau. \tag{7.10}$$

Using $(7.5)_2$ it follows that

$$|\partial_k W(t-\tau) * F_k(\tau)| \le C \int_{\mathbb{R}^n} \frac{|\boldsymbol{F}(y,\tau)|}{\big(|x-y|^2 + (t-\tau)\big)^{(n+1)/2}} dy.$$

Since

$$\frac{1}{(|x|^2 + t)^{(n+1)/2}} \le \frac{c(\beta)}{|x|^{\beta(n+1)} t^{(n+1)(1-\beta)/2}}, \quad \beta \in (0,1),$$

we obtain

$$|\partial_k W(t-\tau) * F_k(\tau)| \le \frac{C}{(t-\tau)^{(n+1)(1-\beta)/2}} \int_{\mathbb{R}^n} \frac{|\boldsymbol{F}(y,\tau)|}{|x-y|^{\beta(n+1)}} dy.$$

Thus, choosing $\beta < n/(n+1)$, and using (7.6) we deduce

$$\|\partial_k W(t-\tau) * F_k(\tau)\|_\sigma \le \frac{\|\boldsymbol{F}\|_{\sigma_1}}{(t-\tau)^{\frac{n}{2}\left(\frac{1}{\sigma_1} - \frac{1}{\sigma}\right) + \frac{1}{2}}}, \quad \frac{1}{\sigma_1} = \frac{1}{\sigma} + 1 - \frac{(n+1)\beta}{n}. \tag{7.11}$$

Since β is arbitrary in $(0, n/(n+1))$, we have that this last relation holds for all $\sigma_1 < \sigma$. Thus, inserting the inequality in (7.11) into (7.10), we conclude

$$\|u(t)\|_\sigma \le \|W(t) * u_0\|_\sigma + \int_0^t \frac{\|\boldsymbol{F}(\tau)\|_{\sigma_1}}{(t-\tau)^{\frac{n}{2}\left(\frac{1}{\sigma_1} - \frac{1}{\sigma}\right) + \frac{1}{2}}} d\tau, \quad 1 < \sigma_1 < \sigma < \infty. \tag{7.12}$$

We next differentiate (7.8) with respect to x_k and take the L^2-norm of both sides of the resulting equation, to get

$$\|\partial_k u(t)\|_2 \leq \|\partial_k(W(t) * u_0)\|_2 + \int_0^t \|\partial_k W(t-\tau) * \operatorname{div} \boldsymbol{F}(\tau)\|_2 d\tau. \tag{7.13}$$

Proceeding as before, one shows that

$$\|\partial_k W(t-\tau) * \operatorname{div} \boldsymbol{F}(\tau)\|_2 \leq C \frac{\|\operatorname{div} \boldsymbol{F}\|_{s_1}}{(t-\tau)^{\frac{1}{2}\left(1+\frac{n}{s}\right)}}$$

where $1/s_1 = 1/s + 1/2$. Replacing this estimate into (7.13), we deduce

$$\|\nabla u(t)\|_2 \leq \|\nabla(W(t) * u_0)\|_2 + \int_0^t \frac{\|\operatorname{div} \boldsymbol{F}\|_{s_1}}{(t-\tau)^{\frac{1}{2}\left(1+\frac{n}{s}\right)}} d\tau, \quad \frac{1}{s_1} = \frac{1}{s} + \frac{1}{2}. \tag{7.14}$$

We now choose in (7.12) $\sigma = s$, $\sigma_1 = s/2$ to obtain

$$\|u(t)\|_s \leq \|W(t) * u_0\|_s + \int_0^t \frac{\|\boldsymbol{F}(\tau)\|_{s/2}}{(t-\tau)^{\frac{1}{2}\left(1+\frac{n}{s}\right)}} d\tau. \tag{7.15}$$

We take the L^r-norm in time of both sides of this relation. If $s > n$, we may apply inequality (7.6) with $n = 1$, $p = r$ and $\lambda = (1 - n/s)/2$ to the integral in (7.15) to show the validity of (7.7_1). To show (7.7_2), we multiply both sides of (7.16) by t^α, $\alpha = (1 - n/s)/2$, and notice that

$$t^\gamma \alpha \int_0^t \frac{d\tau}{(t-\tau)^{\frac{1}{2}\left(1+\frac{n}{s}\right)} \tau^{(1+\gamma)\alpha}} = B_\gamma = \text{const}, \quad \gamma = 0, 1. \tag{7.16}$$

To show (7.7_3), we take in (7.12) $\sigma = 2$, $\sigma_1 = s_1$ and notice that, by (7.17),

$$\int_0^t \frac{\|\boldsymbol{F}(\tau)\|_{s_1}}{(t-\tau)^{\frac{1}{2}\left(1+\frac{n}{s}\right)}} d\tau \leq \langle\!\langle \boldsymbol{F} \rangle\!\rangle_{s_1,\alpha,T} \int_0^t \frac{d\tau}{(t-\tau)^{\frac{1}{2}\left(1+\frac{n}{s}\right)} \tau^\alpha} = B_0 \langle\!\langle \boldsymbol{F} \rangle\!\rangle_{s_1,\alpha,T}.$$

Finally, to prove of $(7.7)_4$, we take the L^2-norm in time of (7.14), apply (7.6) with $n = 1$, $p = 2$, $\lambda = (1 - n/2s)/2$, and notice that, for the solution $W(t) * u_0$ of the Cauchy problem for the heat equation it is

$$\|\nabla(W(t) * u_0)\|_2^2 \leq \frac{1}{2\nu} \|u_0\|_2^2.$$

The lemma is thus proved.

Before proving the main result of this section, we need a further preliminary lemma. The first part is a simple consequence of the Young inequality for convolutions while the second is due to Giga (1986, Lemma p. 196). [34]

Lemma 7.3 *Let* $1 < \sigma, s \leq \infty$,

$$\frac{1}{r} = \frac{n}{2}\left(\frac{1}{\sigma} - \frac{1}{s}\right),$$

[34] Actually, Giga's lemma applies to more general situations than the Cauchy problem for the heat equation described in Lemma 7.3.

and let $a \in L^\sigma(\mathbb{R}^n)$. Then, there exists $C = C(\nu, n, s, \sigma)$ such that the following properties hold, for all $t \in (0, T]$ and all $T > 0$:

(i) $\|W * u_0\|_s \le C\, t^{-1/r}\|u_0\|_\sigma, \quad \sigma \le s$

(ii) $\|W * u_0\|_{s,r,T} \le C\,\|u_0\|_\sigma, \quad \sigma < s.$

We shall now prove the main result of this section.

Theorem 7.1 *Let $n \le \sigma < s < \infty$, and let*

$$\frac{2}{r} + \frac{n}{s} = \frac{n}{\sigma}.$$

Then, for any $\boldsymbol{v}_0 \in H(\mathbb{R}^n) \cap \boldsymbol{L}^\sigma(\mathbb{R}^n)$, there exists $T > 0$ and a unique weak solution $\boldsymbol{v}$ to the Navier-Stokes equations in Ω_T such that $\boldsymbol{v} \in L^{s,r}(\mathbb{R}^n_T)$. Moreover, denoting by f_η the (spatial) mollifier of the function f, we have that the number T is estimated as follows:

(i) *If $\sigma > n$:*

$$T \ge \frac{C}{\|\boldsymbol{v}_0\|_\sigma^{1/\beta_1}}, \quad \beta_1 = \frac{1}{2}\left(1 - \frac{n}{\sigma}\right);$$

(ii) *If $\sigma = n$:*

$$T \ge \left(\frac{C - \|\boldsymbol{v}_0 - \boldsymbol{v}_{0\eta}\|_n}{\|\boldsymbol{v}_{0\eta}\|_q}\right)^{1/\beta_2}, \quad \beta_2 = \frac{1}{2}\left(1 - \frac{n}{q}\right),$$

where $C = C(\nu, n, s, \sigma) > 0$, q is arbitrary in (n, s), and η is taken as small as to satisfy the condition $\|\boldsymbol{v}_0 - \boldsymbol{v}_{0\eta}\|_n < C$.

Proof. We use the method of successive approximations. We set

$$\boldsymbol{v}_1(x,t) \equiv W(t) * \boldsymbol{v}_0,$$

and, for $k = 1, 2, \dots$, $\boldsymbol{v}_{k+1}$ solves the following Stokes-like problem

$$\begin{aligned}\int_0^\infty \left\{\left(\boldsymbol{v}_{k+1}, \frac{\partial \boldsymbol{\varphi}}{\partial t}\right) - \nu(\nabla \boldsymbol{v}_{k+1}, \nabla \boldsymbol{\varphi})\right\} dt \\ = \int_0^\infty (\boldsymbol{v}_k \otimes \boldsymbol{v}_k, \nabla \boldsymbol{\varphi}) dt - (\boldsymbol{v}_0, \boldsymbol{\varphi}(0)), \quad \text{for all } \boldsymbol{\varphi} \in \mathcal{D}_T.\end{aligned} \tag{7.17}$$

Using (7.7) and the Hölder inequality, we find that [35]

$$\begin{aligned}\|\boldsymbol{v}_{k+1}\|_{s,r,T} &\le \|\boldsymbol{v}_1\|_{s,r,T} + C\|\boldsymbol{v}_k\|^2_{s,q_1,T} \le \|\boldsymbol{v}_1\|_{s,r,T} + CT^{\beta_1}\|\boldsymbol{v}_k|^{\,2}_{s,r,T} \\ \langle\langle \boldsymbol{v}_{k+1}\rangle\rangle_{s,\alpha,T} &\le \langle\langle \boldsymbol{v}_1\rangle\rangle_{s,\alpha,T} + C\langle\langle \boldsymbol{v}_k\rangle\rangle^2_{s,\alpha,T} \\ \|\boldsymbol{v}_{k+1}\|_{2,\infty,T} &\le \|\boldsymbol{v}_0\|_2 + C\|\boldsymbol{v}_k\|_{2,\infty,T}\langle\langle \boldsymbol{v}_k\rangle\rangle_{s,\alpha,T} \\ \|\nabla \boldsymbol{v}_{k+1}\|_{2,2,T} &\le C(\|\boldsymbol{v}_0\|_2 + \|\boldsymbol{v}_k\|_{s,\frac{2s}{(s-n)},T}\|\nabla \boldsymbol{v}_k\|_{2,2,T}) \\ &\le C(\|\boldsymbol{v}_0\|_2 + T^{\beta_1}\|\boldsymbol{v}_k\|_{s,r,T}\|\nabla \boldsymbol{v}_k\|_{2,2,T}).\end{aligned} \tag{7.18}$$

[35] Throughout the proof of this theorem, we denote by C a generic constant which depends, at most, on $n, \sigma, s, \nu,$.

Denote by $K_0^{(1)} = K_0^{(1)}(T)$ and $K_0^{(2)} = K_0^{(2)}(T)$ two majorants for $\|\boldsymbol{v}_1\|_{s,r,T}$ and $\langle\!\langle \boldsymbol{v}_1 \rangle\!\rangle_{s,\alpha,T}$, respectively. We want to show that there exist $T > 0$ and $K_0^{(i)}$, $i = 1, 2$, such that

$$\begin{aligned} \|\boldsymbol{v}_{k+1}\|_{s,r,T} &\leq 2K_0^{(1)} \\ \langle\!\langle \boldsymbol{v}_{k+1} \rangle\!\rangle_{s,\alpha,T} &\leq 2K_0^{(2)} \end{aligned} \qquad \text{for all } k = 1, 2, \dots . \tag{7.19}$$

We proceed by induction. From $(7.18)_{1,2}$ we obtain

$$\begin{aligned} \|\boldsymbol{v}_{k+1}\|_{s,r,T} &\leq K_0^{(1)} \left(1 + CT^{\beta_1} K_0^{(1)}\right) \\ \langle\!\langle \boldsymbol{v}_{k+1} \rangle\!\rangle_{s,\alpha,T} &\leq K_0^{(2)} \left(1 + CK_0^{(2)}\right). \end{aligned}$$

Thus, (7.19) follows whenever the following conditions are met

$$\begin{aligned} CT^{\beta_1} K_0^{(1)} &< 1 \\ CK_0^{(2)} &< 1. \end{aligned} \tag{7.20}$$

Let us first consider the case $\sigma > n$. From Lemma 7.3 we find

$$\|\boldsymbol{v}_1\|_{s,r,T} \leq C\|\boldsymbol{v}_0\|_\sigma.$$

Thus, we choose

$$K_0^{(1)} = C\|\boldsymbol{v}_0\|_\sigma$$

and condition $(7.19)_1$ is certainly satisfied for those T such that

$$T^{\beta_1}\|\boldsymbol{v}_0\|_\sigma < C. \tag{7.21}$$

Moreover, again from Lemma 7.3, we find for $t \in [0, T]$

$$t^{\frac{1}{2}(1-\frac{n}{s})}\|\boldsymbol{v}_1(t)\|_s \leq Ct^{-\frac{1}{r}+\frac{1}{2}(1-\frac{n}{s})}\|\boldsymbol{v}_0\|_\sigma \leq CT^{\beta_1}\|\boldsymbol{v}_0\|_\sigma$$

and so, choosing

$$K_0^{(2)} = CT^{\beta_1}\|\boldsymbol{v}_0\|_\sigma,$$

condition (7.19_2) is satisfied again for those T verifying (7.21). In the case $\sigma = n$, observing that $\boldsymbol{v}_{0\eta} \in \boldsymbol{L}^q(\mathbb{R}^n)$, for all $q \in (1, \infty]$, from Lemma 7.3 we deduce for any $q \in (n, s)$

$$\|\boldsymbol{v}_0\|_{s,r,T} \leq \|\boldsymbol{v}_0 - \boldsymbol{v}_{0\eta}\|_n + CT^{\beta_2}\|\boldsymbol{v}_{0\eta}\|_q.$$

Thus, choosing

$$K_0^{(1)} = \|\boldsymbol{v}_0 - \boldsymbol{v}_{0\eta}\|_n + CT^{\beta_2}\|\boldsymbol{v}_{0\eta}\|_q,$$

we see that $(7.19)_1$ is satisfied if we select η sufficiently small and T such that

$$T^{\beta_2}\|\boldsymbol{v}_{0\eta}\|_q \leq C - \|\boldsymbol{v}_0 - \boldsymbol{v}_{0\eta}\|_n. \tag{7.22}$$

Likewise, we show that if we take $K_0^{(2)}$ of the same form as $K_0^{(1)}$, condition $(7.19)_2$ is satisfied for a choice of T of the type (7.22). Using (7.19) and (7.20), into $(7.18)_{3,4}$ we also find that

$$\begin{aligned} \|\boldsymbol{v}_{k+1}\|_{2,\infty,T} &\le \|\boldsymbol{v}_0\|_2 \\ \|\nabla \boldsymbol{v}_{k+1}\|_{2,2,T} &\le C\|\boldsymbol{v}_0\|_2. \end{aligned} \tag{7.23}$$

Let us now show that the sequence $\{\boldsymbol{v}_k\}$ is converging to a weak solution belonging to the space $L^{s,r}$. To this end, we write (7.17) for $\boldsymbol{v}_{k+1}$ and for $\boldsymbol{v}_k$, then subtract the two resulting equations and apply the estimates of Lemma 7.2 to find ($k \ge 1$, $\boldsymbol{v}_0 \equiv 0$)

$$\|\boldsymbol{v}_{k+1} - \boldsymbol{v}_k\|_{s,r,T} \le CT^{\beta_1}\left(\|\boldsymbol{v}_k\|_{s,r,T} + \|\boldsymbol{v}_{k-1}\|_{s,r,T}\right)\|\boldsymbol{v}_k - \boldsymbol{v}_{k-1}\|_{s,r,T}.$$

If we employ $(7.19)_1$ and $(7.20)_1$ into this inequality, we end up with an estimate of the following form

$$\|\boldsymbol{v}_{k+1} - \boldsymbol{v}_k\|_{s,r,T} \le \alpha\|\boldsymbol{v}_k - \boldsymbol{v}_{k-1}\|_{s,r,T}, \tag{7.24}$$

where α is a constant strictly less than one and independent of k. From (7.24) it easy to show that $\{\boldsymbol{v}_k\}$ is a Cauchy sequence in the space $L^{s,r}(\mathbb{R}^n_T)$. In fact, (7.24) implies

$$\|\boldsymbol{v}_{k+1} - \boldsymbol{v}_k\|_{s,r,T} \le K_0^{(1)}\alpha^k,$$

and so, for all $k' = k + l$, $l > 0$,

$$\|\boldsymbol{v}_k - \boldsymbol{v}_{k'}\|_{s,r,T} \le \sum_{i=1}^{l}\|\boldsymbol{v}_{k+i} - \boldsymbol{v}_{k+i-1}\|_{r,s,T} \le \alpha^k\sum_{i=1}^{l}\alpha^i \le \frac{\alpha^{k+1}}{1-\alpha} \to 0 \text{ as } k \to \infty.$$

Denoting by $\boldsymbol{v}$ the limit field, from (7.23) we also deduce that $\boldsymbol{v} \in V_T$ and, by a simple calculation which uses (7.17) and the convergence properties of $\{\boldsymbol{v}_k\}$, that $\boldsymbol{v}$ satisfies (2.8) (with $\boldsymbol{f} \equiv 0$). The existence proof is thus completed. Since uniqueness is a consequence of Theorem 4.2 and Remark 4.7, the theorem is completely proved.

We shall now analyze some consequences of Theorem 7.1. We begin with the following result which improves Theorem 4.2(ii), see also Remark 4.5

Theorem 7.2
Let $\boldsymbol{v}, \boldsymbol{u}$ be two weak solutions in Ω_T corresponding to the same data $\boldsymbol{v}_0$. Assume that $\boldsymbol{u}$ satisfies the energy inequality (EI) *and that $\boldsymbol{v} \in L^\infty(0,T;\boldsymbol{L}^n(\mathbb{R}^n))$. Then $\boldsymbol{v} = \boldsymbol{u}$ a.e. in $\mathbb{R}^n_T$.*

Proof. As in the proof of Theorem 4.2, we establish (4.23). Let $\mathcal{T}$ and τ_0 be defined as in that proof, and assume $\tau_0 < T$. Thus, (4.23) implies

$$\|\boldsymbol{w}(t)\|_2^2 + 2\nu\int_{\tau_0}^{t}\|\nabla \boldsymbol{w}\|_2^2 d\tau \le 2\int_{\tau_0}^{t}(\boldsymbol{w}\cdot\nabla\boldsymbol{w}, \boldsymbol{v})d\tau, \quad t \in (\tau_0, \mathcal{T}). \tag{7.25}$$

We shall show that

$$\boldsymbol{v}(t) \in \boldsymbol{L}^n(\mathbb{R}^n), \quad \text{for all } t \in [0,T], \tag{7.26}$$

and so, in particular, that $\boldsymbol{v}(\tau_0) \in \boldsymbol{L}^n(\mathbb{R}^n)$. In fact, denote by $E \subset [0,T]$ the set where possibly (7.26) does not hold. Clearly, E is of zero Lebesgue measure. Let $t_* \in E$ and let $\{t_k\} \subset [0,T] - E$ be a sequence converging to t_*. By assumption, it follows that there exists $\boldsymbol{U} \in \boldsymbol{L}^n(\Omega)$ such that

$$\lim_{k\to\infty} (\boldsymbol{v}(t_k), \boldsymbol{\psi}) = (\boldsymbol{U}, \boldsymbol{\psi}), \quad \text{for all } \boldsymbol{\psi} \in \boldsymbol{C}_0^\infty(\Omega).$$

On the other hand, by the weak L^2 continuity, we have

$$\lim_{k\to\infty} (\boldsymbol{v}(t_k), \boldsymbol{\psi}) = (\boldsymbol{v}(t_*), \boldsymbol{\psi}), \quad \text{for all } \boldsymbol{\psi} \in \boldsymbol{C}_0^\infty(\Omega),$$

and (7.26) follows. Now, by Theorem 7.1, we infer that there exists a weak solution $\widetilde{\boldsymbol{v}}$, say, assuming the initial data $\boldsymbol{v}(\tau_0)$ and belonging to the space $L^r(\tau_0, \tau_0 + T(\tau_0); \boldsymbol{L}^s(\mathbb{R}^n))$, $2/r + n/s = 1$, $s > n$. In view of Theorem 4.1 and Remark 4.3, $\boldsymbol{v}$ satisfies the energy equality in $[\tau_0, \tau_0 + T(\tau_0))$ and so, from Theorem 4.2(i), we conclude $\boldsymbol{v} = \widetilde{\boldsymbol{v}}$ in $[\tau_0, \tau_0 + T(\tau_0))$. We then use (7.25), and reason as in the proof of Theorem 4.2(i) to show $\boldsymbol{v} = \boldsymbol{u}$ in $[\tau_0, \tau_0 + T(\tau_0))$, contradicting the fact that τ_0 is a maximum.

Another consequence of Theorem 7.1 is contained in the following one, which extends the results of Theorem 6.4(i) to the case $\Omega = \mathbb{R}^n$. [36]

Theorem 7.3 *Let $\boldsymbol{v}$ be a weak solution in $\mathbb{R}^n_T$, for all $T > 0$, corresponding to the initial data $\boldsymbol{v}_0 \in H(\mathbb{R}^n)$, and satisfying the strong energy inequality (4.1) and let t_1 be an epoch of irregularity for $\boldsymbol{v}$. Then, $\|\boldsymbol{v}(t)\|_\sigma$ diverges as $t \to t_1^-$, for all $n < \sigma < \infty$, in such a way that*

$$\|\boldsymbol{v}(t)\|_\sigma \geq \frac{C}{(t_1 - t)^{(\sigma-n)/2\sigma}}, \quad t < t_1,$$

with $C = C(n, \sigma, \nu) > 0$;

Proof. Reasoning as in the proof of Theorem 6.4(i), we show that there can not exist a sequence $\{\tau_k\}$, say, tending to t_1, along which $\|\boldsymbol{v}(\tau_k)\|_\sigma$ stays bounded. [37] In fact, otherwise, in view of Theorem 7.2, we could construct a solution $\widetilde{\boldsymbol{v}}$, having $\boldsymbol{v}(\tau_k)$ as initial data and belonging to $L^r(\tau_k, \tau_k + T_k; \boldsymbol{L}^s(\mathbb{R}^n))$, for some $r = 2s/(s-n)$, $s > n$, [38] and $\tau_k + T_k > t_1$. By Theorem 5.2(i), $\boldsymbol{v} \in C^\infty(\overline{\mathbb{R}_n} \times (\tau_k, \tau_k + T_k))$ and by the uniqueness Theorem 4.2(i), $\boldsymbol{v} = \widetilde{\boldsymbol{v}}$ on $(\tau_k, \tau_k + T_k)$, contradicting the assumption that t_1 is an epoch of irregularity. From Theorem 7.2(i), we then have

$$(t_1 - t) \geq C/\|\boldsymbol{v}(t)\|_\sigma^{2\sigma/(\sigma-n)}, \quad t < t_1,$$

and the result is proved.

Remark 7.1 From Theorem 7.3, we reobtain the sufficient condition for the absence of epochs of irregularity given in Theorem 5.2. The estimate of Theorem 7.3 was first obtained for $n = 3$ by Leray (1934b, pp. 227). Actually, following the work

[36] We refer to Giga (1986), for the more general case when Ω has a compact boundary.

[37] Recall that, from the Definition 6.1 of epoch of irregularity, it follows that $\boldsymbol{v}(t) \in L^q(\mathbb{R}^n)$ for all $q \geq 2$.

[38] See Remark 4.7.

of Leray, *loc. cit.* pp. 222–224, we could show that this estimate also holds in the case $\sigma = \infty$.

As we have noticed in Remark 5.4, one important point which is left out in Theorem 5.4 is to show that a weak solution $\boldsymbol{v}$ which in addition satisfies

$$\boldsymbol{v} \in L^\infty(0,T;\boldsymbol{L}^n(\Omega)) \tag{7.27}$$

is in fact regular. So far, it is not known whether this property is true or not. The last part of this section will be devoted to investigate the kind of regularity achieved by weak solution satisfying (7.27). This will be obtained by means of Theorem 7.1.

We begin to show the following result.

Lemma 7.4 *Let $\boldsymbol{v}$ be a weak solution in $\mathbb{R}^n_T$, verifying (7.27). Then, for any $t_0 \in [0,T)$, there exists $\delta(t_0) > 0$ such that $\boldsymbol{v} \in C([t_0, t_0+\delta(t_0)); \boldsymbol{L}^n(\mathbb{R}^n))$. In particular, $\boldsymbol{v}(t)$ is right continuous in the $\boldsymbol{L}^n$-norm, at each $t \in [0,T)$.*

Proof. We already know that $\boldsymbol{v}(t) \in \boldsymbol{L}^n(\mathbb{R}^n)$, for all $t \in [0,T)$, see (7.26). Therefore, for any fixed $t_0 \in [0,T)$, by Theorem 7.1 we know that there exists $\delta(t_0) > 0$ such that

$$\boldsymbol{v} \in L^r(t_0, t_0+\delta(t_0); \boldsymbol{L}^n(\mathbb{R}^n)), \quad \text{for all } s > n \text{ and } r = 2s/(s-n), \tag{7.28}$$

and thus $\boldsymbol{v}$ is regular in $I = (t_0, t_0+\delta(t_0))$. We may then multiply the Navier-Stokes equations (0.1) – written in $\mathbb{R}^n \times I$, with $\boldsymbol{f} \equiv 0$ – by $|\boldsymbol{v}|^{n-2}\boldsymbol{v}$, and integrate by parts over $\mathbb{R}^n$, to obtain [39]

$$\frac{1}{n}\frac{d}{dt}\|\boldsymbol{v}\|_n^n + \nu D_1(\boldsymbol{v}) + 4\nu\frac{n-2}{n^2}D_2(\boldsymbol{v}) = -(n-2)\int_{\mathbb{R}^n} p|\boldsymbol{v}|^{n-4}\boldsymbol{v}\cdot\nabla\boldsymbol{v}\cdot\boldsymbol{v}\,dx, \tag{7.29}$$

where

$$D_1(\boldsymbol{v}) = \int_{\mathbb{R}^n} |\boldsymbol{v}|^{n-2}|\nabla\boldsymbol{v}|^2 dx$$

$$D_2(\boldsymbol{v}) = \int_{\mathbb{R}^n} |\nabla|\boldsymbol{v}|^{n/2}|^2 dx.$$

We now apply the Cauchy-Schwarz inequality in the integral at the right-hand side of (7.29) to deduce

$$\frac{1}{n}\frac{d}{dt}\|\boldsymbol{v}\|_n^n + \tfrac{1}{2}\nu D_1(\boldsymbol{v}) + 4\nu\frac{n-2}{n^2}D_2(\boldsymbol{v}) \le C\int_{\mathbb{R}^n} p^2|\boldsymbol{v}|^{n-2}dx. \tag{7.30}$$

Since

$$\Delta p = \partial_i\partial_j(v_i v_j),$$

from the Calderon-Zygmund theorem on singular integrals we obtain

$$\|p\|_{(n+2)/2} \le C\|\boldsymbol{v}\|_{n+2}^2.$$

[39]For this type of technique, see Rionero and Galdi (1979), and Beirão da Veiga (1987).

Using this inequality at the right-hand side of (7.30), we conclude

$$\frac{1}{n}\frac{d}{dt}\|\boldsymbol{v}\|_n^n + \tfrac{1}{2}\nu D_1(\boldsymbol{v}) + 4\nu\frac{n-2}{n^2}D_2(\boldsymbol{v}) \le C\|\boldsymbol{v}\|_{n+2}^{n+2}. \tag{7.31}$$

By the same procedure, one also shows

$$\left|\frac{d}{dt}\|\boldsymbol{v}\|_n^n\right| \le C\left(D_1(\boldsymbol{v}) + D_2(\boldsymbol{v}) + \|\boldsymbol{v}\|_{n+2}^{n+2}\right). \tag{7.32}$$

From (7.28) we know that $\boldsymbol{v} \in L^{n+2}(t_0, t_0+\delta(t_0); \boldsymbol{L}^{n+2}(\mathbb{R}^n))$ and, by assumption, that $\boldsymbol{v} \in L^\infty(0,T;\boldsymbol{L}^n(\mathbb{R}^n))$. Therefore, (7.31) gives

$$\int_{t_0}^{t_0+\delta(t_0)} \left[D_1(\boldsymbol{v}) + D_2(\boldsymbol{v})\right] dt \le M,$$

which, in turn, once replaced into (7.32), allows us to infer

$$\int_{t_0}^{t_0+\delta(t_0)} \left|\frac{d}{dt}\|\boldsymbol{v}\|_n^n\right| \le M.$$

Thus, $\|\boldsymbol{v}(t)\|_n$ is continuous in $[t_0, t_0+\delta(t_0))$. On the other hand, the weak continuity of $\boldsymbol{v}$ in $\boldsymbol{L}^2$, along with the uniform boundedness in $\boldsymbol{L}^n$, implies that $\boldsymbol{v}$ is weakly continuous in $\boldsymbol{L}^n$ and we conclude the continuity of $\boldsymbol{v}$ in $\boldsymbol{L}^n(\mathbb{R}^n)$.

We are now able to prove the following partial regularity result (Sohr and von Wahl, 1984, Theorem III.4).

Theorem 7.4
Let $\boldsymbol{v}$ be a weak solution in $\mathbb{R}^n_T$ verifying the condition $\boldsymbol{v} \in L^\infty(0,T;\boldsymbol{L}^n(\mathbb{R}^n))$. Then, there exists a set $E \subset [0,T]$ with the following properties

(i) $\boldsymbol{v} \in C^\infty\left(\overline{\mathbb{R}_n} \times E\right)$;
(ii) *The set $S \equiv [0,T] - E$ is at most countable;*
(iii) *For every epoch of irregularity t_1 ($\in S$) we have*

$$\limsup_{t\to t_1^-}\|\boldsymbol{v}(t)\|_n > \lim_{t\to t_1^+}\|\boldsymbol{v}(t)\|_n.$$

Proof. Point (i) is already known from Theorem 6.3. [40] For $t_0 \in S$, by Theorem 7.1, we may construct a regular solution in $(t_0, t_0+\delta(t_0))$. We can take a rational number in $(t_0, t_0+\delta(t_0))$ to show that S is countable. Let now t_1 be an epoch of irregularity. Then, by Theorem 7.3, $\boldsymbol{v}$ is right continuous at t_1 in the $\boldsymbol{L}^n$-norm, that is,

$$\lim_{t\to t_1^+}\|\boldsymbol{v}(t)\|_n = \|\boldsymbol{v}(t_1)\|_n.$$

Since $\boldsymbol{v}$ is weakly continuous at t_1, we also have that

$$\limsup_{t\to t_1^-}\|\boldsymbol{v}(t)\|_n \text{ exists.}$$

[40] We recall that, by assumption and by Theorem 4.1, $\boldsymbol{v}$ satisfies the energy *equality* in $[0,T)$.

If

$$\limsup_{t \to t_1^-} \|\boldsymbol{v}(t)\|_n \leq \|\boldsymbol{v}(t_1)\|_n = \lim_{t \to t_1^+} \|\boldsymbol{v}(t)\|_n,$$

we would then have that $\boldsymbol{v}(t)$ is strongly continuous in $\boldsymbol{L}^n$ at t_1. From Theorem 5.2(ii) it then follows that

$$\boldsymbol{v} \in C\left([t_1 - \eta_1, t_1]; \boldsymbol{L}^n(\mathbb{R}^n)\right), \quad \eta_1 > 0,$$

On the other hand, by Lemma 7.1, we also have that

$$\boldsymbol{v} \in C\left([t_1, t_1 + \eta_2]; \boldsymbol{L}^n(\mathbb{R}^n)\right), \quad \eta_2 > 0,$$

and so,

$$\boldsymbol{v} \in C\left([t_1 - \eta, t_1 + \eta]; \boldsymbol{L}^n(\mathbb{R}^n)\right), \quad \eta > 0,$$

and, by Theorem 5.2(ii), t_1 can not be an epoch of irregularity.

Remark 7.2 Condition (ii) in Theorem 7.4 can be refined in the following way, see Kozono and Sohr (1996b), Beirão da Veiga (1996). Let the assumption of that theorem be satisfied and let t_1 be any instant of time. Set

$$a \equiv \limsup_{t \to t_1^-} \|\boldsymbol{v}(t)\|_n - \lim_{t \to t_1^+} \|\boldsymbol{v}(t)\|_n.$$

Then, there exists a constant C independent of the particular solution $\boldsymbol{v}$ such that, if $a < C$, then necessarily $a = 0$, that is, $\boldsymbol{v}$ is strongly continuous in $\boldsymbol{L}^n$ at t_1 and, therefore, smooth at t_1. Further investigation on the structure of the possible irregular points of a solution satisfying the assumption of Theorem 7.4, has been more recently carried out by J. Neustupa (1999).

Remark 7.3 The estimate from below for the time T of existence of a solution with data in $\boldsymbol{L}^n$ may play a crucial role in the theory of regularity. Though it is very unlikely that we can give for T a bound of the type

$$T \geq C\|\boldsymbol{v}_0\|_n^{-\beta}, \quad \beta > 0,$$

we may still conjecture the following estimate

$$T \geq f(\|\boldsymbol{v}_0\|_n) \tag{7.33}$$

where $f(\lambda)$ is a positive, strictly decreasing function of λ. The following two possibilities may then arise

i) $\lim_{\lambda \to \infty} f(\lambda) = f_0 > 0;$

ii) $\lim_{\lambda \to \infty} f(\lambda) = 0.$

In the case i), no epoch of irregularity can exist. In fact, we have $T \geq f_0$. Let t_1 be an epoch of irregularity. Then, we could choose t_0 such that $t_1 - t_0 < f_0/2$ (say), and we would conclude, by Theorem 7.1, that $\boldsymbol{v}$ is regular in $(t_0, t_0 + f_0)$, contradicting the fact that t_1 is an epoch of irregularity. In case ii), we distinguish again the following two possibilities:

ii)′ $\limsup_{t\to t_1^-}\|\boldsymbol{v}(t)\|_n = \infty$;

ii)″ $\boldsymbol{v} \in L^\infty(0,t_1;\boldsymbol{L}^n(\Omega))$.

In case ii)′, for t very close to t_1, we would have

$$\|\boldsymbol{v}(t)\|_n \ge f^{-1}(t_1 - t), \quad t < t_1,$$

and, therefore, since

$$\lim_{\lambda\to 0} f^{-1}(\lambda) = \infty,$$

a condition even weaker than (7.27) – depending on f – would imply regularity. In case ii)″, setting

$$M = \operatorname{ess}\sup_{t\in[0,t_1]} \|\boldsymbol{v}(t)\|_n.$$

we would have $T \ge f(M)$, and so, reasoning as in case i), we would deduce that t_1 can not be an epoch of irregularity. From all the above, we then conclude that, if an estimate of the type (7.33) holds for T, then a condition weaker than (7.27) and depending on f, would suffice to ensure regularity of a weak solution. However, we only have for T the estimate of Theorem 7.1(ii).

Remark 7.4 In view of Theorem 6.4(i) and Theorem 7.3, we deduce that a weak solution $\boldsymbol{v}$ in dimension 3 will never go through an epoch of irregularity t_1, provided that the condition $\boldsymbol{v} \in V_T$ is incompatible with the following ones:

$$\left.\begin{aligned} \|\nabla\boldsymbol{v}(t)\|_2 &\ge \frac{C}{(t_1-t)^{1/4}} \\ \|\boldsymbol{v}(t)\|_\sigma &\ge \frac{C}{(t_1-t)^{(\sigma-n)/2\sigma}} \end{aligned}\right. \quad t < t_1, \ \ \sigma > n. \tag{7.34}$$

With this in mind, J.Leray (1934b, p.225) proposed a *possible* counter example to the existence of a global regular solution. This counter example *would* lead to a weak solution possessing just one epoch of irregularity. Even though the existence of such a solution has been recently ruled out by Nečas, Růžička and Šverák (1996) (see also Tsai (1998)), we deem it interesting to reproduce and discuss it here. This solution is constructed as follows. *Assume* that the following system of equations

$$\left.\begin{aligned} \nu\Delta\boldsymbol{U}(y) - \alpha\left[\boldsymbol{U}(y) + \boldsymbol{y}\cdot\nabla\boldsymbol{U}(y)\right] + \nabla P(y) &= \boldsymbol{U}(y)\cdot\nabla\boldsymbol{U}(y) \\ \operatorname{div}\boldsymbol{U}(y) &= 0 \end{aligned}\right. \quad y \in \mathbb{R}^n,$$

admits a non-zero solution $\boldsymbol{U} \in \boldsymbol{W}^{1,2}(\mathbb{R}^n)$, for some $\alpha > 0$, and set

$$\lambda(t) = (2\alpha(t_1 - t))^{-1/2}, \quad t < t_1.$$

Then, the function

$$\boldsymbol{u}(x,t) = \begin{cases} \lambda(t)\boldsymbol{U}(\lambda(t)x) & \text{if } t < t_1 \\ 0 & \text{if } t \ge t_1 \end{cases} \tag{7.35}$$

is a weak solution to the Navier-Stokes problem in $\mathbb{R}^n_T$. By a simple calculation which uses (7.35) we show that

$$\begin{aligned} \|\boldsymbol{u}(t)\|_s &= C\,(\lambda(t))^{1-n/s}, \quad s \in [2,\infty), \quad t < t_1 \\ \|\nabla \boldsymbol{u}(t)\|_2 &= C(\lambda(t))^{1/2}, \quad t < t_1. \end{aligned} \tag{7.36}$$

From (7.36) it is clear that $\boldsymbol{u}$ satisfies all requirements of a weak solution and that, in fact, it possesses even more regularity, such as strong L^2-continuity in time. However, $\boldsymbol{u}$ blows up at t_1 exactly in the way prescribed by (7.34), so that t_1 is the only epoch of irregularity. Moreover, $\|\boldsymbol{u}(t)\|_n \le C$, uniformly in t, and $\|\boldsymbol{u}(t)\|_n$ becomes irregular at t_1 just in the way predicted by Theorem 7.4(iii). As we mentioned, such a solution does not exist, since Nečas, Růžička and Šverák, *loc. cit.*, have shown that $\boldsymbol{U} \equiv 0$. This result gives more weight to the conjecture that the class $L^{n,\infty}(\mathbb{R}^n_T)$ is a regularity class.

References

AMANN, H., 1999, On the Strong Solvability of the Navier-Stokes Equations, *J. Math. Fluid Mech.*, in press

BEIRÃO DA VEIGA, H., 1987, Existence and Asymptotic Behavior for Strong Solutions of the Navier-Stokes Equations in the Whole Space, *Indiana Univ. Math. J.*, **36**, 149

BEIRÃO DA VEIGA, H., 1995a, Concerning the Regularity Problem for the Solutions of the Navier-Stokes Equations, *Compte Rend. Acad. Sci. Paris*, **321**, 405

BEIRÃO DA VEIGA, H., 1995b, A New Regularity Class for the Navier-Stokes Equations in $\mathbb{R}^n$, *Chin. Ann. of Math*, **16B**, 407

BEIRÃO DA VEIGA, H., 1996, Remarks on the Smoothness of the $L^\infty(0,T;L^3)$ Solutions of the 3-D Navier-Stokes Equations, Preprint, University of Pisa

BERSELLI L.C., 1999, Sufficient Conditions for the Regularity of the Solutions of the Navier-Stokes Equations, *Math. Meth. Appl. Sci.*, **22**, 1079

BORCHERS, W. AND MIYAKAWA, T., 1988, Decay for the Navier-Stokes Flow in Halfspaces, *Math. Ann.*, **282**, 139

CAFFARELLI, L., KOHN, R. AND NIRENBERG, L., 1982, Partial Regularity of Suitable Weak Solutions of the Navier-Stokes Equations, *Comm. Pure Appl. Math.*, **35**, 771

CALDERON, P.C., 1990a, Existence of Weak Solutions for the Navier-Stokes Equations with Initial Data in L^p, *Trans. American Math. Soc.*, **318**, 179

CALDERON, P.C., 1990b, Addendum to the paper "Existence of Weak Solutions for the Navier-Stokes Equations with Initial Data in L^p", *Trans. American Math. Soc.*, **318**, 201

CALDERON, P.C., 1993, Initial Values of Solutions of the Navier-Stokes Equations, *Proc. American Math. Soc.*, **117**, 761

CANNONE M., 1997, A generalization of a theorem by Kato on Navier-Stokes equations, *Rev. Mat. Iberoamer.*, **13** 515.

CHEMIN J.Y., 1999, Theorems of Unicity for the Threedimensional Navier-Stokes System, *J. Anal. Math.*, **77** 27

FABES, E.B., JONES, B.F. AND RIVIÈRE, N.M., 1972, The Initial Value Problem for the Navier-Stokes Equations with Data in L^p, *Arch. Ratl Mech. Anal.*, **45**, 222

FABES, E.B., LEWIS, J. AND RIVIÈRE, N.M., 1977a, Singular Integrals and Hydrodynamical Potentials, *American J. Math.*, **99**, 601

FABES, E.B., LEWIS, J. AND RIVIÈRE, N.M., 1977b, Boundary Value Problems for the Navier-Stokes Equations, *American J. Math.*, **99**, 626

FOIAŞ, C. , 1961, Une Remarque sur l'Unicité des Solutions des Equations de Navier-Stokes en dimension n, *Bull. Soc. Math. France*, **89**, 1

FOIAŞ, C. AND TEMAM, R., 1979, Some Analytical and Geometric Properties of the Solutions of the Navier-Stokes Equations, J. Math. Pure Appl., **58**, 339

GALDI, G.P., 1994, *An Introduction to the Mathematical Theory of the Navier-Stokes Equations, Vol. I: Linearized Theory*, Springer Tracts in Natural Phylosophy, Vol.38. Springer-Verlag.

GALDI, G.P. AND MAREMONTI, P., 1986, Monotonic Decreasing and Asymptotic Behavior of the Kinetic Energy for Weak Solutions of the Navier-Stokes Equations in Exterior Domains, *Arch. Ratl Mech. Anal.*, **94**, 253

GALDI, G.P. AND MAREMONTI, P., 1988, Sulla Regolarità delle Soluzioni Deboli al Sistema di Navier-Stokes in Domini Arbitrari, *Ann. Univ. Ferrara, Sci. Mat.*, **34**, 59

GIGA, Y., 1986, Solutions for Semilinear Parabolic Equations in L^p and Regularity of Weak Solutions of the Navier-Stokes System, *J. Differential Eq.*, **62**, 182

GOLOVKIN, K.K., 1964, On the Non-Uniqueness of Solutions of Certain Boundary-Value Problems for the Equations of Hydromechanics, *Ž. Vyčisl. Mat. i. Mat. Fiz.*, **4**, 773 (in Russian)

HEYWOOD, J.G., 1980, The Navier-Stokes Equations: On the Existence, Regularity and Decay of Solutions, *Indiana Univ. Math. J.*, **29**, 639

HEYWOOD, J.G., 1988, Epochs of Irregularity for Weak Solutions of the Navier-Stokes Equations in Unbounded Domains, *Tôhoku Math. J.*, **40**, 293

HILLE E. AND PHILLIPS, R.S., 1957, *Functional Analysis and Semigroups*, American Math. Soc. Colloq. Publ. **31**

HOPF, E., 1950/1951, Über die Anfangswertaufgabe für die Hydrodynamischen Grundgleichungen, *Math. Nachr.*, **4**, 213

KANIEL, S., 1969, A Sufficient Condition for Smoothness of Solutions of Navier-Stokes Equations, *Israel J. Math.*, **6**, 354

KANIEL, S. AND SHINBROT M., 1967, Smoothness of Weak Solutions of the Navier-Stokes Equations, *Arch. Ratl Mech. Anal.*, **24**, 302

KATO, H., 1977/78, On a Regularity of E. Hopf's Weak Solutions for the Navier-Stokes Equations, *Math. Rep. Kyushu Univ.*, **1**, 15

KATO, H., 1986, Regularity of Weak Solutions of the Navier-Stokes Equations, *Math. Rep. Kyushu Univ.*, **15**, 1

KATO, H., 1989, Weak Solutions of the Navier-Stokes Initial Value Problem, *Math. Rep. Kyushu Univ.*, **17**, 57

KATO, H., 1993, On the Regularity and Uniqueness of Weak Solutions for the Navier-Stokes Equations, *Memoirs of the Faculty Sci. Kyushu Univ. Ser. A*, **47**, 183

KATO T., 1984, Strong L^p-Solutions to the Navier-Stokes Equations in $\mathbb{R}^n$, with Applications to Weak Solutions, *Math. Z.*, **187**, 471

KISELEV, A.A. AND LADYZHENSKAYA, O.A., 1957, On Existence and Uniqueness of the Solution of the Nonstationary Problem for a Viscous Incompressible Fluid, *Izv. Akad. Nauk SSSR*, **21**, 655

KOZONO H., 1998, Uniqueness and regularity of weak solutions to the Navier-Stokes equations, *Lecture Notes in Num. and Appl. Anal.* **16**, 161

KOZONO, H. AND SOHR, H., 1996a, Remark on Uniqueness of Weak Solutions to the Navier-Stokes Equations, *Analysis*, **16**, 255

KOZONO, H. AND SOHR, H., 1996b, Regularity Criterion on Weak Solutions to the Navier-Stokes Equations, Preprint, Universität Paderborn

KOZONO, H. AND YAMAZAKI, M., 1998, Exterior problem for the stationary Navier-Stokes equations in the Lorentz space, *Mathematische Annalen*, **310**, 279

LADYZHENSKAYA, O.A., 1958, Solution "in the Large" of the Nonstationary Boundary-Value Problem for the Navier-Stokes System for the Case of Two Space Variables, *Doklady Akad. Nauk. SSSR*, **123**, 427 (in Russian)

LADYZHENSKAYA, 0.A., 1967, Uniqueness and Smoothness of Generalized Solutions of the Navier-Stokes Equations, *Zap. Naucn Sem. Leningrad Otdel Mat. Inst. Steklov*, **5**, 169 (in Russian)

LADYZHENSKAYA, 0.A., 1969, Example of Nonuniqueness in the Hopf Class of Weak Solutions for the Navier-Stokes Equations, *Izv. Akad. Nauk SSSR., Ser. Mat.*, **33** (1), 229 (in Russian)

LADYZHENSKAYA, O.A., AND SEREGIN, G.A., 1999, On Partial Regularity of Suitable Weak Solutions to the Three-Dimensional Navier-Stokes Equations, *J. Math. Fluid Mech.*, in press

LADYZHENSKAYA, O.A., URAL´CEVA, N.N. AND SOLONNIKOV V.A., 1968, *Linear and Quasilinear Equations of Parabolic Type*, Transl. Math. Monographs, **23**, American Math. Soc., Providence R.I.

LERAY, J., 1934a, Essai sur les Mouvements Plans d'un Liquide Visqueux que Limitent des Parois, *J. Math. Pures Appl.*, **13**, 331

LERAY, J., 1934b, Sur les Mouvements d'un Liquide Visqueux Emplissant l'Espace, *Acta Math.*, **63**, 193

LIGHTHILL, J., 1956, *Nature*, **178**, 343

LIN, F.H., 1998, A new proof of the Caffarelli-Kohn-Nirenberg theorem, *Comm. Pure Appl. Math.*, **51**, 241.

LIONS, J.L., 1960, Sur la Regularité et l'Unicité des Solutions Turbulentes des Équations de Navier-Stokes, *Rend. Sem. Mat. Università di Padova*, **30**, 1

LIONS, J.L., 1969, *Quelques Méthodes de Résolution des Problèmes aux Limites Non Linéaires.* Paris: Gauthier-Villars

LIONS, J.L. AND PRODI, G., 1959, Un Théorème d'Existence et Unicité dans les Équations de Navier-Stokes en Dimension 2, *Compte Rend. Acad. Sci. Paris.*, **248**, 3519

MAREMONTI, P., 1987, Partial Regularity of a Generalized Solution to the Navier-Stokes Equations in Exterior Domains, *Commun. Math. Phys.*, **110**, 75

MASUDA, K., 1984, Weak Solutions of the Navier-Stokes Equations, *Tohoku Math. J.*, **36**, 623

MONNIAUX S., 1999, Uniqueness of Mild Solutions of the Navier-Stokes Equation and Maximal L^p-regularity, *C.R. Acad. Sci-Math*, **328**, 663

MIYAKAWA, T., AND SOHR, H., 1988, On Energy inequality, Smoothness and Large Time Behavior in L^2 for Weak Solutions of the Navier-Stokes Equations in Exterior Domains, *Math. Z.*, **199**, 455

NAVIER, C.L.M.H., 1827, Mémoire sur les Lois du Mouvement des Fluides, *Mem. Acad. Sci. Inst. de France*, **6** (2), 389

NEUSTUPA, J., 1999, Partial Regularity of Weak Solutions to the Navier-Stokes Equations in the Class $L^\infty(0,T;\, L^3(\Omega)^3)$, *J. Math. Fluid Mech.*, in press

NEČAS, J., RŮŽIČKA, M. AND ŠVERÁK, V., 1996, On Leray's Self-similar Solutions of the Navier-Stokes Equations, *Acta Math.*, **176**, 283

OHYAMA, T., 1960, Interior Regularity of Weak Solutions of the Time Dependent Navier-Stokes Equation, *Proc. Japan Acad.*, **36**, 273

PRODI, G., 1959, Un Teorema di Unicità per le Equazioni di Navier-Stokes, *Ann. Mat. Pura Appl.*, **48**, 173

RAUTMANN, R., 1983, On Optimum Regularity of Navier-Stokes Solution at Time $t = 0$, *Math. Z.*, **184**, 141

RIONERO, S., AND GALDI, G.P., 1979, The Weight Function Approach to Uniqueness of Viscous Flows in Unbounded Domains, *Arch. Ratl Mech. Anal.*, **69**, 37

SCHEFFER, V., 1976a, Turbulence and Hausdorff Dimension, *Springer Lecture Notes in Mathematics*, **565**, 174

SCHEFFER, V., 1976b, Partial Regularity of Solutions to the Navier-Stokes Equations, *Pacific J. Math.*, **66**, 535

SCHEFFER, V., 1977, Hausdorff Measure and the Navier-Stokes Equations, *Commun Math. Phys.*, **55**, 97

SCHEFFER, V., 1978, The Navier-Stokes Equations in Space Dimension Four, *Commun Math. Phys.*, **61**, 41

SCHEFFER, V., 1980, The Navier-Stokes Equations on a Bounded Domain, *Commun Math. Phys.*, **73**, 1

SCHEFFER, V., 1982, Boundary Regularity for the Navier-Stokes Equations in Half-Space, *Commun Math. Phys.*, **80**, 275

SCHEFFER, V., 1985, A Solution to the Navier-Stokes Equations with an Internal Singularity, *Commun Math. Phys.*, **101**, 47

SERRIN, J., 1962, On the Interior Regularity of Weak Solutions of the Navier-Stokes Equations, *Arch. Ratl Mech. Anal.*, **9**, 187

SERRIN, J., 1963, The Initial Value Problem for the Navier-Stokes Equations, *Nonlinear Problems.*, R.E. Langer Ed., Madison: University of Wisconsin Press **9**, 69

SHINBROT, M., 1973, *Lectures on Fluid Mechanics*, Gordon and Breach, New York

SHINBROT, M., 1974, The Energy Equation for the Navier-Stokes System, *SIAM J. Math. Anal.*, **5**, 948

SIMON, J., 1999, On the existence of the pressure for solutions of the variational Navier-Stokes equations, *J. Math. Fluid Mech.*, in press

SIMON, L., 1983, *Lectures on Geometric Measure Theory*, University of Canberra, Australia

SOHR, H., 1983, Zur Regularitätstheorie der Instationären Gleichungen von Navier-Stokes, *Math. Z.*, **184**, 359

SOHR, H., 1984, Erweiterung eines Satzes von Serrin über die Gleichungen von Navier-Stokes, *J. Reine Angew. Math.*, **352**, 81

SOHR, H. AND VON WAHL, W., 1984, On the Singular Set and the Uniqueness of Weak Solutions of the Navier-Stokes Equations, *Manuscripta Math.*, **49**, 27

SOHR, H. AND VON WAHL, W., 1985, A New Proof of Leray's Structure Theorem and the Smoothness of Weak Solutions Navier-Stokes Equations for large $|x|$, *Bayreuth. Math. Schr.*, **20**, 153

SOHR, H. AND VON WAHL, W., 1986, On the Regularity of the Pressure of Weak Solutions of Navier-Stokes Equations, *Arch. Math.*, **46**, 428

SOHR, H., VON WAHL, W. AND WIEGNER, W., 1986, Zur Asymptotik der Gleichungen von Navier-Stokes, *Nachr. Akad. Wiss. Göttingen.*, **146**, 1

SOLONNIKOV, V.A., 1964, Estimates of Solutions of Nonstationary Linearized Systems of Navier-Stokes equations, *Trudy Mat. Inst. Steklov*, **70**, 213; English Transl.: A.M.S. Transl., **75**, 1968, 1

STEIN, E.M., 1970, *Singular Integrals and Differentiability Properties of Functions*, Princeton University Press, Princeton

STOKES, G.H., 1845, On the Theories of the Internal Friction of Fluids in Motion *Trans. Cambridge Phil. Soc.*, **8**, 287

STRUWE, M., 1988, On Partial Regularity Results for the Navier-Stokes Equations, *Commun. Pure Appl. Math.*, **41**, 437

TAKAHASHI, S., 1990, On Interior Regularity Criteria for Weak Solutions of the Navier-Stokes Equations, *Manuscripta Math.*, **69**, 237

TAKAHASHI, S., 1992, On a Regularity Criterion up to the Boundary for Weak Solutions of the Navier-Stokes Equations, *Comm. Partial Diff. Eq.*, **17**, 261; Corrigendum: *ibidem*, **19**, 1994, 1015

TANAKA, A., 1987, Régularité par Rapport au Temps des Solutions Faibles de l'Équation de Navier-Stokes, *J. Math. Kyoto Univ.*, **27**, 217

TANIUCHI Y., 1997, On generalized energy equality of the Navier-Stokes equations, *Manuscripta Math*, **94** 365

TEMAM, R., 1977, *Navier-Stokes Equations*, North-Holland

TEMAM, R., 1980, Behavior at $t = 0$ of the Solutions of Semi-Linear Evolution Equations, MRC Technical Summary Report **2162**, Madison: University of Wisconsin

TITCHMARSH, M.A., 1964, *The Theory of Functions*, Oxford University Press

TSAI, T.P., 1998, On Leray's Self-Similar Solutions of the Navier-Stokes Equations Satisfying Local Energy Estimates, *Arch. Ratl Mech. Anal.*, **143**, 29; Corrigendum: *ibidem*, **147**, 1999, 363

VON WAHL, W., 1983, Über das Behlten für $t \to 0$ der Lösungen Nichtlinearer Parabolischer Gleichungen, insbesondere der Gleichungen von Navier-Stokes, Sonderforschungsbereich, Universität Bonn

VON WAHL, W., 1985, *The Equations of Navier-Stokes and Abstract Parabolic Equations*, Vieweg Verlag

VON WAHL, W., 1986, Regularity of Weak Solutions of the Navier-Stokes Equations, *Proc. Symp. Pure Appl. Math.*, **45**, 497

WU, L., 1991, Partial Regularity Problems for Navier-Stokes Equations in Three-Dimensional Space, *Acta Math. Sinica*, **34**, 541

Giovanni P. Galdi
Department of Mechanical Engineering
University of Pittsburgh, USA

Spectral Approximation of Navier-Stokes Equations*

P. Gervasio, A. Quarteroni and F. Saleri

Abstract. We review some basic aspects of spectral methods and their application to the numerical solution of Navier-Stokes equations for viscous incompressible flows.

Introduction

Spectral methods are today widely used for the approximation of partial differential equations. They are intrinsically very accurate, and they become computationally effective also for non-cartesian domains, provided they are used in a domain decomposition framework in which case the domain is split into subdomains that can be mapped into a reference square (or cube). The spectral element method provides an instance of domain decomposition approach.

Spectral methods were initially used for the approximation of differential problems with periodic boundary conditions, with a Galerkin approach and Fourier basis functions. In 1965 the introduction of the FFT algorithm, that allows to *travel* from the physical space and the frequency space by $\mathcal{O}(N \log_2 N)$ floating point operations versus $\mathcal{O}(N^2)$ that would be required otherwise (say using a matrix-vector product), gave a great impulse to the development of the method.

Earliest theoretical results, concerning stability and convergence of spectral methods are due to Gottlieb and Orszag. Afterwards, in the seventhies, spectral methods have been extended to differential problems with non-periodic boundary conditions, by the use of Chebyshev and Legendre systems, for two- and three-dimensional domains. The first estimates of the approximation errors in Sobolev norms, obtained by functional analysis arguments, are due to Canuto and Quarteroni (see [19]).

From the computational viewpoint, the use of a tensorial basis has limited this type of methods to tensorial shaped domains (i.e. deformations of parallelotop domains) until spectral methods were adapted to domains with more general geometries in the eighties: Patera introduced spectral elements ([62], [51]),

* This research has been carried out with the support of M.U.R.S.T. through "Fondi 40%" and C.N.R. "Prog. Speciale Fluidodinamica e Dinamica Molecolare".

Morchoisne proposed the overlapping domain decomposition methods ([56]) and Quarteroni the non-overlapping domain decomposition methods ([68]).

An historical account on the "early stage" of spectral methods is given e.g. in the books [36], [16], [10].

Since their origin spectral methods were applied to the mechanics of viscous incompressible flows. This paper pays special attention to the approximation of the Navier Stokes equations in the primitive variable formulation. In this case, the velocity and the pressure cannot be chosen independently; indeed, a compatibility condition (the so-called Ladyzenskaya-Brezzi-Babuška or inf-sup condition) has to be satisfied in order to avoid spurious modes on the pressure. Moreover, for the Navier Stokes equations for high Reynolds number flows, the dominating convection terms produce another type of instability that generates oscillations on the velocity field.

In the spectral context, the most widely used method to avoid the spurious modes on the pressure is the so-called ($\mathbb{Q}_N - \mathbb{Q}_{N-2}$) approach. It was proposed by Maday, Meiron, Patera and Ronquist ([50]) and it consists of choosing the space of polynomials of degree N (in each variable) for the velocity and of degree $N-2$ for the pressure. The LBB condition id satisfied with a constant β which behaves like $N^{-1/2}$. Another approach, which is less used, consists in using polynomials with the same degree for both the velocity and the pressure with a-posteriori filtering of parasitic modes for the pressure ([5]).

Otherwise, one has to relax the incompressibility condition, in which case the LBB condition doesn't necessarily have to be fulfilled. An example is provided by the adaption to the spectral context (see [34], [17]) of the stabilization techniques that have been originally proposed for the finite element approximation of the Stokes problem (see Brezzi and Pitkaranta [13] and Hughes, Franca and Balestra [39]) and of the incompressible Navier-Stokes equations (see Brooks and Hughes [14] and Franca and Frey [30]).

About the approximation of the time-dependent Navier-Stokes equations we give an account of those numerical methods which can be defined through a separation between temporal and spatial discretization. We deal with semi-implicit finite-difference and fractional step schemes, largely used in the spectral framework. The former avoid the resolution of non-linear systems (contrary to implicit methods) and retain reasonably good stability properties.

The latter are based on the splitting of the differential operator in two or more suboperators such that the original problem can be reduced to the subsequent resolution of simpler differential problems (such as elliptic or Stokes equations). Inside the spectral method context the fractional step schemes (and, in particular, the projection methods) have been used, among others, by Shen ([77]), Pinelli et al. ([66]), Karniadakis et al. ([42]). Splitting methods using Adams-Bashforth or Adams-Moulton multistep schemes were considered by Karniadakis et al. ([42]), Orszag and Kells ([57]) and Marcus ([53]). Operator factor splitting was considered by Maday, Patera and Ronquist ([52]).

We observe that, with the exception of the stabilized approach, all the previous schemes lead to the successive resolution of subproblems that may be either elliptic problems (linear or non-linear) or a generalized form of the Stokes problem. Therefore, in order to achieve an overall globally efficient algorithm, it is mandatory to tackle effectively these computational kernels.

Domain decomposition methods offer a chance in this respect. We review here the domain decomposition formulation of linear elliptic, generalized Stokes and Navier-Stokes problems. In particular, on the Poisson problem we describe different iterative procedure to implement the domain decomposition techniques: the classical iterative substructuring methods such as Dirichlet-Neumann ([68]), Neumann-Neumann ([9]) and Robin ([48]); and the projection decomposition method associated to spectral collocation ([33]). A brief review on the overlapping Schwarz method is done in connection with the method proposed by Pavarino ([63]) (see also Pahl ([60]) and Casarin ([21])).

The Dirichlet-Neumann scheme is also presented for the Stokes problem, while the Navier-Stokes equations are discretized by stabilized spectral elements. The paper ends with some numerical results obtained by the stabilized spectral element approximation of the Navier-Stokes system, which are obtained by the BiCGStab iterations preconditioned by the finite element matrices.

1. Mathematical foundation and different paradigms of spectral methods

Spectral methods are among the most commonly used methods for the approximation of partial differential equations. Historically, spectral methods have been associated with Fourier expansion and they have been applied to approximate periodic functions. However, nowadays they are used indifferently for periodic as well as general boundary-value problems. For the latter, algebraic polynomial expansions (especially Chebyshev's and Legendre's) are used in lieu of Fourier trigonometric polynomials.

The basic idea behind the classical single-domain spectral approach consists of approximating the solution of a differential problem, $u \in W$, by an infinitely differentiable function $u_N \in W_N$, with W_N a subspace of W. The function u_N is a trigonometric polynomial in the Fourier approach or an algebraic polynomial in the Legendre or Chebyshev system. It follows that the accuracy of the approximation is inherently high and actually depends solely on the regularity of the exact solution.

The discrete spectral problem is obtained by the projection of the differential equations to be solved on a finite-dimensional space V_N (possibly coinciding with W_N) by a suitable projection operator P_N and, if $W_N \neq V_N$, it is required that $\dim V_N \equiv \dim W_N$. We refer to W_N and V_N as the space of the trial and test functions, respectively. The choice of $\{P_N, W_N, V_N\}$ characterises completely the type of spectral approximation. Three different spectral schemes, *Galerkin, tau* and *collocation*, are typically used.

The Galerkin and collocation schemes can be applicable to the Fourier or the algebraic approximation, on the contrary the tau approach is applicable solely to problems with non-periodic boundary conditions.

In order to briefly describe the spectral schemes we consider the following linear differential problem:

$$\begin{cases} Lu = f & \text{in } \Omega, \\ Bu = 0 & \text{on } \partial\Omega, \end{cases} \tag{1}$$

where Ω is an open bounded domain in $\mathbb{R}^d$ $(d = 1,2,3)$ of boundary $\partial\Omega$, L is a linear differential operator, B is a set of linear boundary differential operators on $\partial\Omega$ and f a suitable function in $L^2(\Omega)$. The corresponding variational formulation reads:

$$\text{find } u \in W: \ a(u,v) = \mathcal{F}(v) \qquad \forall v \in V, \tag{2}$$

where $a : W \times V \to \mathbb{R}$ is the bilinear form associated to L and $\mathcal{F}$ is a linear continuous functional associated to f and to the boundary differential operator B.

The *Galerkin* approach is characterised by choosing the same space for the test and trial functions (i.e. $W_N = V_N$) and P_N as the classical $L^2(\Omega)$ projection operator from V to V_N, i.e.

$$(P_N u, v_N)_{L^2(\Omega)} = (u, v_N)_{L^2(\Omega)}, \qquad \forall v_N \in V_N. \tag{3}$$

The Galerkin approximation to (2) reads:

$$\text{find } u_N \in V_N: \ a(u_N, v_N) = \mathcal{F}(v_N) \qquad \forall v_N \in V_N. \tag{4}$$

The spectral *tau* approach can be viewed as a special case of the Petrov-Galerkin that, applied to (2) reads:

$$\text{find } u_N \in W_N: \ a(u_N, v_N) = \mathcal{F}(v_N) \qquad \forall v_N \in V_N. \tag{5}$$

The *tau* method was proposed by Lanczos in 1938 ([16]), and it differs from the Galerkin one for the treatment of boundary conditions. In particular, the test functions are not required to satisfy the boundary conditions, since the latter are enforced through a set of supplementary equations. The projection operator P_N is defined as for the Galerkin scheme. We refer to Sec. 1.2 for a more detailed description of tau approach.

Finally, the *collocation* approach is characterised by choosing $V_N = W_N$ as the space of lagrangian polynomials on a set of Gaussian quadrature nodes and by letting P_N to be the interpolation operator with respect to the Gaussian quadrature formula.

The collocation form of (1) reads:

$$\text{find } u_N \in V_N: \ \begin{cases} L_N u_N = f & \text{at } \mathbf{x}_i \in \mathring{\Omega} \\ B_N u_N = 0 & \text{at } \mathbf{x}_i \in \partial\Omega, \end{cases} \tag{6}$$

where L_N and B_N are suitable approximations of L and B, respectively, and $\mathbf{x}_i$ are the Gaussian nodes in $\overline{\Omega}$.

The use of a spectral scheme is also characterised by the way the solution $u_N \in W_N$ is represented, whether in *transform* or in *physical* space.

In transform space a function v is described through its coefficients with respect to a complete orthogonal system (Fourier, Chebyshev or Legendre); in physical space by its nodal values in Ω. For Galerkin or tau approaches, the solution is represented through the transform space, while for the collocation approach it will be represented in the physical space.

1.1. Fourier approximation

For a smooth and periodic complex function $u(x)$, defined on the interval $(0, 2\pi)$, its truncated Fourier series of order N reads

$$P_N u(x) = \sum_{k=-N/2}^{N/2-1} \hat{u}_k \varphi_k(x), \tag{7}$$

with $\varphi_k(x) = e^{ikx}$.

For a suitable N, if $u \in C^\infty(0, 2\pi)$ and it is periodic with all its derivatives in $(0, 2\pi)$, the truncated series (7) is a good approximation of u, thanks to the property that the k-th coefficient $\hat{u}_k$ of the expansion decays faster than any negative power of k. This property is usually referred to as "spectral accuracy" or "exponential accuracy" of the Fourier method.

More generally, given a periodic complex function u defined on the domain $\Omega = (0, 2\pi)^d$ $(d = 2, 3)$, its truncated Fourier series will be:

$$P_N u(\mathbf{x}) = \sum_{k_1,\dots,k_d=-N/2}^{N/2-1} \hat{u}_{\mathbf{k}} \varphi_{\mathbf{k}}(\mathbf{x}), \tag{8}$$

where: $\mathbf{x} = (x_1, \dots, x_d)$ will be an element of $\mathbb{R}^d$, $\mathbf{k} = (k_1, \dots, k_d)$ a multi-index and $\varphi_{\mathbf{j}}(\mathbf{x}) = e^{i\mathbf{k}\cdot\mathbf{x}}$, with $\mathbf{k} \cdot \mathbf{x} = k_1 x_1 + \dots + k_d x_d$.

The following result provides an estimate in all Sobolev norms for the remainder of the Fourier series of u:

$$\|u - P_N u\|_{H^s(\Omega)} \le C N^{s-m} \|u\|_{H^m(\Omega)} \qquad \text{for } 0 \le s \le m. \tag{9}$$

Let us consider the differential problem

$$Lu = f \quad \text{in } \Omega = (0, 2\pi)^d, \qquad u \text{ periodic in } \Omega, \tag{10}$$

with L being a differential operator and f a suitable function defined in $\Omega = (0, 2\pi)^d$. We look for an approximate solution u_N of u by the Galerkin method (4) as follows. We define the space

$$V_N = span\{\varphi_{\mathbf{k}}(\mathbf{x}),\ -N/2 \le k_j \le N/2 - 1,\ j = 1, \dots, d\} \tag{11}$$

and seek a function of the following form

$$u_N(\mathbf{x}) = \sum_{k_1,\dots,k_d=-N/2}^{N/2-1} \hat{a}_{\mathbf{k}} \varphi_{\mathbf{k}}(\mathbf{x}) \tag{12}$$

that satisfies the equations

$$\int_\Omega Lu_N(\mathbf{x})\overline{\varphi}_{\mathbf{k}}(\mathbf{x})d\Omega = \int_\Omega f(\mathbf{x})\overline{\varphi}_{\mathbf{k}}(\mathbf{x})d\Omega \qquad -N/2 \leq k_1,\ldots,k_d \leq N/2-1. \tag{13}$$

By the orthogonality of the Fourier system $\{\varphi_{\mathbf{k}}(\mathbf{x})\}$ in V_N, we obtain the following equations:

$$\widehat{(Lu_N)}_{\mathbf{k}} = \hat{f}_{\mathbf{k}} \qquad -N/2 \leq k_1,\ldots,k_d \leq N/2-1, \tag{14}$$

where $\hat{f}_{\mathbf{k}}$ and $\widehat{(Lu_N)}_{\mathbf{k}}$ are the coefficients of $f(\mathbf{x})$ and $Lu_N(\mathbf{x})$ respectively, with respect to the Fourier expansion.

We observe that 1-D differentiation of $P_N u$ in transform space consists of multiplying each Fourier coefficient $\hat{u}_k$ by the imaginary unity times the corresponding wavenumber, i.e.:

$$(P_N u)'(x) = P_N u'(x) = \sum_{k=-N/2}^{N/2-1} \hat{u}_k ik\varphi_k(x). \tag{15}$$

Similarly for partial derivatives we have:

$$\frac{\partial(P_N u)}{\partial x_j}(\mathbf{x}) = P_N\frac{\partial u}{\partial x_j}(\mathbf{x}) = \sum_{k_1,\ldots,k_d=-N/2}^{N/2-1} \hat{u}_{\mathbf{k}} ik_j\varphi_{\mathbf{k}}(\mathbf{x}), \qquad j=1,\ldots,d. \tag{16}$$

This implies that, if the operator L is linear, the system (14) can be easily reduced to a diagonal system on the coefficients $\hat{a}_{\mathbf{k}}$ of (12). Otherwise convolution sums appear in the system which is not yet diagonal.

As an example of application of the Fourier-Galerkin method we consider the Helmholtz equation

$$\begin{cases} -\Delta u + \lambda u = f & \text{in } \Omega = (0,2\pi)^2 \\ u \text{ periodic} & \text{in } \Omega \end{cases} \tag{17}$$

with $\lambda \geq 0$ and $f \in L^2(\Omega)$. The Fourier-Galerkin solution $u_N \in V_N$ satisfies the equation

$$\begin{gathered}\int_\Omega (-\Delta u_N(\mathbf{x}) + \lambda u_N(\mathbf{x}))e^{-i(k_1x_1+k_2x_2)}\, d\Omega = \int_\Omega f(\mathbf{x})e^{-i(k_1x_1+k_2x_2)}\, d\Omega \\ -N/2 \leq k_1,k_2 \leq N/2-1,\end{gathered} \tag{18}$$

by (12), (16) and by the orthogonality of the basis in V_N we obtain a diagonal system for the $(N+1)^2$ discrete Fourier coefficients $\{\hat{a}_{\mathbf{k}}\}$:

$$(k_1^2+k_2^2+\lambda)\hat{a}_{\mathbf{k}} = \hat{f}_{\mathbf{k}} \qquad -N/2 \leq k_1,k_2 \leq N/2-1, \tag{19}$$

whose solution is very cheap. The coefficients $\hat{a}_{\mathbf{k}}$ are often named "frequency unknowns".

The Fourier Galerkin approach has pros and cons. Among the pros we note:

- its spectral accuracy (that we mentioned above), which means that, if $u \in H^s(\Omega)$, there exists a positive constant C independent of N such that (see [16])
$$\|u - u_N\|_{H^m(\Omega)} \leq CN^{m-s}\|u\|_{H^s(\Omega)} \qquad 0 \leq m \leq s,\ s \geq 0 \tag{20}$$
(this makes spectral methods much more accurate than finite difference or finite element methods); ($H^r(\Omega)$ is the space of functions of $L^2(\Omega)$ whose distributional derivatives of order up to r belong to $L^2(\Omega)$ (see [47]);
- the phase error is minimised; actually Fourier spectral methods are free of phase errors as $\varphi_k'(x) = ik\varphi_k(x),\ \forall k$;
- if L is a linear differential operator with constant coefficients, then the system (14) is diagonal and only $\mathcal{O}(N)$ floating point operations are needed to compute the coefficients $\hat{a}_{\mathbf{k}}$.

On the other hand, the Fourier-Galerkin approach has several drawbacks:

- only periodic solutions can be treated;
- if the differential problem is non-linear and/or it has non-constant coefficients, convolution sums appear in the system (14), with a consequently high computational effort.

The first drawback is overcome by using algebraic polynomials which preserve spectral accuracy (see Sec. 1.2), while for the latter, one resorts to the *pseudo-spectral Fourier method*, whose theoretical properties have been earlier pointed out in Gottlieb and Orszag ([36]) and in Kreiss and Oliger ([44]).

For simplicity of exposition we take now $d = 1$, a similar approach can be followed if $d > 1$ by building the set of nodes in Ω as a cartesian product of the one-dimensional sets of nodes.

We denote by
$$\Xi_N^G = \{x_j = \frac{2\pi}{N}j,\ j = 0, \dots, N-1\} \tag{21}$$
the set of Gaussian nodes on the interval $[0, 2\pi]$, and by $I_N u \in V_N$ the trigonometric interpolant of u on the set Ξ_N^G
$$I_N u(x) = \sum_{k=-N/2}^{N/2-1} \tilde{u}_k \varphi_k(x). \tag{22}$$

The coefficients
$$\tilde{u}_k = \frac{1}{N}\sum_{j=0}^{N-1} u(x_j)e^{-ikx_j} \qquad -N/2 \leq k \leq N/2 - 1 \tag{23}$$
are the "discrete Fourier coefficients" of the complex function u in $[0, 2\pi]$ with respect to the nodes in Ξ_N^G. The polynomial (22) is also known as the discrete

Fourier series of u and (23) is the *discrete Fourier transform* from N nodal values $\{u(x_j)\}$ to N discrete coefficients $\{\tilde{u}_k\}$, (the latter being an approximation of $\{\hat{u}_k\}$). The interpolant (22) is preferred to the truncated series (7) since the coefficients $\hat{u}_k$ of an arbitrary function are not known in closed form, so that they must be approximated starting from the information in the physical space.

Due to the orthogonality of the Fourier system, we have the inversion formula or inverse discrete Fourier transform

$$u(x_j) = \sum_{k=-N/2}^{N/2-1} \tilde{u}_k e^{ikx_j} \qquad j = 0, \ldots, N-1. \tag{24}$$

The convergence properties of the interpolation polynomials $\{I_N u\}_{N\geq 0}$ are similar to those of the sequence of truncated Fourier series $\{P_N u\}_{N\geq 0}$, i.e.

$$\|u - I_N u\|_{H^s(\Omega)} \leq C N^{s-m} \|u\|_{H^m(\Omega)} \qquad \text{for } 0 \leq s \leq m,\ s \geq 0. \tag{25}$$

Furthermore, the continuous and the discrete Fourier coefficients ($\hat{u}_k$ and $\tilde{u}_k$) have the same asymptotic behaviour.

We define the *pseudo-spectral derivative* of $u(x)$ as the exact derivative of its interpolant polynomial $I_N u(x)$, i.e.

$$\mathcal{D}_N u := (I_N u)'. \tag{26}$$

If $u \notin V_N$ then $(I_N u)' \neq I_N(u')$, but the difference $\|(I_N u)' - I_N(u')\|$ is of the same order as $\|u' - P_N u'\|$ so that the pseudo-spectral derivative is spectrally accurate.

The Fourier pseudo-spectral derivative $\mathcal{D}_N u$ can be represented in the Gaussian nodes by a matrix-vector product, i.e.:

$$(\mathcal{D}_N u)(x_l) = \sum_{j=0}^{N-1} (D_N)_{lj} u(x_j) \qquad l = 0, \ldots, N-1 \tag{27}$$

with $D_N \in \mathbb{R}^{N\times N}$ and

$$(D_N)_{lj} = \begin{cases} \frac{1}{2}(-1)^{l+j} cot \left[\frac{(l-j)\pi}{N}\right] & l \neq j \\ 0 & l = j \end{cases} \qquad l, j = 0, \ldots, N-1. \tag{28}$$

The pseudo-spectral counterpart L_N of a differential operator L is obtained by substituting each derivative in L by a pseudo-spectral derivative; for example, if $Lu := -(a\ u')' + (bu)' + cu$, with $a = a(x)$, $b = b(x)$ and $c = c(x)$, then the pseudo-spectral operator L_N reads:

$$L_N u = -(I_N(a\ u_N'))' + (I_N(bu_N))' + cu_N. \tag{29}$$

In vector form (29) reads:

$$-D_N(\mathbf{a} \otimes (D_N \mathbf{u})) + D_N(\mathbf{b} \otimes \mathbf{u}) + \mathbf{c} \otimes \mathbf{u}, \tag{30}$$

where: $\mathbf{u} = (u_N(x_0), \ldots, u_N(x_{N-1}))^T$, $\mathbf{a} = (a(x_0), \ldots, a(x_{N-1}))^T$, $\mathbf{b} = (b(x_0), \ldots, b(x_{N-1}))^T$, $\mathbf{c} = (c(x_0), \ldots, c(x_{N-1}))^T$ and $\mathbf{a} \otimes \mathbf{b}$ denotes the point-wise product of two vectors $\mathbf{a}$ and $\mathbf{b}$.

We can now approximate the differential problem (1) at Gaussian nodes by a collocation approach by solving the system (not-diagonal in general):

$$L_N u_N(x_j) = f(x_j) \qquad \text{at } x_j \in \Xi_N^G. \tag{31}$$

The unknowns $\{u_N(x_j)\}_{j=0}^{N-1}$ of the system are said "physical unknowns", when they are available, an approximation $\tilde{a}_k$ of the frequency unknowns $\hat{a}_k$ (the coefficients of u_N in (43)) can be computed via a discrete Fourier transform as follows:

$$\tilde{a}_k = \frac{1}{N}\sum_{j=0}^{N-1} u_N(x_j)e^{-ikx_j} \qquad -N/2 \le k \le N/2-1. \tag{32}$$

We observe that the discrete Fourier transform (23) can be used also to easily compute the pseudo-spectral derivative of a function u that is known by its nodal values $u(x_j)$. In fact, in order to compute $\mathcal{D}_N u(x_j)$ one can "travel" from the nodal values $\{u(x_j),\ j = 0,\dots,N-1\}$ to the discrete coefficients $\{\tilde{u}_k,\ -N/2 \le k \le N/2-1\}$ by a Discrete Fourier Transform (23), compute $\tilde{u}_k^{(1)} = ik\tilde{u}_k$ (which are the discrete coefficients of $\mathcal{D}_N u$) and then, go back to the physical space, by an inverse Discrete Fourier Transform (24).

A great impulse to the development of the pseudo-spectral Fourier approach was given by Cooley and Tukey ([23]) in 1965 by the introduction of the Fast Fourier Transform (FFT) algorithm that permits to compute the discrete Fourier coefficients $\{\tilde{u}_k\}$ from the nodal values $\{u_N(x_j)\}$ in $\mathcal{O}(N\ \log_2 N)$ floating point operations (provided that N is a power of 2).

In order to treat the case of general boundary conditions, we introduce now the algebraic polynomial systems.

1.2. Galerkin spectral methods for problems with non-periodic boundary conditions

The property of spectral accuracy (20) is also attainable for smooth but non-periodic functions, provided that the expansion functions are properly chosen. Both Legendre and Chebyshev systems (see [16]) allow spectral accuracy in the expansion of smooth functions, they are algebraic polynomials and satisfy the orthogonality property with respect to a suitable weighted inner product in $L^2(-1,1)$.

From now on we denote by $\{T_k(x)\}_{k\ge0}$ the set of Chebyshev polynomials:

$$\begin{aligned} &T_0(x) \equiv 1, \qquad T_1(x) = x,\\ &T_{k+1}(x) = 2x\, T_k(x) - T_{k-1}(x) \quad k \ge 1 \end{aligned} \tag{33}$$

and by $\{L_k(x)\}_{k\ge0}$ the set of Legendre polynomials:

$$\begin{aligned} &L_0(x) \equiv 1, \qquad L_1(x) = x,\\ &L_{k+1}(x) = \frac{2k+1}{k+1}xL_k(x) - \frac{k}{k+1}L_{k-1}(x) \quad k \ge 1. \end{aligned} \tag{34}$$

They are orthogonal with respect to the weighted inner product

$$(u,v)_w := \int_{-1}^{1} u(x)v(x)w(x)dx, \tag{35}$$

where the weight is

$$w(x) = \begin{cases} \dfrac{1}{\sqrt{(1-x^2)}} & \text{for the Chebyshev system} \\ 1 & \text{for the Legendre system.} \end{cases} \tag{36}$$

The same notation used in (35) will be considered with functions defined on a two- or three-dimensional domain. In this case the weight will be the product of the weight functions given in (36) (see [16], Sec. 2.4 for more details).

1.2.1. THE GALERKIN APPROACH To deal with the weak formulation of (1) we denote by $H^1(\Omega)$ the space of functions $v \in L^2(\Omega)$ whose first-order distributional derivatives belong to $L^2(\Omega)$, endowed with the norm

$$\|v\|_{H^1(\Omega)} = \{\|v\|^2_{L^2(\Omega)} + \|\nabla v\|^2_{L^2(\Omega)}\}^{1/2}. \tag{37}$$

The symbol $H^1_0(\Omega)$ is used to denote the subspace of $H^1(\Omega)$ of the functions whose trace at the boundary is zero, i.e. $H^1_0(\Omega) = \{v \in H^1(\Omega) : \ v|_{\partial\Omega} = 0\}$.

Consider the Poisson equation in the unit square $\Omega = (-1,1)^2$ with homogeneous Dirichlet boundary condition:

$$-\Delta u = f \text{ in } \Omega, \qquad u = 0 \text{ on } \partial\Omega \tag{38}$$

and $f \in L^2(\Omega)$. This problem admits the variational form (2) with

$$V = H^1_0(\Omega), \quad a(u,v) = \int_\Omega \nabla u \cdot \nabla v d\Omega \ \text{ and } \ \mathcal{F}(v) = \int_\Omega fv \, d\Omega. \tag{39}$$

We denote by $\mathbb{P}_N(I)$ the set of the algebraic polynomials of degree less than or equal to N on the interval $I \subset \mathbb{R}$, and by $\mathbb{Q}_N(\Omega)$ the set of algebraic polynomials of degree less than or equal to N in each variable in Ω. Then we look for the solution of the Galerkin method taking

$$V_N = \{v \in \mathbb{Q}_N(\Omega) : v = 0 \text{ on } \partial\Omega\} \tag{40}$$

We choose in V_N the following basis functions $\varphi_{\mathbf{k}}(\mathbf{x}) = \varphi_{k_1}(x_1)\varphi_{k_2}(x_2)$, $2 \leq k_1, k_2 \leq N$ with

$$\varphi_k(x) = \begin{cases} T_k(x) - T_0(x) & k \text{ even} \\ T_k(x) - T_1(x) & k \text{ odd} \end{cases} \tag{41}$$

or

$$\varphi_k(x) = \begin{cases} L_k(x) - L_0(x) & k \text{ even} \\ L_k(x) - L_1(x) & k \text{ odd.} \end{cases} \tag{42}$$

Then we expand the numerical solution u_N with respect to this basis:

$$u_N(\mathbf{x}) = \sum_{k_1,k_2=2}^{N} a_{\mathbf{k}}\varphi_{\mathbf{k}}(\mathbf{x}). \tag{43}$$

Since the basis functions $\varphi_{\mathbf{k}}(\mathbf{x})$ satisfy the homogeneous Dirichlet boundary conditions given in (38) no explicit boundary conditions will be imposed on u_N.

The Galerkin formulation of (38) reads:

$$\text{find } u_N \in V_N : \qquad (\nabla u_N, \nabla\varphi_{\mathbf{j}})_w = (f, \varphi_{\mathbf{j}})_w \qquad 2 \le j_1, j_2 \le N \tag{44}$$

and owing to (43), this yields the following system for the $\{a_{\mathbf{k}}\}$:

$$\sum_{k_1,k_2=2}^{N} a_{\mathbf{k}}(\nabla\varphi_{\mathbf{k}}, \nabla\varphi_{\mathbf{j}})_w = (f, \varphi_{\mathbf{j}})_w \qquad 2 \le j_1, j_2 \le N. \tag{45}$$

The matrix form of (45) reads:

$$S\mathbf{a} = \mathbf{f} \tag{46}$$

where $S = [(\nabla\varphi_{\mathbf{k}}, \nabla\varphi_{\mathbf{j}})_w]$ is the stiffness matrix, $\mathbf{a} = [a_{\mathbf{k}}]$ is the vector of unknowns and $\mathbf{f} = [(f, \varphi_{\mathbf{j}})_w]$ is the right-hand side.

The matrix S is a structured sparse matrix and it is symmetric only if the Legendre system is used.

We observe that, in general, this approach is scarcely used in the spectral context since, for large values of N, a large computational effort is required to evaluate both the bilinear form and the right-hand side by sufficiently accurate quadrature formulas.

1.2.2. THE SPECTRAL TAU APPROACH The main feature of this approach is that the test space V_N and the trial space W_N do not coincide. As for the Galerkin approach the operator P_N is the orthogonal projection operator from W to V_N relative to the inner product of V. Let Ξ_N^{CGL} be the set of nodes of the Gauss-Lobatto quadrature formulas of Chebyshev type:

$$\Xi_N^{CGL} = \{\xi_0 = -1,\ \xi_i = -cos(i\pi/N),\ i = 1, \dots, N-1,\ \ \xi_N = 1\} \tag{47}$$

and Ξ_N^{LGL} the set of Legendre type:

$$\Xi_N^{LGL} = \{\xi_0 = -1,\ \ \xi_N = 1,\ \xi_i, i = 1, \dots, N-1, \text{ are the zeros of } L'_N(\xi)\}. \tag{48}$$

Then we define $\boldsymbol{\Xi}_N^{CGL} = \Xi_N^{CGL} \times \Xi_N^{CGL}$ and denote by $\mathbf{x}_{ij} = (\xi_i, \xi_j)$ its elements. (The same notation holds for the Legendre set of nodes.) We consider the problem (1) on the computational domain $\Omega \subset \mathbb{R}^2$ and the Petrov-Galerkin approach (5), where we define the trial space as

$$W_N = \{v \in \mathbb{Q}_N(\Omega) :\ Bv(\mathbf{x}_{ij}) = 0 \text{ at } \mathbf{x}_{ij} \in \partial\Omega\}, \tag{49}$$

while the choice of V_N depends on the number of boundary conditions which are imposed on each side of $\partial\Omega$.

As an example we consider the Neumann problem for a second-order elliptic operator in the interval $(-1, 1)$:

$$\begin{cases} Lu := -u'' + u = f & -1 < x < 1 \\ u'(-1) = u'(1) = 0, \end{cases} \tag{50}$$

and we look for the tau solution

$$u_N \in W_N = \{v \in \mathbb{P}_N(-1,1) : \; v'(-1) = v'(1) = 0\} \tag{51}$$

expanded in Chebyshev polynomials, i.e.: $u_N(x) = \sum_{k=0}^{N} \tilde{a}_k T_k(x)$.
Here $V_N = \mathbb{P}_{N-2}(-1,1) = span\{T_k(x), \; k = 0, \ldots, N-2\}$ and the solution u_N is determined by the system:

$$\begin{cases} \displaystyle\int_{-1}^{1} (-u_N'' + u_N)(x) T_k(x) w(x) \, dx = \int_{-1}^{1} f(x) T_k(x) w(x) \, dx \\ \qquad\qquad\qquad\qquad\qquad\qquad \text{for } k = 0, \ldots, N-2 \\ u_N'(-1) = u_N'(1) = 0. \end{cases} \tag{52}$$

Since the derivatives of u_N can be expanded with respect to the Chebyshev system as follows (see [16], Sec. 2.4.2):

$$u_N'(x) = \sum_{k=0}^{N-1} \tilde{a}_k^{(1)} T_k(x) \tag{53}$$

with

$$\tilde{a}_k^{(1)} = \frac{2}{c_k} \sum_{\substack{p=k+1 \\ p+k \; odd}}^{N} p\tilde{a}_p \qquad c_0 = 2, \; c_k = 1 \text{ for } k \geq 1, \tag{54}$$

and

$$u_N''(x) = \sum_{k=0}^{N-2} \tilde{a}_k^{(2)} T_k(x) \tag{55}$$

with

$$\tilde{a}_k^{(2)} = \frac{1}{c_k} \sum_{\substack{p=k+2 \\ p+k \; even}}^{N} p(p^2 - k^2)\tilde{a}_p, \tag{56}$$

we can transform the system (52) in terms of the unknown coefficients $\{\tilde{a}_k\}$ and obtain:

$$\begin{cases} -\dfrac{1}{c_k} \displaystyle\sum_{\substack{p=k+2 \\ p+k \ even}}^{N} p(p^2-k^2)\tilde{a}_p + \tilde{a}_k = \hat{f}_k & k = 0,\dots,N-2 \\ \displaystyle\sum_{k=0}^{N-1} \frac{(-1)^k}{c_k} \sum_{\substack{p=k+1 \\ p+k \ odd}}^{N} p\tilde{a}_p = \sum_{k=0}^{N-1} \frac{1}{c_k} \sum_{\substack{p=k+1 \\ p+k \ odd}}^{N} p\tilde{a}_p = 0. \end{cases} \tag{57}$$

Following [16] (Sec. 5.1.2), the Neumann boundary conditions in (57) can be rewritten as

$$\sum_{\substack{k=1 \\ k \ odd}}^{N} k^2\tilde{a}_k = \sum_{\substack{k=2 \\ k \ even}}^{N} k^2\tilde{a}_k = 0, \tag{58}$$

so that the even and odd coefficients in (57) can be decoupled and a block diagonal linear system has to be solved.

The matrix associated to the linear system for the even (or odd) coefficients has an upper Hessenberg structure and a Gauss elimination without pivoting can be used with a computational effort of the order of $N^2/4$ operations.

In general the accuracy of this approach is less than the one from Galerkin's approach if the same number of degrees of freedom is used.

We have seen that both Galerkin and tau spectral methods are formulated in terms of the frequency unknowns $\{\tilde{a}_k\}$ of the numerical solution u_N. As such, they are cumbersome to handle for general problems with variables coefficients and/or nonlinearities, due to the presence of convolution sums. As already seen in the periodic case, this leads to the use of spectral methods with nodal variables, i.e. to collocation or generalized Galerkin spectral methods.

1.3. Collocation spectral methods

These collocation methods are in general amenable to Galerkin spectral methods since they make use of Gaussian nodes for the numerical evaluation of integrals.

Consider the two-dimensional elliptic problem with homogeneous Dirichlet boundary conditions:

$$\begin{cases} Lu := -\nabla\cdot(\nu\nabla u) + \gamma u = f & \text{in } \Omega = (-1,1)^2 \\ u = 0 & \text{on } \partial\Omega \end{cases} \tag{59}$$

with $f \in L^2(\Omega)$, $\nu,\ \gamma \in L^\infty(\Omega)$ non-negative.

Given $N \geq 2$, we consider the 2-D Chebyshev Gauss-Lobatto set Ξ_N^{CGL} (or the Legendre Gauss-Lobatto set Ξ_N^{LGL}) and we look for a function $u_N \in \mathbb{Q}_N(\Omega)$ such that:

$$\begin{cases} -\nabla\cdot(\nu(\mathbf{x}_{ij})\nabla u_N(\mathbf{x}_{ij})) + \gamma(\mathbf{x}_{ij})u_N(\mathbf{x}_{ij}) = f(\mathbf{x}_{ij}) & \forall \mathbf{x}_{ij} \in \mathring{\Omega} \\ u_N(\mathbf{x}_{ij}) = 0 & \forall \mathbf{x}_{ij} \in \partial\Omega. \end{cases} \tag{60}$$

The system (60) is referred to as the *point-wise form of the spectral collocation method.*

The primary unknowns of the collocation problem (60) are the nodal values $u_N(\mathbf{x}_{ij})$. They are computed by solving the linear system that arises from the pseudo-spectral approximation of (60), i.e. by substituting each derivative in (59) by a pseudo-spectral derivative as done for the Fourier expansion.

In order to derive the algebraic form of (60) we introduce the Lagrangian basis functions associated to the nodes Ξ_N^{CGL} or Ξ_N^{LGL}:

$$\psi_i \in \mathbb{P}_N(-1,1),\ \psi_i(\xi_j) = \delta_{ij},\ i,j = 0,\dots,N, \tag{61}$$

and the 2-D Lagrangian basis associated to the sets $\mathbf{\Xi}_N^{CGL}$ or $\mathbf{\Xi}_N^{LGL}$ defined by a tensorial product as:

$$\varphi_{ij}(\mathbf{x}) = \psi_i(x_1)\psi_j(x_2) \in \mathbb{Q}_N(\Omega),\ i,j = 0,\dots,N. \tag{62}$$

Clearly $\varphi_{ij}(\mathbf{x}_{kl}) = \delta_{ik}\delta jl$. The Lagrange interpolant of $u \in \mathcal{C}^0(\overline{\Omega})$ on the nodes of $\mathbf{\Xi}_N^{LGL}$ or $\mathbf{\Xi}_N^{CGL}$ is

$$I_N u(\mathbf{x}) = \sum_{i,j=0}^{N} u(\mathbf{x}_{ij})\varphi_{ij}(\mathbf{x}), \tag{63}$$

and the solution $u_N(\mathbf{x})$ of (60) reads:

$$u_N(\mathbf{x}) = \sum_{i,j=0}^{N} u_N(\mathbf{x}_{ij})\varphi_{ij}(\mathbf{x}). \tag{64}$$

The pseudo-spectral partial derivatives of a regular function u are defined as follows:

$$(\mathcal{D}_{N,x_m} u)(\mathbf{x}) = \frac{\partial I_N u}{\partial x_m}(\mathbf{x}) = \sum_{i,j=0}^{N} u(\mathbf{x}_{ij}) \frac{\partial \varphi_{ij}}{\partial x_m}(\mathbf{x}), \quad m = 1,2. \tag{65}$$

Setting

$$(D_{N,1})_{kl,ij} = \frac{\partial \varphi_{ij}}{\partial x_1}(\mathbf{x}_{kl}), \qquad (D_{N,2})_{kl,ij} = \frac{\partial \varphi_{ij}}{\partial x_2}(\mathbf{x}_{kl}), \tag{66}$$

for $i,j,k,l = 0,\dots,N$, they can be represented by the matrix-vector form:

$$(\mathcal{D}_{N,x_m})u(\mathbf{x}_{kl}) = \sum_{i,j=0}^{N} (D_{N,m})_{kl,ij}\ u(\mathbf{x}_{ij}), \qquad m = 1,2. \tag{67}$$

The squared matrices in (66) of order $(N+1)^2$, are the "pseudo-spectral derivative matrices". They have a sparse structure, with $(N+1)^3$ non-zero entries.

As for the Fourier system, an alternative way to compute the pseudo-spectral derivative of u consists in passing from the physical space to the frequency space, differentiating and then coming back to the physical space. This procedure is efficient for Chebyshev expansions for large N, as a fast transform is still available.

When Ω is a rectangular domain, the matrices (66) can be obtained from their one-dimensional counterpart, the pseudo-spectral derivative matrix D_N, of order $(N+1)$. In fact, the x_1 (resp. x_2) pseudo-spectral derivative is the same on all rows (resp. columns) of the collocation grid, i.e.

$$(\mathcal{D}_{N,x_1})(\mathbf{x}_{kl}) = \psi_i'((x_1)_k)\psi_j((x_2)_l) = \psi_i'((x_1)_k)\delta_{jl} \tag{68}$$

and

$$(\mathcal{D}_{N,x_2})(\mathbf{x}_{kl}) = \psi_i((x_1)_k)\psi_j'((x_2)_l) = \delta_{ik}\psi_j'((x_2)_l). \tag{69}$$

If we consider the set of Chebyshev Gauss-Lobatto nodes (47), and we write the Lagrange polynomials $\psi_i(x_1)$, $\psi_j(x_2)$ in terms of the Chebyshev functions, the pseudo-spectral derivative matrix in one dimension has the following entries:

$$(D_N)_{lj} = \begin{cases} \dfrac{\bar{c}_l}{\bar{c}_j}\dfrac{(-1)^{l+j}}{\xi_l-\xi_j} & l \neq j \\ \dfrac{-\xi_j}{2(1-\xi_j^2)} & 1 \leq l = j \leq N-1 \\ -\dfrac{2N^2+1}{6} & l = j = 0 \\ \dfrac{2N^2+1}{6} & l = j = N \end{cases} \tag{70}$$

with

$$\bar{c}_l = \begin{cases} 2 & l = 0, N \\ 1 & 1 \leq l \leq N-1. \end{cases} \tag{71}$$

Similarly, for the Legendre system (48) we have:

$$(D_N)_{lj} = \begin{cases} \dfrac{L_N(x_l)}{L_N(x_l)}\dfrac{1}{\xi_l-\xi_j} & l \neq j \\ 0 & 1 \leq l = j \leq N-1 \\ -\dfrac{(N+1)N}{4} & l = j = 0 \\ \dfrac{(N+1)N}{4} & l = j = N. \end{cases} \tag{72}$$

The system (60) can be written in matrix form through the following matrix

$$A_{sp}^c = -D_{N,1}VD_{N,1} - D_{N,2}VD_{N,2} + G \tag{73}$$

with $V_{ij,kl} = \nu(\mathbf{x}_{ij})\delta_{ik}\delta_{jl}$ and $G_{ij,kl} = \gamma(\mathbf{x}_{ij})\delta_{ik}\delta_{jl}$ (for $i, j, k, l, = 0, \ldots, N$).

The matrix A_{sp}^c is the algebraic counterpart of the differential operator of (60). Its structure is represented in Fig. 1. By substituting the rows of A_{sp}^c associated to the boundary nodes with the rows of the identity matrix (corresponding to Dirichlet boundary conditions), the algebraic form of (60) reads

$$A_{sp}^c\mathbf{u} = \mathbf{f} \tag{74}$$

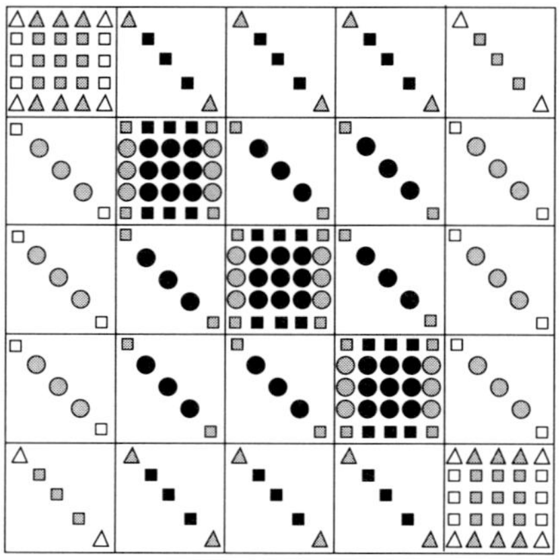

FIGURE 1. The pattern of matrix S

where

$$\mathbf{u}_{ij} = u_N(\mathbf{x}_{ij}), \text{ and } \mathbf{f}_{ij} = \begin{cases} f(\mathbf{x}_{ij}) & \forall \mathbf{x}_{ij} \in \mathring{\Omega} \\ 0 & \forall \mathbf{x}_{ij} \in \partial\Omega. \end{cases} \tag{75}$$

We observe that the system can be reduced by eliminating the unknowns associated to the nodes on the boundary.

The condition number of A^c_{sp} is $\chi_{sp}(A^c_{sp}) = \mathcal{O}(N^4)$ so that, when using iterative methods, the use of preconditioners is mandatory.

In 1980 Orszag introduced the centred finite difference preconditioner, i.e. the matrix obtained by the approximation of the same differential operator by finite difference on the Gauss-Lobatto grid. This is an optimal preconditioner for the spectral collocation since the eigenvalues of the centred finite difference derivative matrix on the Gauss-Lobatto grid and the pseudo-spectral matrix D_N have the same asymptotic behaviour.

Later, in 1985, the finite element matrix was proposed by Deville and Mund ([25]) and by Canuto and Quarteroni ([20]), using bilinear finite elements on the grid induced by the Gauss-Lobatto nodes. Effective generalizations of the finite element preconditioner have been carried out later by Canuto and Pietra ([18]), Deville and Mund ([26]) and by Quarteroni and Zampieri ([73]). For elliptic equations, both finite differences and finite elements matrices lead to a preconditioned matrix whose condition number is independent of N, see Haldenwang, Labrosse, Abboudi and Deville ([38]).

1.4. Generalized Galerkin method

The Generalized Galerkin (GG) method is obtained from a Galerkin method in which every integral is replaced by suitable quadrature formulas.

In particular for the spectral methods, the Gauss-Lobatto quadrature formulas are used. For any function $u \in \mathcal{C}^0([-1,1])$ the Gauss Lobatto quadrature formulas read as follows:

$$\int_{-1}^{1} u(\xi)w(\xi)d\xi \simeq \sum_{i=0}^{N} u(\xi_i)w_i, \tag{76}$$

where the values ξ_i and w_i are said respectively nodes and weights of the quadrature formulas. The nodes of the Chebyshev Gauss-Lobatto quadrature formulas are those given in (47) and the weights are

$$w_i = \begin{cases} \dfrac{\pi}{2N} & i = 0, N \\ \dfrac{\pi}{N} & i = 1, \dots, N-1, \end{cases} \tag{77}$$

while the Legendre Gauss-Lobatto nodes are those given in (48) and the weights are:

$$w_i = \frac{2}{N(N+1)} \frac{1}{L_N^2(\xi_j)} \qquad i = 0, \dots, N. \tag{78}$$

The Gauss-Lobatto quadrature formulas are exact for polynomials of degree less than or equal to $2N-1$. For any fixed $N \geq 1$, the nodes for 2-D quadrature formulas are those given in Ξ_N^{LGL} or Ξ_N^{CGL} and the weights are $w_{ij} = w_i w_j$ for $i, j = 0, \dots, N$. Then, $\forall u, v \in \mathcal{C}^0(\overline{\Omega})$, we define the discrete scalar product:

$$(u, v)_{N,\Omega} = \sum_{i,j=0}^{N} u(\mathbf{x}_{ij})v(\mathbf{x}_{ij})w_{ij} \tag{79}$$

and we note that

$$(u, 1)_{N,\Omega} = \int_{\Omega} u(\mathbf{x})w(\mathbf{x})d\Omega \qquad \forall u \in \mathbb{Q}_{2N-1}(\Omega). \tag{80}$$

The interpolation on the Legendre Gauss-Lobatto nodes satisfies the following bounds (see [19]): if $u \in H^s(\Omega)$, for some $s \geq 2$

$$\|u - I_N u\|_{H^k(\Omega)} \leq CN^{k-s}\|u\|_{H^s(\Omega)} \qquad \text{for } k = 0, 1. \tag{81}$$

Thanks to (81) it holds that: if $u \in H^s(\Omega)$, for some $s \geq 2$ and $v_N \in \mathbb{Q}_N(\Omega)$, then there exists $C > 0$ independent of N such that

$$|(u, v_N)_{L^2(\Omega)} - (u, v_N)_{N,\Omega}| \leq CN^{-s}\|u\|_{H^s(\Omega)}\|v_N\|_{L^2(\Omega)}. \tag{82}$$

Similar estimates hold for the interpolation on the Chebyshev Gauss-Lobatto nodes, by replacing the classical Sobolev norm in H^k with the weighted Sobolev norm.

By setting $V_N = \mathbb{Q}_N(\Omega) \cap H_0^1(\Omega)$, and

$$a_N(u_N, v_N) := (\nu\nabla u_N, \nabla v_N)_{N,\Omega} + (\gamma u_N, v_N)_{N,\Omega} \qquad \forall u_N, v_N \in V_N, \tag{83}$$

the GG formulation of problem (59) reads:

$$\text{find } u_N \in V_N: \qquad a_N(u_N, v_N) = (f, v_N)_{N,\Omega} \qquad \forall v_N \in V_N. \tag{84}$$

In the case of Dirichlet boundary condition, the Generalized Galerkin approach is equivalent to the collocation approach introduced in the previous section.

For all $u_N, v_N \in \mathbb{Q}_N(\Omega)$ we start noticing that

$$(\nu \nabla u_N, \nabla v_N)_{N,\Omega} = -(\nabla \cdot I_N(\nu \nabla u_N), v_N)_{N,\Omega} + \sum_{k=1}^{4} \left(\nu \frac{\partial u_N}{\partial \mathbf{n}}, v_N \right)_{N,S_k} \tag{85}$$

where S_k denotes the k-th side of the boundary $\partial\Omega$ and $\mathbf{n}$ the outward normal vector to $\partial\Omega$. To obtain the system (60) starting from (84), it is sufficient to put $v_N(\mathbf{x}) = \varphi_{ij}$ (for $i, j = 1, \dots, N-1$) in (84), use the identity (85) and to eliminate the sum in (85) thanks to the homogeneous Dirichlet boundary conditions.

By the linearity of the problem and by (43), the generalized Galerkin formulation (84) also reads:

$$\sum_{k,l=0}^{N} a_N(\varphi_{kl}, \varphi_{ij}) u_N(\mathbf{x}_{kl}) = (f, \varphi_{ij})_{N,\Omega} \qquad i, j = 0, \dots, N, \tag{86}$$

or again

$$A_{sp}\mathbf{u} = \mathbf{f} \tag{87}$$

with $(A_{sp})_{ij,kl} = a_N(\varphi_{kl}, \varphi_{ij})$ and $(\mathbf{f})_{ij} = (f, \varphi_{ij})_{N,\Omega}$ for the internal nodes, $(A_{sp})_{ij,kl} = \delta_{ki}\delta_{lj}$ and $(\mathbf{f})_{ij} = 0$ for the Dirichlet boundary nodes.

The matrix A_{sp} differs from the collocation matrix A^c_{sp} on the rows associated to internal nodes, for the presence of the quadrature weights, more precisely:

$$(A_{sp})_{ij,kl} = w_{ij}(A^c_{sp})_{ij,kl} \qquad \forall i, j, k, l = 1, \dots, N-1. \tag{88}$$

Even if the collocation and the GG approaches are equivalent, the latter formulation is often preferred, in fact:

1. the condition number of A_{sp} is $\chi(A_{sp}) = \mathcal{O}(N^3)$ versus $\chi(A^c_{sp}) = \mathcal{O}(N^4)$;
2. by eliminating the rows and the columns associated to the Dirichlet nodes, the matrix A_{sp} becomes symmetric.

Collocation and GG approaches differ on the treatment of Neumann data. Set $\mathbf{n} = (n_x, n_y)$ and assign the Neumann boundary condition $\partial u/\partial \mathbf{n} = g$ on an open subset $\partial\Omega_N$ of $\partial\Omega$, with $g \in L^2(\partial\Omega_N)$.

When using a collocation approach, the Neumann boundary condition is strongly enforced on the boundary nodes and it reads:

$$\begin{aligned} \nu(\mathbf{x}_{ij}) \frac{\partial u_N}{\partial \mathbf{n}}(\mathbf{x}_{ij}) &= \nu(\mathbf{x}_{ij}) \frac{\partial u_N}{\partial x}(\mathbf{x}_{ij}) n_x(\mathbf{x}_{ij}) + \nu(\mathbf{x}_{ij}) \frac{\partial u_N}{\partial y}(\mathbf{x}_{ij}) n_y(\mathbf{x}_{ij}) \\ &= \nu(\mathbf{x}_{ij}) n_x(\mathbf{x}_{ij}) (D_{N,1}\mathbf{u})_{ij} + \nu(\mathbf{x}_{ij}) n_y(\mathbf{x}_{ij}) (D_{N,2}\mathbf{u})_{ij} = g(\mathbf{x}_{ij}). \end{aligned} \tag{89}$$

The matrix A^c_{sp} is updated on the rows associated to Neumann boundary nodes with the terms $(A^c_{sp})_{ij,kl} = \nu(\mathbf{x}_{ij}) n_x(\mathbf{x}_{ij}) (D_{N,1})_{ij,kl} + \nu(\mathbf{x}_{ij}) n_y(\mathbf{x}_{ij}) (D_{N,2})_{ij,kl}$.

Otherwise, thanks to the equivalence (85), inside the GG approach the Neumann condition is expressed in weak form by the equations

$$a_N(u_N, \varphi_{ij}) = (f, \varphi_{ij})_{N,\Omega} + \sum_{\substack{k=1 \\ S_k \subset \partial\Omega_N}}^{4} (g, \varphi_{ij}|_{S_k})_{N,S_k}, \qquad \forall i,j:\ \mathbf{x}_{ij} \in \partial\Omega. \quad (90)$$

The matrix A_{sp} is not changed, while now the component $\mathbf{f}_{ij}$ of the right-hand side will contain the right-hand side of (90). It is worthwhile observing that Neumann boundary conditions expressed in weak form do not affect the symmetry of the matrix A_{sp}, contrary to what happens for the strong form (89).

For the convergence analysis, the Strang lemma ([71]) can be advocated. For that consider the linear differential problem (1) and the following GG approximation:

$$\text{find } u_N \in V_N : a_N(u_N, v_N) = \mathcal{F}_N(v_N) \qquad \forall v_N \in V_N. \quad (91)$$

Theorem 1.1. (Strang lemma) *Let V be a Hilbert space endowed with the norm $\|\cdot\|$, $a : V \times V \to \mathbb{R}$ a bilinear and continuous form, i.e.:*

$$\exists \gamma > 0 :\ |a(w,v)| \le \gamma \|w\|\,\|v\| \qquad \forall w, v \in V \quad (92)$$

and $\mathcal{F} : V \to \mathbb{R}$ a linear continuous functional.

Suppose further that $\mathcal{F}_N(\cdot)$ is a linear map and the bilinear form $a_N(\cdot,\cdot)$ is uniformly coercive over $V_N \times V_N$. This means that there exists a positive constant α^* such that for all $N \ge 1$ it holds

$$a_N(v_N, v_N) \ge \alpha^* \|v_N\|^2 \qquad \forall v_N \in V_N. \quad (93)$$

Then there exists a unique solution u_N of (91), which satisfies

$$\|u_N\| \le \frac{1}{\alpha^*} \sup_{\substack{v_N \in V_N \\ v_N \neq 0}} \frac{\mathcal{F}_N(v_N)}{\|v_N\|}. \quad (94)$$

Moreover, if u is the solution to (2), then

$$\begin{aligned} \|u - u_N\| \le \inf_{w_N \in V_N} \Bigg[\left(1 + \frac{\gamma}{\alpha^*}\right) \|u - w_N\| \\ + \frac{1}{\alpha^*} \sup_{\substack{v_N \in V_N \\ v_N \neq 0}} \frac{|a(w_N, v_N) - a_N(w_N, v_N)|}{\|v_N\|} \Bigg] \\ + \frac{1}{\alpha^*} \sup_{\substack{v_N \in V_N \\ v_N \neq 0}} \frac{|\mathcal{F}(v_N) - \mathcal{F}_N(v_N)|}{\|v_N\|}. \end{aligned} \quad (95)$$

□

If we take $V = H_0^1(\Omega)$, if the bilinear form a_N is that given in (83), $\mathcal{F}_N(v_N) := (f, v_N)_{N,\Omega}$ and u_N is the GG solution of (84), we have a more detailed result. In fact, by taking the interpolant of u of degree $N-1$ in each variable in place of w_N in (1.1), the term $\|u - w_N\|_{H^1(\Omega)}$ is bounded by $C(N-1)^{1-s}\|u\|_{H^s(\Omega)}$ (by the inequality (81)) and the difference $|a(w_N, v_N) - a_N(w_N, v_N)|$ is equal to zero. Finally, for some

$r > 1$, the term $|\mathcal{F}(v_N) - \mathcal{F}_N(v_N)|$ is bounded by $CN^{-r}\|f\|_{H^r(\Omega)}\|v_N\|_{L^2(\Omega)}$ (by (82)), so that

$$\|u - u_N\|_{H^1(\Omega)} \le C(N^{1-s}\|u\|_{H^s(\Omega)} + N^{-r}\|f\|_{H^r(\Omega)}). \tag{96}$$

1.5. The spectral element method

Spectral Element Method (SEM) combines the high accuracy of spectral approximation to (at some extent) the geometric versatility of the Finite Element Method. The computational domain is partitioned into macro "spectral" elements, either quadrilateral or triangular then, a Generalized Galerkin formulation based on Gaussian quadrature formulas and relatively high degree polynomials are used.

SEM are very similar to some $h-p$ type Finite Element (see [3]), the essential difference between the two approaches lies in the choice of the basis for the trial and the test functions of the variational formulation.

SEM were proposed by Patera in 1984 ([62]); initially, conforming partitions in rectangular domains were considered. Later, Korczak and Patera ([43]) adopted the isoparametric mappings (largely used in FEM) in order to discretize more complex geometries. In 1989 Bernardi, Maday and Patera ([6]) introduced a new non-conforming approach to domain decomposition: the mortar element method, and Mavripilis, Maday and Patera ([49]) applied it to the spectral approximation. The latter has become very competitive in the treatment of complex geometries with non-uniform distribution of the computational nodes (see [6], [49]).

In 1991 Dubiner ([29]) and later Karnadiakis ([78]) gave a great impulse to triangular spectral elements, by the use of orthogonal polynomial basis on triangular domains.

Now we confine to continuous and conforming spectral elements for ease of exposition. In order to briefly illustrate the conforming SEM we introduce in Ω a decomposition $\mathcal{T}_H$ of quadrilaterals T_k, with $k = 1, \ldots, Ne$, we set

$$H = \max_{T_k \in \mathcal{T}_H} \mathrm{diam}(T_k) \qquad \mathrm{diam}(T_k) = \max \mathrm{diag}(T_k) \tag{97}$$

and we use the set Ξ_N^{LGL} of Legendre Gauss-Lobatto nodes (48) on each element T_k. Then we set:

$$V_\mathcal{H} = \{u_\mathcal{H} \in \mathcal{C}^0(\overline{\Omega}) : \; u_\mathcal{H}|_{T_k} \in \mathbb{Q}_N(T_k), \; \forall T_k \in \mathcal{T}_H\} \tag{98}$$

and $u_{N,k} = u_\mathcal{H}|_{T_k}$, $v_{N,k} = v_\mathcal{H}|_{T_k}$. The discretization of (2) by the SEM reads:

$$\text{find } u_\mathcal{H} \in V_\mathcal{H} : \qquad \sum_{k=1}^{Ne} a_{N,k}(u_{N,k}, v_{N,k}) = \sum_{k=1}^{Ne} (f, v_{N,k})_{N,k} \qquad \forall v_\mathcal{H} \in V_\mathcal{H} \tag{99}$$

where $a_{N,k}(u_{N,k}, v_{N,k})$ is obtained from $a|_{T_k}(u_\mathcal{H}, v_\mathcal{H})$ by replacing each integral with the Legendre Gauss-Lobatto quadrature formulas, and $(f, v_{N,k})_{N,k}$ is the Legendre Gauss-Lobatto numerical integration of $\int_{T_k} f v_{N,k} \mathrm{d}T_k$.

The following interpolation and quadrature error formulas hold (see [34]):

Theorem 1.2. *For all $u \in H^s(\Omega)$ with $s \geq 2$, there exists a constant $C_1 > 0$ independent of H_k and N such that $\forall T_k \in \mathcal{T}_H$*

$$\|u - I_N u\|_{H^m(T_k)} \leq C_1 H_k^{\min(N+1,s)-m} N^{m-s} \|u\|_{H^s(T_k)} \qquad m = 0, 1. \tag{100}$$

Moreover, there exists a constant $C_2 > 0$ independent of H and N such that

$$\|u - I_{\mathcal{H}} u\|_{H^m(\Omega)} \leq C_2 H^{\min(N+1,s)-m} N^{m-s} \|u\|_{H^s(\Omega)} \qquad m = 0, 1. \tag{101}$$

□

Lemma 1.3. *If $u \in H^s(T_k)$ with $s \geq 2$, then $\exists C > 0$ independent of N and H such that $\forall v_N \in \mathbb{Q}_N(T_k)$*

$$|(u, v_N)_{L^2(T_k)} - (u, v_N)_{N,T_k}| \leq C(s) N^{-s} H^{\min(N+1,s)} \|u\|_{H^s(T_k)} \|v_N\|_{L^2(T_k)}. \tag{102}$$

□

On the basis of these results, if u is the solution to (59), $u_{\mathcal{H}}$ to (99) and $f \in H^r(\Omega)$ with $r > 1$, then there exists $C > 0$ independent of N and H such that

$$\|u - u_{\mathcal{H}}\|_{H^1(\Omega)} \leq C \left[H^{\min(N+1,s)-1} N^{1-s} \|u\|_{H^s(\Omega)} + H^{\min(N+1,r)} N^{-r} \|f\|_{H^r(\Omega)} \right]. \tag{103}$$

The proof follows by the application of theorem 1.2, lemma 1.3 and by the following approximation result of Babuska and Suri [4]:

Theorem 1.4. *Let $u \in H^s(\Omega) \cap H_0^1(\Omega)$, with $s > 3/2$, then for any $N \geq 1$ and $H > 0$ there exists $z_{\mathcal{H}} \in V_{\mathcal{H}}$ such that*

$$\|u - z_{\mathcal{H}}\|_{H^k(\Omega)} \leq C H^{\min(N+1,s)-k} N^{k-s} \|u\|_{H^s(\Omega)} \qquad k = 0, 1 \tag{104}$$

where C is independent of u, N, H and $\mathcal{T}_H$. □

Consider now the algebraic form of problem (99). The matrix $A_{\mathcal{H}}$ associated to the bilinear form

$$a_{\mathcal{H}}(u_{\mathcal{H}}, v_{\mathcal{H}}) = \sum_{k=1}^{Ne} a_{N,k}(u_{N,k}, v_{N,k}) \tag{105}$$

has a sparse structure. An example of $A_{\mathcal{H}}$ for a domain which is split into 4×4 elements with polynomials of degree $N = 4$ is shown in Fig. 2.

From numerical experiments we obtained that the condition number of the Spectral Element matrix is

$$\chi(A_{\mathcal{H}}) \leq C N^3 H^{-2}. \tag{106}$$

The sparsity of the matrix and the high dependence of the condition number by the discretization parameters N and H suggest to solve the linear system $A_{\mathcal{H}} \mathbf{u} = \mathbf{f}$ associated to (99) by preconditioned Krylov Methods. An optimal preconditioner for this system can be provided by the global bilinear finite element matrix (see the Remark in Sec. 4.1). Another is given by the inexact additive Schwarz method with coarse correction ([27], [28], [60], [32]), that we briefly describe hereafter.

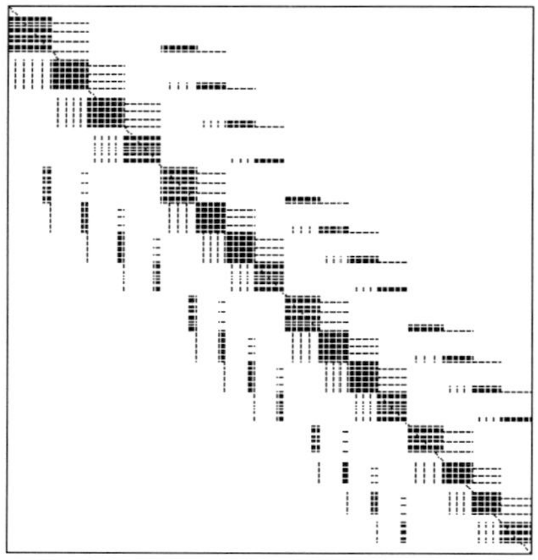

FIGURE 2. The pattern of the matrix $A_{\mathcal{H}}$ on a decomposition in 4×4 spectral elements with $N = 4$.

Consider a decomposition of the computational domain Ω in Ne conformal non-overlapping spectral elements T_k and a coarse mesh $\mathcal{M}_0$ given by the vertices of all the elements T_k. Then, for any $T_k \in \mathcal{T}_H$, consider an extension $T_{k,e}$ overlapping the adjacents elements of one or two nodes in each direction and the fine mesh $\mathcal{M}_{k,e}$ of all Legendre Gauss-Lobatto nodes in $T_{k,e}$ (see Fig. 3).

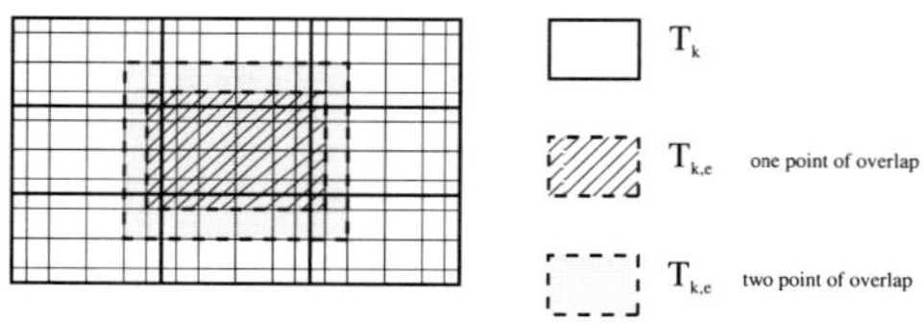

FIGURE 3. The spectral elements T_k and the extended elements $T_{k,e}$ with one or two nodes of overlap.

Let A_0 (resp. $A_{k,e}$) be the bilinear finite (or spectral) element matrix associated to the discretization of the problem (2) on the mesh $\mathcal{M}_0$ (resp. $\mathcal{M}_{k,e}$). Let R_0^T (resp. $R_{k,e}^T$) be the extension matrix whose action extends to zero a vector of nodal values in $\mathcal{M}_0$ (resp. in $\mathcal{M}_{k,e}$). The transposes R_0 and $R_{k,e}$ are restrictions matrices whose action restricts a full vector defined on all the nodes in Ω to a local vector on $\mathcal{M}_0$ or $\mathcal{M}_{k,e}$.

The inexact additive Schwarz preconditioner P_{as} for $A_{\mathcal{H}}$ with coarse refinement is defined as follows:

$$P_{as}^{-1} := R_0^T A_0^{-1} R_0 + \sum_{k=1}^{Ne} R_{k,e}^T A_{k,e}^{-1} R_{k,e}. \tag{107}$$

The attribute "inexact" to the preconditioner refers to the fact that the local matrices $A_{k,e}$ and the coarse matrix A_0 arise from bilinear finite (or spectral) element discretization rather than on the primary spectral element discretization directly. If finite element methods were used for both the primary discretization and the preconditioner, Driya and Widlund ([28], [27]) proved that $\exists C > 0$ independent of H and h (the size of the fine meshes on $T_{k,e}$), but possibly depending on ν such that:

$$\chi(P_{as}^{-1}A) \leq C(\nu)(1+\beta^{-1}), \tag{108}$$

A being the finite element matrix associated to the primary discretization and βH the maximum extension of the elements $T_{k,e}$ along the x_1 or x_2 direction. For large jumps on the coefficient ν at the interface, the convergence rate can deteriorate and precisely $\exists C > 0$ independent of H, h and ν, but depending on β such that

$$\chi(P_{as}^{-1}A) \leq C(\beta)\left(1+\log\frac{H}{h}\right). \tag{109}$$

We come back now to spectral element approximation. An estimate similar to (108) holds also if we consider the matrix $A_{\mathcal{H}}$ instead of A, thanks to the equivalence of the two discretizations (see [15]). If we consider a decomposition in squared elements with $H_k = H$ for any $T_k \in \mathcal{T}_H$ and if we measure the extension of an element T_k by mesh nodes and the extension amounts of one node in each direction, it follows that the overlap is $\beta H \simeq H/N^2$, i.e. the distance between the Legendre Gauss-Lobatto endpoint and the nearest one. The immediate consequence is that the condition number of the precondioned matrix $P_{as}^{-1}A_{\mathcal{H}}$ is bounded uniformly with respect to H but not with respect to the polynomial degree N. In particular:

$$\chi(P_{as}^{-1}A_{\mathcal{H}}) \leq C(\nu)(1+N^2), \tag{110}$$

for mild variations of ν, and

$$\chi(P_{as}^{-1}A_{\mathcal{H}}) \leq C(\beta)(1+\log(HN^2)), \tag{111}$$

in the case of severe discontinuities across the interfaces.

Since iterative methods are used, the matrices $A_{\mathcal{H}}$ and P_{as}^{-1} are never explicitly assembled; a matrix-vector product involving $A_{\mathcal{H}}$ is performed by local matrix-vector products. Then (107) yields only local systems (for the matrices $A_{k,e}$), or the coarse system (for A_0), and all of them have relatively little dimension.

Let us show the number of conjugate gradient iterations preconditioned by the matrix P_{as}^{-1} to solve with a tolerance $\epsilon = 10^{-6}$ the following problem:

$$\begin{cases} -\Delta u = f & \text{in } \Omega = (0,1)^2 \\ u = g & \text{on } \partial\Omega, \end{cases} \tag{112}$$

where f and g are fixed in a way that the exact solution is $u(\mathbf{x}) = e^{x_1+x_2}$. We have considered a decomposition of Ω into $Ne = M_1 \times M_1$ equal subdomains. The results refer to the overlap involving only one grid point outside each subdomain.

		M_1									
		2	3	4	5	6	7	8	9	10	11
	3	9	13	16	18	17	19	19	19	19	19
	4	9	13	15	16	15	16	16	16	16	16
	5	11	15	17	17	17	18	17	17	17	17
N	6	12	15	17	17	17	18	18	18	18	18
	7	14	16	15	16	16	16	16	16	16	16
	8	14	15	17	17	15	15	15	16	16	16
	9	16	17	17	16	16	16	16	18	18	18

TABLE 1. The number of conjugate gradient iterations for the additive Schwarz preconditioner with coarse correction for the problem (112) with a decomposition of Ω in $M_1 \times M_1$ equal subdomains and one point of overlap.

		M_1									
		2	3	4	5	6	7	8	9	10	11
	2	5	15	18	21	24	26	26	27	27	27
	3	10	16	23	25	27	27	27	28	28	28
	4	12	16	22	23	24	24	24	23	24	23
	5	13	15	19	21	21	21	21	21	21	21
N	6	12	16	19	19	20	20	19	20	20	20
	7	12	15	16	18	18	18	18	18	17	17
	8	13	16	15	17	16	16	16	16	16	16
	9	14	16	16	16	16	16	16	16	16	16

TABLE 2. The number of conjugate gradient iterations for the additive Schwarz preconditioner with coarse correction for the problem (112) with a decomposition of Ω in $M_1 \times M_1$ equal subdomains and two points of overlap.

2. Stokes and Navier-Stokes equations

Given an open bounded domain $\Omega \subset \mathbb{R}^2$ and a time interval $(0, T)$, the Navier Stokes equations for Newtonian, viscous incompressible fluids read:

$$\begin{cases} \dfrac{\partial \mathbf{u}}{\partial t} - \nu \Delta \mathbf{u} + \nabla p = \mathbf{f} - (\mathbf{u} \cdot \nabla)\mathbf{u} & \text{in } \Omega \times (0, T) \\ \nabla \cdot \mathbf{u} = 0 & \text{in } \Omega \times (0, T) \\ \mathbf{u} = \mathbf{0} & \text{on } \partial\Omega \times (0, T) \\ \mathbf{u} = \mathbf{u}_0 & \text{in } \Omega \times \{0\}, \end{cases} \tag{113}$$

where ν is the kinematic viscosity, $\mathbf{u} = \mathbf{u}(\mathbf{x},t)$ is the velocity field, $p = p(\mathbf{x},t)$ is the kinematic pressure, $\mathbf{u}_0 = \mathbf{u}_0(\mathbf{x})$ and $\mathbf{f} = \mathbf{f}(\mathbf{x},t)$ are the initial data and the external force field, respectively.

Before analysing the Navier-Stokes system and its discretization in time, we want to face up the discretization of the generalized Stokes problem, since many time-differentiation algorithms for the Navier-Stokes system can be reduced to a successive resolution of generalized Stokes systems.

2.1. Analysis of the generalized Stokes problem

The generalized Stokes equations read:

$$\begin{cases} \alpha\mathbf{u} - \nu\Delta\mathbf{u} + \nabla p = \mathbf{f} & \text{in } \Omega \\ \nabla\cdot\mathbf{u} = 0 & \text{in } \Omega \\ \mathbf{u} = 0 & \text{on } \partial\Omega, \end{cases} \tag{114}$$

with $\mathbf{f} \in [L^2(\Omega)]^2$ and α is a non-negative constant. Given

$$\mathbf{V} = \left[H_0^1(\Omega)\right]^2 \text{ and } \qquad Q = L_0^2(\Omega) = \left\{ q \in L^2(\Omega : \int_\Omega q d\Omega = 0 \right\}, \tag{115}$$

the weak formulation of (114) reads: find $\mathbf{u} \in \mathbf{V},\ p \in Q$:

$$\begin{cases} a(\mathbf{u},\mathbf{v}) + b(\mathbf{v},p) = (\mathbf{f},\mathbf{v})_{L^2(\Omega)} & \forall \mathbf{v} \in \mathbf{V} \\ b(\mathbf{u},q) = 0 & \forall q \in Q, \end{cases} \tag{116}$$

with

$$\begin{aligned} a(\mathbf{u},\mathbf{v}) &= \int_\Omega \nu\nabla\mathbf{u}\cdot\nabla\mathbf{v}d\Omega + \int_\Omega \alpha\mathbf{u}\cdot\mathbf{v}d\Omega, \\ b(\mathbf{v},q) &= -\int_\Omega q\nabla\cdot\mathbf{v}d\Omega. \end{aligned} \tag{117}$$

When the Stokes (or Navier-Stokes) equations in the primitive variable formulation are approximated by Galerkin methods, a compatibility condition between the finite-dimensional spaces for velocity and pressure has to be satisfied.

As a matter of fact, when the continuity equation is discretized directly, some parasitic modes on the pressure can appear with a deterioration of the accuracy on the numerical solution. The parasitic modes have been characterised mathematically for equal order polynomial approximation of velocity and pressure. Various techniques have been developed in order to filter them out. We give a short account hereafter.

We consider the generalized Galerkin approach (see Sec. 1.4) in order to approximate (116): find $\mathbf{u}_N \in \mathbf{V}_N \subset \mathbf{V},\ p_N \in Q_N \subset Q$:

$$\begin{cases} a_N(\mathbf{u}_N,\mathbf{v}_N) + b_N(\mathbf{v}_N,p_N) = (\mathbf{f},\mathbf{v}_N)_{N,\Omega} & \forall \mathbf{v}_N \in \mathbf{V}_N \\ b_N(\mathbf{u}_N,q_N) = 0 & \forall q_N \in Q_N, \end{cases} \tag{118}$$

with

$$\begin{aligned} a_N(\mathbf{u}_N, \mathbf{v}_N) &= (\nu\nabla\mathbf{u}_N, \nabla\mathbf{v}_N)_{N,\Omega} + (\alpha\mathbf{u}_N, \mathbf{v}_N)_{N,\Omega} \\ b_N(\mathbf{v}_N, q_N) &= -(q_N, \nabla\cdot\mathbf{v}_N)_{N,\Omega}; \end{aligned} \tag{119}$$

and where $\mathbf{V}_N$ and Q_N are finite-dimensional polynomial subspaces of $\mathbf{V}$ and Q respectively.

The space of parasitic modes for the pressure, i.e. the elements of the following space:

$$Z_N = \{q_N \in Q_N : \; b_N(\mathbf{v}_N, q_N) = 0, \; \forall \mathbf{v}_N \in \mathbf{V}_N\}, \tag{120}$$

is empty if and only if the following *inf-sup* or *Ladyzenskaya-Brezzi-Babuška (LBB)* condition is satisfied (see [2], [12]):

$$\exists\beta > 0 : \; \forall q_N \in Q_N \; \exists \mathbf{v}_N \in V_N : \quad b_N(\mathbf{v}_N, q_N) \geq \beta \|\mathbf{v}_N\|_{H^1(\Omega)} \|q_N\|_{L^2(\Omega)}. \tag{121}$$

The a priori most obvious choice

$$\mathbf{V}_N = \left[\mathbb{Q}_N(\Omega) \cap H_0^1(\Omega)\right]^2 \text{ and } Q_N = \mathbb{Q}_N(\Omega) \cap L_0^2(\Omega) \tag{122}$$

gives $\dim Z_N = 7$, with

$$Z_N = span\{L_N(x), \; L_N(y), \; L_N(x)L_N(y), \; L_N'(x)L_N'(y)(1 \pm x)(1 \pm y)\}. \tag{123}$$

Should the Chebyshev (rather than Legendre) Gauss-Lobatto quadrature be used, the space Z_N would have a similar characterisation, with the Chebyshev polynomials T_N instead of L_N. See [5].

The parasitic modes can be filtered out, for example, by a transform from the physical space to the frequency space, projecting p_N upon the space $Z_N^{\perp}$, and then by anti-transforming to the physical space.

Another strategy consists of choosing $\mathbf{V}_N$ as before and Q_N satisfying (see [5]):

$$\begin{aligned} &Q_N \oplus Z_N = \mathbb{Q}_N(\Omega) \cap L_0^2(\Omega), \qquad\qquad Q_N \cap Z_N = \{0\}, \\ &\dim Q_N + \dim Z_N = \dim(\mathbb{Q}_N(\Omega) \cap L_0^2(\Omega)). \end{aligned} \tag{124}$$

In this case, the (*LBB*) condition (121) is satisfied with a constant $\beta = \beta_N = \mathcal{O}(N^{-1})$ and, if $\mathbf{u} \in H^{s+1}(\Omega)$, $p \in H^s(\Omega)$, the following error estimate holds:

$$\|\mathbf{u} - \mathbf{u}_N\|_{H^1(\Omega)} + N^{-1}\|p - p_N\|_{L^2(\Omega)} \leq CN^{-s}\left(\|\mathbf{u}\|_{H^{s+1}(\Omega)} + \|p\|_{H^s(\Omega)}\right). \tag{125}$$

Due to the difficulty to build the space Q_N, the interest of this choice is more theoretical than practical.

A more practical choice from the computational viewpoint is to take

$$Q_N = \mathbb{Q}_{N-2}(\Omega) \cap L_0^2(\Omega). \tag{126}$$

The (*LBB*) condition (121) is satisfied with $\beta = \beta_N = \mathcal{O}(N^{-1/2})$, and (see [50])

$$\begin{aligned} &\|\mathbf{u} - \mathbf{u}_N\|_{H^1(\Omega)} + N^{-1/2}\|p - p_N\|_{L^2(\Omega)} \leq \\ &\qquad CN^{-s}\left(\|\mathbf{u}\|_{H^{s+1}(\Omega)} + \|p\|_{H^s(\Omega)}\right). \end{aligned} \tag{127}$$

Two ways can be followed in order to implement this approach. The former consists in using a staggered grid, a Legendre Gauss-Lobatto grid for the velocity and a Legendre Gauss grid for the pressure (as in Fig. 4, left); in this way an interpolation process is needed in order to compute the values for ∇p_N on the velocity grid. Alternatively, one can use for the pressure the $(N-1)^2$ Legendre Gauss-Lobatto quadrature nodes internal to the computational domain (see Fig. 4, right); the incompressibility condition is enforced by a discrete quadrature formula which is no more a Gaussian quadrature formula ([5]).

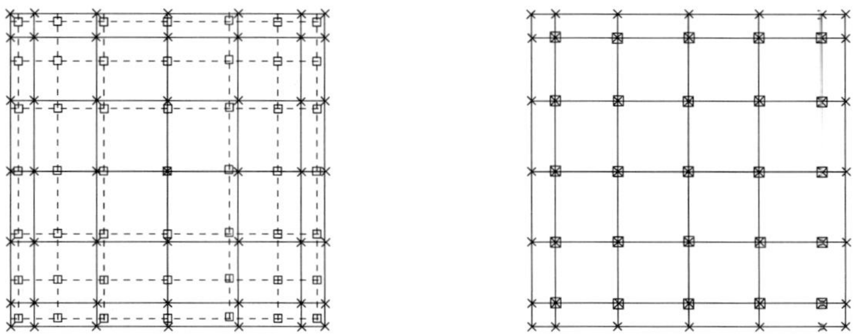

FIGURE 4. At left the two staggered grids: Legendre Gauss-Lobatto ($\times$) for the velocity and Legendre Gauss ($\Box$) for the pressure. At right the two grids: Legendre Gauss-Lobatto ($\times$) for the velocity and internal Legendre Gauss-Lobatto ($\Box$) for the pressure.

An alternative approach to those mentioned above is to use equal order interpolation spaces (122) and consequently stabilize the Stokes equations *à la* Galerkin Least Squares, or *à la* Douglas-Wang (see [40], [39]).

The stabilization techniques, initially proposed in the finite element context, have then been extended to spectral methods (see [17], [61]) and the spectral element methods (see [34]) with a suitable design of the stabilization parameters.

We briefly describe the stabilization *à la* Douglas-Wang on the spectral element approximation of (118). It consists of finding $\mathbf{u}_N$ and p_N in the spaces (122), such that:

$$\begin{aligned} a_N(\mathbf{u}_N, \mathbf{v}_N) + b_N(\mathbf{v}_N, p_N) - b_N(\mathbf{u}_N, q_N) & \\ +(\alpha\mathbf{u}_N - \nu\Delta\mathbf{u}_N + \nabla p_N, \tau(\alpha\mathbf{v}_N - \nu\Delta\mathbf{v}_N + \nabla q_N))_{N,\Omega} & \\ = (\mathbf{f}, \tau(\alpha\mathbf{v}_N - \nu\Delta\mathbf{v}_N + \nabla q_N))_{N,\Omega} & \\ \forall \mathbf{v}_N \in \mathbf{V}_N, \ \forall q_N \in Q_N, & \end{aligned} \tag{128}$$

where τ is the stabilization parameter which depends on the viscosity ν, on the domain size, and on the polynomial degree. A particular choice for τ is

$$\tau = \frac{m \ \text{meas}(\Omega)}{N^4}, \tag{129}$$

The bilinear form

$$\begin{aligned}\mathcal{B}_N(\mathbf{u}_N, p_N; \mathbf{v}_N, q_N) :=& a_N(\mathbf{u}_N, \mathbf{v}_N) + b_N(\mathbf{v}_N, p_N) - b_N(\mathbf{u}_N, q_N) \\ &+ (\alpha\mathbf{u}_N - \nu\Delta\mathbf{u}_N + \nabla p_N, \tau(\alpha\mathbf{v}_N - \nu\Delta\mathbf{v}_N + \nabla q_N)),\end{aligned} \tag{130}$$

associated to the approximation (128), is coercive with a coercivity constant equal to one. Consequently the solution satisfy the following stability inequality: $\exists\beta > 0$ independent of N such that

$$|||[\mathbf{u}_N, p_N]|||_\Omega \le \beta\|\mathbf{f}\|_{N,\Omega} \tag{131}$$

where

$$\begin{aligned}|||[\mathbf{u}_N, p_N]|||_\Omega := & \Big(\alpha\|\mathbf{u}_N\|^2_{L^2(\Omega)} + \nu\|\nabla\mathbf{u}_N\|^2_{L^2(\Omega)} \\ & + \|\tau^{1/2}(\alpha\mathbf{u}_N - \nu\Delta\mathbf{u}_N + \nabla p_N)\|^2_{L^2(\Omega)}\Big)^{1/2}.\end{aligned} \tag{132}$$

Furthermore, the solution of (128) satisfies the following error estimate: if the solution of (116) is such that $\mathbf{u} \in H^s(\Omega)$, $s \ge 2$, $p \in H^l(\Omega)$, $l \ge 1$ and $\mathbf{f} \in H^r(\Omega)$, $r > 1$, then there exists a positive constant C independent of N such that

$$|||[\mathbf{u} - \mathbf{u}_N, p - p_N]|||_\Omega \le C\left[N^{1-s}\|\mathbf{u}\|_{H^s(\Omega)} + N^{-l}\|p\|_{H^l(\Omega)} + N^{-r}\|\mathbf{f}\|_{H^r(\Omega)}\right]. \tag{133}$$

Other stabilization approaches are possible, for instance based on *streamline upwind Galerkin* methods, or else on the *bubble stabilized spectral* method proposed by Canuto and Van-Kemenade ([17]). Stabilization methods for the so-called *hp*-FEM are analyzed in [75] and [76].

3. Time-differentiation of Navier Stokes equations

We focus now our attention on the discretization of the temporal derivative in (113) and review some of the schemes which have been more successfully used in the past years.

3.1. Finite difference schemes

The most classical approach consists in advancing in time by suitable explicit, implicit or semi-implicit finite difference schemes.

The explicit schemes are easy to implement, the basic computational kernel required is a matrix-vector product, but an obvious restriction on the choice of the time step is needed to ensure stability to the scheme.

On the other hand, a fully implicit scheme requires the solution of a non-linear system at every time step, but no bounds on the time step are needed.

One way to avoid the resolution of a non-linear system is to use semi-implicit methods in which the nonlinear convective term is replaced by a linearised one. An example is provided by the Euler semi-implicit scheme that reads as follows. Given $\Delta t \in (0, T)$ we set $t^0 = 0$ and $t^n = t^0 + n \cdot \Delta t$ with $n = 1, \ldots, \mathcal{N}$ and

$\mathcal{N} = \left[\dfrac{T}{\Delta t}\right]$ (where $[a]$ stands for integral part of a). Then, for $n = 0, \ldots, \mathcal{N} - 1$, find $[\mathbf{u}^{n+1}, p^{n+1}] \in \mathbf{V} \times Q$:

$$\begin{cases} \dfrac{\mathbf{u}^{n+1} - \mathbf{u}^n}{\Delta t} - \nu\Delta\mathbf{u}^{n+1} + \nabla p^{n+1} = \mathbf{f}^{n+1} - (\mathbf{u}^n \cdot \nabla)\mathbf{u}^{n+1} & \text{in } \Omega \\ \nabla \cdot \mathbf{u}^{n+1} = 0 & \text{in } \Omega. \end{cases} \tag{134}$$

Otherwise the non-linear terms can be treated fully explicitly, and the other ones fully implicitly as in the Crank-Nicolson/Adams-Bashforth (CN-AB) second order scheme, that, on system (113), reads as follows:

$$\begin{cases} \dfrac{\mathbf{u}^{n+1} - \mathbf{u}^n}{\Delta t} - \dfrac{\nu}{2}\Delta(\mathbf{u}^{n+1} + \mathbf{u}^n) + \dfrac{1}{2}\nabla(p^{n+1} + p^n) = \\ \dfrac{1}{2}(\mathbf{f}^{n+1} + \mathbf{f}^n) - \dfrac{1}{2}\left[3(\mathbf{u}^n \cdot \nabla)\mathbf{u}^n - (\mathbf{u}^{n-1} \cdot \nabla)\mathbf{u}^{n-1}\right] & \text{in } \Omega \\ \nabla \cdot \mathbf{u}^{n+1} = 0 & \text{in } \Omega \end{cases} \tag{135}$$

with a suitable second order approximation of $\mathbf{u}^1$. At each time step a generalized Stokes problem as (114) has to be solved. If the spectral collocation method is used for space discretization, this method is stable under the condition that $\Delta t \leq CN^{-2}$.

Another semi-implicit scheme is based on the second-order backward finite-difference formula (see [37]):

$$\begin{cases} \dfrac{3\mathbf{u}^{n+1} - 4\mathbf{u}^n + \mathbf{u}^{n-1}}{2\Delta t} - \nu\Delta\mathbf{u}^{n+1} + \nabla p^{n+1} = \\ \mathbf{f}^{n+1} - \left((2\mathbf{u}^n - \mathbf{u}^{n-1}) \cdot \nabla\right)(2\mathbf{u}^n - \mathbf{u}^{n-1}) & \text{in } \Omega \\ \nabla \cdot \mathbf{u}^{n+1} = 0 & \text{in } \Omega. \end{cases} \tag{136}$$

Also in this case at each time step a generalized Stokes problem has to be solved.

We refer also to Simo and Armero (see [79]) for other finite difference schemes of various order, explicit, implicit or semi-implicit, with very-good stability properties in time, that may be applied to spectral approximation. In particular, a second order accurate algorithm that retains the key property of unconditional stability and linearity within the time step, reads

$$\begin{cases} \dfrac{\mathbf{u}^{n+1} - \mathbf{u}^n}{\Delta t} - \dfrac{\nu}{2}\Delta(\mathbf{u}^{n+1} + \mathbf{u}^n) + \nabla p^{n+1} = \\ \dfrac{1}{2}(\mathbf{f}^{n+1} + \mathbf{f}^n) - \dfrac{1}{4}\left[(3\mathbf{u}^n - \mathbf{u}^{n-1}) \cdot \nabla(\mathbf{u}^{n+1} + \mathbf{u}^n)\right] & \text{in } \Omega \\ \nabla \cdot (\mathbf{u}^{n+1} + \mathbf{u}^n) = 0 & \text{in } \Omega. \end{cases} \tag{137}$$

Now we show how to combine finite difference schemes for time-discretization with space-stabilized Galerkin methods. As an instance, we use the Euler semi-implicit scheme (134) for the Douglas-Wang stabilization method that was formerly used in

(128) on the Stokes problem (114) (we refer to [34] for a more detailed description and analysis). For $n = 0, \dots, \mathcal{N}-1$, find $[\mathbf{u}_N^{n-1}, p_N^{n+1}] \in \mathbf{V}_N \times Q_N$:

$$\begin{aligned} &\frac{1}{\Delta t}(\mathbf{u}_N^{n+1} - \mathbf{u}_N^n, \mathbf{v}_N)_{N,\Omega} + a_N(\mathbf{u}_N^{n+1}, \mathbf{v}_N) + b_N(\mathbf{u}_N^{n+1}, q_N) \\ &\quad + c_N(\mathbf{u}_N^n, \mathbf{u}_N^{n+1}, \mathbf{v}_N) + b_N(\mathbf{v}_N, p_N^{n+1}) - (\nabla \cdot \mathbf{u}_N^{n+1}, \gamma(\mathbf{x}) \nabla \cdot \mathbf{v}_N)_{N,\Omega} \\ &\quad + \left(\frac{\mathbf{u}_N^{n+1}}{\Delta t} + L(\mathbf{u}_N^n, \mathbf{u}_N^{n+1}, p_N^{n+1}), \tau(\mathbf{x}) L(\mathbf{u}_N^n, \mathbf{v}_N, q_N) \right)_{N,\Omega} = \\ &\quad = (\mathbf{f}^{n+1}, \mathbf{v}_N)_{N,\Omega} + \left(\mathbf{f}^{n+1} + \frac{\mathbf{u}_N^n}{\Delta t}, \tau(\mathbf{x}) L(\mathbf{u}_N^n, \mathbf{v}_N, q_N) \right)_{N,\Omega} \\ &\qquad\qquad \forall \mathbf{v}_N \in \mathbf{V}_N,\ q_N \in Q_N \end{aligned} \tag{138}$$

with $\mathbf{u}_N^0 = \mathbf{u}_{0N}$, where

$$L(\mathbf{w}, \mathbf{u}, p) = -\Delta \mathbf{u} + (\mathbf{w} \cdot \nabla)\mathbf{u} + \nabla p, \tag{139}$$

and

$$c_N(\mathbf{w}_N, \mathbf{u}_N, \mathbf{v}_N) = ((\mathbf{w}_N \cdot \nabla)\mathbf{u}_N, \mathbf{v}_N)_{N,\Omega}, \ \forall \mathbf{w}_N, \mathbf{u}_N, \mathbf{v}_N \in \mathbf{V}_N. \tag{140}$$

We observe that if in (138) we put $q_N = 0$ then we obtain the stabilized momentum equation, while, if we put $\mathbf{v}_N = 0$ we obtain the stabilized continuity equation.

The stabilization parameters $\tau(\mathbf{x})$ and $\gamma(\mathbf{x})$ can be defined as follows:

$$\tau(\mathbf{x}) = \frac{\text{meas}(\Omega)}{2|\mathbf{u}_N^n(\mathbf{x})|_p N^2} \xi(Re(\mathbf{x})), \qquad \gamma(\mathbf{x}) = \frac{\lambda |\mathbf{u}_N^n(\mathbf{x})|_p \,\text{meas}(\Omega)}{N^2} \xi(Re(\mathbf{x})), \tag{141}$$

where:

$$Re(\mathbf{x}) = \frac{m|\mathbf{u}_N^n(\mathbf{x})|_p \,\text{meas}(\Omega)}{2\nu N^2}, \qquad \xi(Re(\mathbf{x})) = \begin{cases} Re(\mathbf{x}) & \text{if } 0 \le Re(\mathbf{x}) < 1 \\ 1 & \text{if } 1 \le Re(\mathbf{x}), \end{cases} \tag{142}$$

$$|\mathbf{u}_N^n(\mathbf{x})|_p = \begin{cases} (|(\mathbf{u}_N^n)_1(\mathbf{x})|^p + |(\mathbf{u}_N^n)_2(\mathbf{x})|^p)^{1/p} & \text{if } 1 \le p < \infty \\ \max\limits_{i=1,2} |(\mathbf{u}_N^n)_i(\mathbf{x})| & \text{if } p = \infty. \end{cases} \tag{143}$$

For each $n = 0, \dots, \mathcal{N}$ there exists a unique solution of (138). Moreover, if $\mathbf{u} \in [H_0^1(\Omega) \cap H^{s+1}(\Omega)]^2$ with $s \ge 1$, $p \in L_0^2(\Omega) \cap H^l(\Omega)$ with $l \ge 1$, $\mathbf{w} \in W^{\max(s,l,r),\infty}(\Omega)$, $\mathbf{f} \in H^r(\Omega)$ with $r > 1$, then there exists a positive constant C independent of N such that:

$$|||[\mathbf{u} - \mathbf{u}_N, p - p_N]|||_\Omega \le C\Big(N^{1-s}\|\mathbf{u}\|_{H^s(\Omega)} + N^{-l}\|p\|_{H^l(\Omega)} + N^{-r}\|\mathbf{f}\|_{H^r(\Omega)}\Big). \tag{144}$$

Another possible choice for τ and γ is given in [17], where τ is locally defined on each quadrilateral T_k arising from the Legendre Gauss-Lobatto grid in the following way:

$$\tau_k(\mathbf{x}) = \frac{h_k}{2|\mathbf{u}_N^n(\mathbf{x})|_p} \xi(Re_k(\mathbf{x})), \tag{145}$$

$$\xi(Re_k(\mathbf{x})) = \begin{cases} \dfrac{m h_k |\mathbf{u}_N^n(\mathbf{x})|_p}{4\nu} & \text{if } 0 \le Re_k(\mathbf{x}) \le 1 \\ 1 & \text{if } Re_k(\mathbf{x}) > 1, \end{cases} \tag{146}$$

with

$$h_k = \sqrt{2}\frac{\text{meas}(k)}{\text{diag}(k)} \qquad \text{or} \qquad h_k = h_{k_x}\cos\theta + h_{k_y}\sin\theta \tag{147}$$

and

$$\theta = \arctan(|\mathbf{u}_{N,2}^n|/|\mathbf{u}_{N,1}^n|), \tag{148}$$

while γ is taken equal to zero. This strategy is more sensible to local changes of the ratio $|\mathbf{u}_N|/\nu$ than the former one (141).

We notice that in (138) the velocity components are unfortunally coupled through the term $(\nabla \cdot \mathbf{u}_N^{n+1}, \gamma(\mathbf{x})\nabla \cdot \mathbf{v}_N)_{N,\Omega}$, whose presence is necessary in order to prove the stability of the numerical solution.

3.2. Projection methods

The idea of these schemes, introduced by Chorin (1968) and Temam (1969) consists in decoupling the resolution of the velocity and the pressure in order to overcome the incompressibility constraint $\nabla \cdot \mathbf{u} = 0$.

First, a non-linear elliptic advection-diffusion problem of the following form

$$\begin{cases} \dfrac{\tilde{\mathbf{u}}^{n+1} - \mathbf{u}^n}{\Delta t} - \nu\Delta\tilde{\mathbf{u}}^{n+1} + B(\tilde{\mathbf{u}}^{n+\alpha}, \tilde{\mathbf{u}}^{n+\beta}) = \dfrac{1}{\Delta t}\displaystyle\int_{t_n}^{t_{n+1}} \mathbf{f}(t)dt & \text{in } \Omega \\ \tilde{\mathbf{u}}^{n+1} = 0 & \text{on } \partial\Omega, \end{cases} \tag{149}$$

is solved, where $B(\mathbf{u}, \mathbf{v}) = (\mathbf{u}\cdot\nabla)\mathbf{v} + 1/2(\nabla\cdot\mathbf{u})\mathbf{v}$ is the skew-symmetric form of the non-linear term, α, $\beta \in [0, 1]$. Therefore the numerical solution $\mathbf{u}^{n+1}$ is computed as the projection of $\tilde{\mathbf{u}}^{n+1}$ on the free-divergence space

$$\mathbf{V}_{\text{div}} = \{\mathbf{v} \in \mathbf{V} : \ \nabla \cdot \mathbf{v} = 0\} \tag{150}$$

with respect to the scalar product in $[L^2(\Omega)]^2$, i.e.

$$(\mathbf{u}^{n+1}, \mathbf{w})_{L^2(\Omega)} = (\tilde{\mathbf{u}}^{n+1}, \mathbf{w})_{L^2(\Omega)} \qquad \forall \mathbf{w} \in \mathbf{V}_{\text{div}}. \tag{151}$$

For the projection step (151) the method and all its variants are known as "projection methods".

As a consequence of the so-called Helmholtz decomposition principle ([71]), the computation of $\mathbf{u}^{n+1}$ by (151) is reduced to the resolution of an elliptic problem with Neumann boundary conditions for the pressure p^{n+1}:

$$\begin{cases} \Delta p^{n+1} = \dfrac{1}{\Delta t}\nabla \cdot \tilde{\mathbf{u}}^{n+1} & \text{in } \Omega \\ \dfrac{\partial p^{n+1}}{\partial \mathbf{n}} = 0 & \text{on } \partial\Omega, \end{cases} \tag{152}$$

from which we recover the end-of-step velocity field

$$\mathbf{u}^{n+1} = \tilde{\mathbf{u}}^{n+1} - \Delta t \nabla p^{n+1}. \tag{153}$$

Karniadakis et al. ([42], [78]) proposed a family of high-order time accurate splitting methods with improved boundary conditions of high order in time that minimize the effect of erroneous numerical boundary layers induced by splitting methods. The schemes read:

$$\begin{cases} \dfrac{\hat{\mathbf{u}} - \mathbf{u}^n}{\Delta t} = \displaystyle\sum_{j=0}^{J_e-1} \beta_j (\mathbf{u}^{n-j} \cdot \nabla)\mathbf{u}^{n-j} & \text{in } \Omega \\ \dfrac{\hat{\hat{\mathbf{u}}} - \hat{\mathbf{u}}}{\Delta t} = -\nabla \overline{p}^{n+1} & \text{in } \Omega \\ \dfrac{\mathbf{u}^{n+1} - \hat{\hat{\mathbf{u}}}}{\Delta t} = \nu \displaystyle\sum_{j=0}^{J_p-1} \gamma_j \Delta \mathbf{u}^{n+1-j} & \text{in } \Omega \end{cases} \tag{154}$$

with Dirichlet boundary conditions on $\mathbf{u}^{n+1}$. The incompressibility constraint is satisfied by the intermediate solution $\hat{\hat{\mathbf{u}}}$ and $\overline{p}^{n+1}$ is a scalar field that ensures that final velocity field is incompressible at the end of time level $(n+1)$. The coefficients β_j, for $j = 0, \dots, J_e - 1$, and γ_j, for $j = 0, \dots, J_p - 1$, can be the coefficients of Adams-Bashforth and Adams-Moulton methods (respectively of order J_e and J_p), or more general multistep schemes. For other splitting methods see [16], Sec. 7.3.2.

Rannacher ([74]) has made a thorough analysis of the projection method and of its relation with stabilization methods (penalty methods). He has considered the following form of the projection method: for $n = 0, \dots, \mathcal{N} - 1$:

$$\begin{cases} \dfrac{\mathbf{u}^{n+1/2} - \mathbf{u}^{n-1/2}}{\Delta t} - \nu \Delta \mathbf{u}^{n+1/2} + B(\mathbf{u}^{n+1/2}, \mathbf{u}^{n+1/2}) & \\ \qquad\qquad + \nabla q^n = \dfrac{1}{\Delta t} \displaystyle\int_{t_n}^{t_{n+1}} \mathbf{f}(t) dt & \text{in } \Omega \\ \mathbf{u}^{n+1/2} = 0 & \text{on } \partial\Omega \end{cases} \tag{155}$$

then find q^{n+1}:

$$\begin{cases} \Delta q^{n+1} = \dfrac{1}{\Delta t} \nabla \cdot \mathbf{u}^{n+1/2} & \text{in } \Omega \\ \dfrac{\partial q^{n+1}}{\partial \mathbf{n}} = 0 & \text{on } \partial\Omega, \end{cases} \tag{156}$$

with $\mathbf{u}^{-1/2} = \mathbf{u}_0$, $q^0 = 0$ and $\{q^n\}$ a sequence approximating $\{p^n\}$.

The velocity and the pressure are proven to be convergent at the first order with respect to the norms of $[L^2(\Omega)]^2$ and of the dual space of $H^1(\Omega) \cap L^2_0(\Omega)$, respectively.

The analysis of Rannacher indicates that in the interior of the computational domain Ω the pressure q^n is indeed a reasonable approximation of the exact pressure p at time t_n, as the effects of the non-physical Neumann boundary condition in (156) decay exponentially with respect to $dist(\mathbf{x}, \partial\Omega)\sqrt{\Delta t}$.

Shen ([77]) proposed the following scheme:

$$\begin{cases} \dfrac{\mathbf{u}^{n+1/2}-\mathbf{u}^n}{\Delta t} - \dfrac{\nu}{2}\Delta(\mathbf{u}^{n+1/2}+\mathbf{u}^n) + B\left(\dfrac{\mathbf{u}^{n+1/2}+\mathbf{u}^n}{2}, \dfrac{\mathbf{u}^{n+1/2}+\mathbf{u}^n}{2}\right) & \\ \qquad\qquad +\nabla p^n = \mathbf{f}(t_{n+1/2}) & \text{in } \Omega \\ \mathbf{u}^{n+1/2}+\mathbf{u}^n = 0 & \text{on } \partial\Omega, \end{cases} \tag{157}$$

$$\begin{cases} \Delta(p^{n+1}-p^n) = \dfrac{2}{\Delta t}\nabla\cdot\mathbf{u}^{n+1/2} & \text{in } \Omega \\ \dfrac{\partial p^{n+1}}{\partial \mathbf{n}} = \dfrac{\partial p^n}{\partial \mathbf{n}} & \text{on } \partial\Omega \end{cases} \tag{158}$$

and he proved that the scheme (157)–(158) is second order accurate in time for the velocity and at least first order accurate for the pressure. He proposed to use first-order accurate boundary conditions for the pressure as analysed by Karniadakis et al. ([42]) in the framework of spectral element approximation.

Pinelli et al. ([66]) use the Van Kan scheme ([41]) based on the CN-AB method (135):

$$\begin{cases} \dfrac{\tilde{\mathbf{u}}^{n+1}-\mathbf{u}^n}{\Delta t} - \dfrac{\nu}{2}\Delta(\tilde{\mathbf{u}}^{n+1}+\mathbf{u}^n) + \nabla p^n = \dfrac{1}{2}\left(\mathbf{f}_n+\mathbf{f}_{n+1}\right) & \\ \qquad -\dfrac{1}{2}(3B(\mathbf{u}^n,\mathbf{u}^n) - B(\mathbf{u}^{n-1},\mathbf{u}^{n-1})) & \text{in } \Omega \\ \tilde{\mathbf{u}}^{n+1} = 0 & \text{on } \partial\Omega, \end{cases} \tag{159}$$

$$\begin{cases} \dfrac{1}{\Delta t}(\mathbf{u}^{n+1}-\tilde{\mathbf{u}}^{n+1}) + \dfrac{1}{2}\nabla(p^{n+1}-p^n) = 0 & \text{in } \Omega \\ \nabla\cdot\mathbf{u}^{n+1} = 0 & \text{in } \Omega. \end{cases} \tag{160}$$

Legendre collocation in two directions and Fourier expansion in the third direction are considered in order to simulate a turbulent flow in 3D. At each time level a Projection Decomposition solver (see Sec. 4) on the generalized Stokes problem is used on the directions approximated by Legendre collocation.

4. Domain decomposition methods

The domain decomposition techniques are combined to the spectral methods for their good versatility to treat general geometries, as well as for the good properties that they enjoy as parallel preconditioners for the algebraic system arising from the spectral approximation.

Let us assume that Ω is partitioned into M non-overlapping subdomains Ω_i, $i = 1, \ldots, M$, such that $\overline{\Omega} = \cup_{i=1}^M \overline{\Omega}_i$. We set $\mathring{\Omega} = \cup_{i=1}^M \Omega_i$ and we define the interface between the subdomains as $\Gamma = \Omega \backslash \mathring{\Omega}$.

4.1. The Poisson problem

The first computational kernel used in the approximation of the Navier-Stokes equations is the Poisson problem

$$\begin{cases} -\Delta u = f & \text{in } \Omega \\ u = 0 & \text{on } \partial\Omega. \end{cases} \tag{161}$$

We indicate by u_i the restriction to Ω_i of the solution u to (161) and by $\mathbf{n}_i$ the outward unit normal vector to $\partial\Omega_i \cap \Gamma$.

For simplicity, we consider a rectangular domain $\Omega \subset \mathbb{R}^2$ split in two rectangular subdomains (see Fig. 5). For details and generalisation, see [72]. The mul-

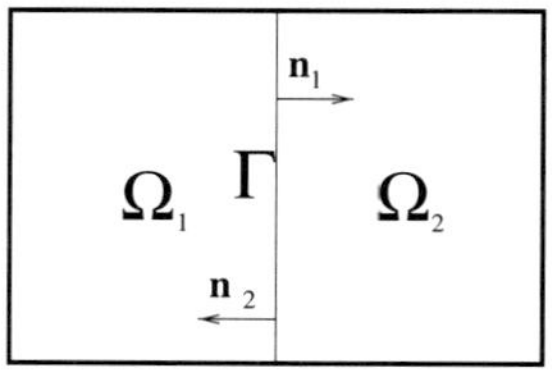

FIGURE 5. A decomposition of Ω in two subdomains.

tidomain formulation of (161) reads:

$$\begin{cases} -\Delta u_1 = f & \text{in } \Omega_1 \\ u_1 = 0 & \text{on } \partial\Omega_1 \cap \partial\Omega \\ u_1 = u_2 & \text{on } \Gamma \\ \dfrac{\partial u_1}{\partial \mathbf{n}_1} = -\dfrac{\partial u_2}{\partial \mathbf{n}_2} & \text{on } \Gamma \\ -\Delta u_2 = f & \text{in } \Omega_2 \\ u_2 = 0 & \text{on } \partial\Omega_2 \cap \partial\Omega. \end{cases} \tag{162}$$

The third and fourth equations of (162) are said transmission conditions for u_1 and u_2 and they ensure the necessary regularity of the multidomain solution.

The equivalence between (161) and (162) is shown by resorting to the weak form of both problems. The variational form of (161) reads:

$$\text{find } u \in V \equiv H_0^1(\Omega): \qquad a(u,v) = (f,v)_{L^2(\Omega)} \qquad \forall v \in V \tag{163}$$

with $a(u,v) = (\nabla u, \nabla v)_{L^2(\Omega)}$, while the variational multidomain form is: find $u_1 \in V_1,\ u_2 \in V_2$:

$$\begin{cases} a_1(u_1,v_1) = (f,v_1)_{L^2(\Omega_1)} & \forall v_1 \in V_1^0 \\ a_2(u_2,v_2) = (f,v_2)_{L^2(\Omega_2)} & \forall v_2 \in V_2^0 \\ u_1 = u_2 & \text{on } \Gamma \\ a_2(u_2, E_2\mu) = (f, E_2\mu)_{L^2(\Omega_2)} - a_1(u_1, E_1\mu) + (f, E_1\mu)_{L^2(\Omega_1)} & \forall \mu \in \Lambda \end{cases} \tag{164}$$

where:

$$V_i = \{v_i \in H^1(\Omega_i) :\ v_i|_{\partial\Omega \cap \partial\Omega_i} = 0\} \qquad i = 1,2, \tag{165}$$

($v_i|_\Gamma$ denotes the trace of v_i on Γ),

$$V_i^0 = H_0^1(\Omega_i), \qquad i = 1,2 \tag{166}$$

$$\Lambda = \{\mu \in H^{1/2}(\Gamma) : \ \mu = v|_\Gamma \text{ for a suitable } v \in V\}, \tag{167}$$

$$a_i(w_i, v_i) = (\nabla w_i, \nabla v_i)_{L^2(\Omega_i)} \qquad i = 1,2 \tag{168}$$

and E_i denotes any possible (continuous) operator from Λ to V_i which satisfies $(E_i\mu)|_\Gamma = \mu$, i.e. E_i is an extension operator from Λ to V_i.

The variational multidomain problem (164) can be equivalently written in terms of the Steklov Poincaré operator (see [1], [70]), where the determination of u, solution of (161) is reduced to find the trace λ of u on Γ and then to solve M independent Dirichlet problems on the subdomains Ω_i.

As a matter of fact, we note that the solution u to (162) can be represented as being $u = u_0 + v$, where u_0 is the solution to the following problems:

$$\begin{cases} -\Delta u_{0,i} = f & \text{in } \Omega_i \\ u_{0,i} = 0 & \text{on } \partial\Omega_i \end{cases} \qquad i = 1,2 \tag{169}$$

(with $u_{0,i} = u_0|_{\Omega_i}$), while v is the solution of

$$\begin{cases} -\Delta v_i = 0 & \text{in } \Omega_i \\ v_i = 0 & \text{on } \partial\Omega_i \cap \partial\Omega \\ v_i = \lambda & \text{on } \partial\Omega_i \cap \Gamma \end{cases} \qquad i = 1,2 \tag{170}$$

(with $v_i = v|_{\Omega_i}$).

The equivalence with the original problem (162) is achieved provided that the following relation holds:

$$\frac{\partial u_1}{\partial \mathbf{n}_1} = -\frac{\partial u_2}{\partial \mathbf{n}_2} \qquad \text{on } \Gamma, \tag{171}$$

or:

$$\frac{\partial v_1}{\partial \mathbf{n}_1} + \frac{\partial v_2}{\partial \mathbf{n}_2} = -\left(\frac{\partial u_{0,1}}{\partial \mathbf{n}_1} + \frac{\partial u_{0,2}}{\partial \mathbf{n}_2}\right) \qquad \text{on } \Gamma. \tag{172}$$

Noticing that the functions $u_{0,i}$ only depend on f (say $u_{0,i} = \mathcal{G}_i(f)$) and v_i on λ (say $v_i = \mathcal{H}_i(\lambda)$), we can formally write the equation (172) as:

$$\mathcal{S}\lambda = \chi \qquad \text{on } \Gamma \tag{173}$$

where

$$\chi = -\left(\frac{\partial u_{0,1}}{\partial \mathbf{n}_1} + \frac{\partial u_{0,2}}{\partial \mathbf{n}_2}\right) = -\left(\frac{\partial \mathcal{G}_1(f)}{\partial \mathbf{n}_1} + \frac{\partial \mathcal{G}_2(f)}{\partial \mathbf{n}_2}\right) \tag{174}$$

and

$$\mathcal{S}\lambda = \frac{\partial v_1}{\partial \mathbf{n}_1} + \frac{\partial v_2}{\partial \mathbf{n}_2} = \frac{\partial \mathcal{H}_1(\lambda)}{\partial \mathbf{n}_1} + \frac{\partial \mathcal{H}_2(\lambda)}{\partial \mathbf{n}_2}. \tag{175}$$

The equation (173) is said *interface equation* and $\mathcal{S}$ is the Steklov-Poincaré operator; $\mathcal{S}$ is self-adjoint and coercive.

To solve the interface equation amounts to solve two Poisson problems as (169) in order to built χ, to solve the interface equation to obtain $\lambda = v|_\Gamma$, then to solve two problems like (170) in order to obtain v_i.

When the differential multidomain problem (162) is approximated by spectral methods with a generalized Galerkin approach (or any other discretization method), the Steklov-Poincaré equation (173) has a finite-dimensional counterpart and it takes the algebraic form $A\mathbf{u} = \mathbf{f}$, or equivalently:

$$\begin{bmatrix} A_{11} & 0 & A_{1\Gamma} \\ 0 & A_{22} & A_{2\Gamma} \\ A_{\Gamma 1} & A_{\Gamma 2} & A_{\Gamma\Gamma} \end{bmatrix} \begin{bmatrix} u_1 \\ u_2 \\ u_\Gamma \end{bmatrix} = \begin{bmatrix} f_1 \\ f_2 \\ f_\Gamma \end{bmatrix}. \tag{176}$$

u_1 (resp. u_2) is the vector of unknowns at the collocation nodes of $\overline{\Omega}_1 \backslash \Gamma$ (resp. $\overline{\Omega}_2 \backslash \Gamma$), u_Γ is the unknowns vector of the nodes of Γ. A block Gauss elimination on the matrix A gives:

$$\tilde{S} u_\Gamma = \tilde{f}_\Gamma \tag{177}$$

with

$$\tilde{S} \equiv (A_{\Gamma\Gamma} - A_{\Gamma 1} A_{11}^{-1} A_{1\Gamma} - A_{\Gamma 2} A_{22}^{-1} A_{2\Gamma}) \tag{178}$$

and

$$\tilde{f}_\Gamma = f_\Gamma - A_{\Gamma 1} A_{11}^{-1} f_1 - A_{\Gamma 2} A_{22}^{-1} f_2. \tag{179}$$

The matrix $\tilde{S}$ is referred to as the Schur complement of $A_{\Gamma\Gamma}$ in A and the equation (177) is a discrete approximation of the interface equation (173).

The matrix $\tilde{S}$ is never explicitly assembled, but it is common practice to solve the system (177) by preconditioned iterative procedures where each matrix-vector product with $\tilde{S}$ involves two (in general M) subdomain solvers. The iterative procedures for solving equation (173) are traditionally referred to as *iterative sub-structuring methods*, since they introduce a sequence of differential subproblems in Ω_i, for which the transmission conditions of (162) provide Dirichlet or Neumann data at the internal boundary Γ.

We note that the Steklov-Poincaré operator $\mathcal{S}$ can be split in $\mathcal{S} = \mathcal{S}_1 + \mathcal{S}_2$ with $\mathcal{S}_i \lambda = \partial \mathcal{H}_i(\lambda)/\partial \mathbf{n}_i$ and, if we consider the Richardson method with preconditioner $\mathcal{P} = \mathcal{S}_1$ we obtain the so-called Dirichlet-Neumann scheme ([8], [54], [11]). Otherwise the so-called Neumann-Neumann scheme ([9]) amounts to solve (173) by the preconditioned conjugate gradient with $\mathcal{P} = (\alpha \mathcal{S}_1^{-1} + (1-\alpha)\mathcal{S}_2^{-1})^{-1}$, $0 < \alpha < 1$.

The *Dirichlet Neumann scheme* was considered, e.g. by Bjorstad and Widlund ([8]), Bramble, Pasciak and Schatz ([11]) and Marini and Quarteroni ([54]) in the framework of finite element method and by Funaro, Quarteroni and Zanolli ([31]) for spectral methods. It reads as follows: given $\lambda^{(0)}$, $\forall k \geq 0$ solve:

$$\begin{cases} -\Delta u_1^{(k+1)} = f & \text{in } \Omega_1 \\ u_1^{(k+1)} = 0 & \text{on } \partial\Omega_1 \cap \partial\Omega \\ u_1^{(k+1)} = \lambda^{(k)} & \text{on } \Gamma, \end{cases} \tag{180}$$

$$\begin{cases} -\Delta u_2^{(k+1)} = f & \text{in } \Omega_2 \\ u_2^{(k+1)} = 0 & \text{on } \partial\Omega_2 \cap \partial\Omega \\ \dfrac{\partial u_2^{(k+1)}}{\partial \mathbf{n}_2} = -\dfrac{\partial u_1^{(k+1)}}{\partial \mathbf{n}_1} & \text{on } \Gamma, \end{cases} \tag{181}$$

then set $\lambda^{(k+1)} = \theta u_2^{(k+1)}|_\Gamma + (1-\theta)\lambda^{(k)}$, θ being a positive acceleration parameter in $(0,1)$ that can be computed dynamically ([54], [68]). The method can be generalized to more complex decompositions, with $M > 2$; in this case the subdomains are divided in two sets as in a chess-board; all the domains of the first set are treated as Ω_1, the others as Ω_2.

If a domain decomposition without internal cross-points is considered (a cross point is a subdomain vertex which is common to almost four subdomains) and the parameter θ is computed in the optimal way by the minimal contraction constant approach ([54], [68]), the condition number of the preconditioned system is independent of the polynomial degree N of the spectral approximation but it linearly depends on the number of subdomains M. If we set $H = \max_i meas(\Omega_i)$ then there exists $C > 0$ independent of N and H such that

$$\chi(P^{-1}\tilde{S}) \leq CH^{-1} \tag{182}$$

where P is the matrix associated to $\mathcal{P}$. Otherwise, in presence of internal cross points, we have

$$\chi(P^{-1}\tilde{S}) \leq CH^{-2}(1 + \log^2(HN^2)) \tag{183}$$

in accordance with ([68], [69]) and the numerical results in [33] (Fig. 2).

The *Neumann-Neumann scheme* was considered by Bourgat, Glowinski, Le Tallec and Vidrascu ([9]) and reads: given $\lambda^{(0)}$, $\forall k \geq 0$ solve:

$$\begin{cases} -\Delta u_i^{(k+1)} = f & \text{in } \Omega_i \\ u_i^{(k+1)} = 0 & \text{on } \partial\Omega_i \cap \partial\Omega \\ u_i^{(k+1)} = \lambda^{(k)} & \text{on } \Gamma, \end{cases} \qquad i = 1, 2, \tag{184}$$

$$\begin{cases} -\Delta v_i^{(k+1)} = f & \text{in } \Omega_i \\ v_i^{(k+1)} = 0 & \text{on } \partial\Omega_i \cap \partial\Omega \\ \dfrac{\partial v_i^{(k+1)}}{\partial \mathbf{n}_i} = \dfrac{\partial u_1^{(k+1)}}{\partial \mathbf{n}_1} + \dfrac{\partial u_2^{(k+1)}}{\partial \mathbf{n}_2} & \text{on } \Gamma, \end{cases} \qquad i = 1, 2, \tag{185}$$

then set $\lambda^{(k+1)} = \lambda^{(k)} - \theta(\sigma_1 v_1^{(k+1)}|_\Gamma - \sigma_2 v_2^{(k+1)}|_\Gamma)$, θ being a positive acceleration parameter in $(0,1)$ and σ_1, σ_2 two positive averaging coefficients.

Also for the Neumann-Neumann scheme a behaviour like (183) holds. Note that the computational cost of each iteration of the Neumann-Neumann method is double with respect to the computational cost of one iteration of the Dirichlet-Neumann scheme. Nevertheless this is balanced by the fact that the Neumann-Neumann scheme is twice faster than the Dirichlet-Neumann scheme in terms of number of iterations to converge with a given accuracy.

Another iterative substructuring approach is the *Robin method* proposed by Agoshkov and Lebedev ([1]) and analysed by P.L. Lions ([48]). It reads: given $u_2^{(0)}$, $\forall k \geq 0$ solve:

$$\begin{cases} -\Delta u_i^{(k+1)} = f & \text{in } \Omega_i \\ u_i^{(k+1)} = 0 & \text{on } \partial\Omega_i \cap \partial\Omega \qquad i = 1,2 \\ \dfrac{\partial u_i^{(k+1)}}{\partial \mathbf{n}_i} + \theta u_i^{(k+1)} = -\dfrac{\partial u_j^{(k)}}{\partial \mathbf{n}_j} + \theta u_j^{(k)} & \text{on } \partial\Omega_i \cap \partial\Omega_j, \qquad j \neq i, \end{cases} \tag{186}$$

where θ, is a non-negative acceleration parameter.

The Dirichlet-Neumann and the Robin methods are particular cases of the Lebedev-Agoshkov method ([46]) that reads: given $u_2^{(0)}$, $\forall k \geq 0$ solve:

$$\begin{cases} -\Delta u_1^{(k+1/2)} = f & \text{in } \Omega_1 \\ u_1^{(k+1/2)} = 0 & \text{on } \partial\Omega_1 \cap \partial\Omega \\ \dfrac{\partial u_1^{(k+1/2)}}{\partial \mathbf{n}_1} + p_k u_1^{(k+1/2)} = -\dfrac{\partial u_2^{(k)}}{\partial \mathbf{n}_2} + p_k u_2^{(k)} & \text{on } \partial\Omega_1 \cap \Gamma \end{cases} \tag{187}$$

then set

$$u_1^{(k+1)} = u_1^{(k)} + \alpha_{k+1}(u_1^{(k+1/2)} - u_1^{(k)}) \qquad \text{in } \overline{\Omega}_1; \tag{188}$$

solve:

$$\begin{cases} -\Delta u_2^{(k+1/2)} = f & \text{in } \Omega_2 \\ u_2^{(k+1/2)} = 0 & \text{on } \partial\Omega_2 \cap \partial\Omega \\ -q_k \dfrac{\partial u_2^{(k+1/2)}}{\partial \mathbf{n}_2} + u_2^{(k+1/2)} = -q_k \dfrac{\partial u_1^{(k+1)}}{\partial \mathbf{n}_1} + u_1^{(k+1)} & \text{on } \partial\Omega_2 \cap \Gamma \end{cases} \tag{189}$$

then set

$$u_2^{(k+1)} = u_2^{(k)} + \beta_{k+1}(u_2^{(k+1/2)} - u_2^{(k)}) \qquad \text{in } \overline{\Omega}_2 \tag{190}$$

with $p_k \geq 0$, $q_k \geq 0$, $\alpha_{k+1}, \beta_{k+1} \in \mathbb{R}$.

For $p_k = q_k = 0$ and $\alpha_k = \beta_k = 1$ we obtain the Dirichlet-Neumann scheme (180)–(181) where the role of Ω_1 and Ω_2 are reversed; for $p_k = \theta$, $q_k = 1/\theta$ and $\alpha_k = \beta_k = 1$ we obtain the Robin method (186).

Another approach to solve the Poisson problem via domain decomposition is the so-called Projection Decomposition Method (PDM) ([58], [33]). The main idea of this method consists of solving the interface equation (173) by the Galerkin projection method. The use of a piecewise-polynomial well-conditioned basis (in the sense of Mikhlin [55]) leads to a well-conditioned linear system for the coordinates of the Galerkin approximation in this basis, which can be solved effectively by the Conjugate Gradient method with the convergence rate independent of the system dimension. The combination of the piecewise-polynomials Galerkin approximation on the interface and the spectral collocation method in the subdomains results in high accuracy of the numerical solutions.

For the details on the construction of the well-conditioned basis we refer to ([58], [33]), here we briefly summarise the algorithm induced by the PDM:

- solve by spectral methods M Poisson problems as (169) to obtain u_0 and compute the right-hand side of the interface equation (written in weak form) with respect to the well-conditioned basis;
- solve by conjugate gradient the interface equation: at each conjugate gradient iteration we have to apply the discrete counterpart of the Steklov-Poincaré operator to a vector, that amounts to solve M Poisson problems by spectral methods;
- solve by spectral methods M Poisson problems as (170) to obtain v.

We note that at each step of this procedure the Poisson problems on the subdomains Ω_i are independent, thus this algorithm can be easily parallelised. The numerical results in [33] show that the rate of convergence of PDM depends on H and N as for the Dirichlet Neumann scheme (see 183) with a constant C which is about 1/10 of the Dirichlet Neumann constant.

A way to improve the performance of domain decomposition methods, in terms of convergence rate, consists of incorporating a mechanism for global coupling, such as through a coarse grid problem based on a coarse mesh inside the computational domain.

For example, the *Bramble, Pasciak and Schatz method* ([11]) consists of preconditioning the Schur complement system by a block Jacobi matrix plus a coarse refinement involving the vertices of the subdomains. The condition number becomes logarithmic in H/h (h is the finite element mesh size), i.e. $\exists C > 0$ independent of H and h such that

$$\chi(P^{-1}\tilde{S}) \leq C\left(1 + \log^2\left(\frac{H}{h}\right)\right). \tag{191}$$

The performance of domain decomposition methods can be improved again by introducing overlapping decompositions as in the *additive Schwarz method* with *coarse correction* ([22], [27]), or the *2D vertex space method* of Smith ([80], [81]). For both of these schemes it has been proved that $\exists C > 0$ independent of H and h, but possibly dependent on the overlap β such that

$$\chi(P^{-1}\tilde{S}) \leq C(\beta)\left(1 + \log\left(\frac{H}{h}\right)\right). \tag{192}$$

The overlapping methods have been largely applied to finite element discretization, but seldom to spectral context for the intrinsic difficulty of this approaches to combine "little" overlapping (i.e. little extensions of the elements) with Gaussian quadrature formulas. As a matter of fact these formulas have a fixed distribution of nodes and the addition of other nodes to the existing Gaussian set, in general causes the loss of precision of the quadrature formula.

Pavarino ([63], [64]) has proposed an overlapping additive Schwarz method for p-type finite elements. This idea amounts to consider quadrilateral finite elements Ω_i (of size H and with high polynomial degree p), and to define the extended elements Ω'_k as the union of four elements $\Omega_{k1}, \ldots, \Omega_{k4}$ which have a common vertex (see Fig. 6). The elements Ω'_k are defined for each vertex internal to the computational domain Ω and they form an overlapping decomposition with an overlap of the same size of the original elements (H). Then an additive Schwarz method is considered to solve the multidomain problem with a coarse refinement involving the vertex of the original decomposition. The rate of convergence of this domain decomposition solver is independent of both the size H of the elements and the polynomial degree p ([63]).

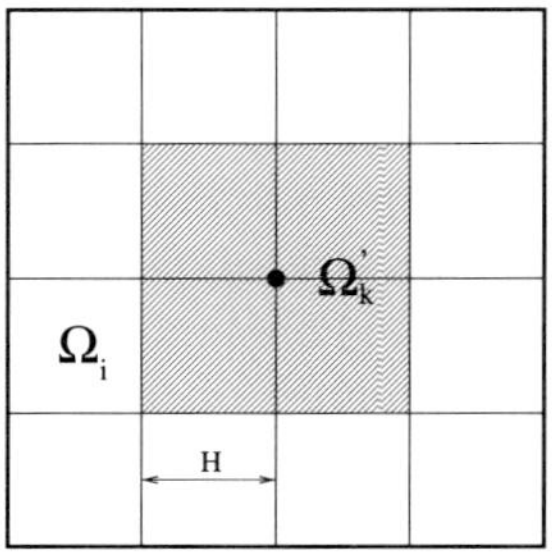

FIGURE 6. The original elements Ω_i and the extended elements of the overlapped decomposition Ω'_k used by Pavarino.

Remark 4.1. *We have seen in Sec. 1 that when a differential problem is approximated by spectral methods, an optimal preconditioner is the matrix obtained by the bilinear finite element approximation of the same problem on the Legendre Gauss-Lobatto grid. This preconditioner can be also used inside a multidomain approach. The system (162), approximated by spectral methods, can be faced by a preconditioned iterative method (e.g. the Richardson or conjugate gradient method) where the preconditioner is referred to the global grid of collocation nodes, i.e. the union of the Legendre Gauss-Lobatto nodes of all the subdomains Ω_i $(i = 1, \ldots, M)$ (see* [18], [26], [73]*). Each step of the iterative solver consists of evaluating the residual of the spectral multidomain system and then solving a finite element problem in the global domain. The method converges with a rate that is substantially independent of the number of gridpoints (*[73], [82]*).* □

Another way to use the bilinear finite element preconditioner consists of building the inexact additive Schwarz preconditioner as described in Sec. 1.5 for the spectral element method.

4.2. The Stokes problem

Consider now the generalized Stokes system (114), and the notations introduced in the previous subsection. The multidomain form of the Stokes system reads: for

$i = 1, \ldots, M$ find $(\mathbf{u}_i, p_i)$ such that:

$$\begin{cases} \alpha \mathbf{u}_i - \nu \Delta \mathbf{u}_i + \nabla p_i = \mathbf{f} & \text{in } \Omega_i \\ \mathbf{u}_i = \mathbf{u}_j & \text{on } \Gamma_{ij} = \partial\Omega_i \cap \partial\Omega_j \\ \nu \dfrac{\partial \mathbf{u}_i}{\partial \mathbf{n}_i} - p_i \mathbf{n}_i = -\nu \dfrac{\partial \mathbf{u}_j}{\partial \mathbf{n}_j} - p_j \mathbf{n}_j & \text{on } \Gamma_{ij} \\ \mathbf{u}_i = \mathbf{0} & \text{on } \partial\Omega_i \cap \partial\Omega \\ \nabla \cdot \mathbf{u}_i = 0 & \text{in } \Omega_i. \end{cases} \tag{193}$$

The transmission conditions on Γ_{ij} enforce the continuity of the solution and of the normal component of the stress tensor.

As for the Poisson problem, one can prove the equivalence between the systems (114) and (193) and introduce the Steklov-Poincaré operator, the interface equation and the iterative methods based on transmission conditions.

We take into account the variational form (116)–(117) of (114); the variational multidomain form of (193) with $M = 2$ reads:

find $(\mathbf{u}_1, p_1) \in \mathbf{V}_1 \times L^2(\Omega_1)$, $(\mathbf{u}_2, p_2) \in \mathbf{V}_2 \times L^2(\Omega_2)$:

$$\begin{cases} a_1(\mathbf{u}_1, \mathbf{v}_1) + b_1(\mathbf{v}_1, p_1) = (\mathbf{f}, \mathbf{v}_1)_{L^2(\Omega_1)} & \forall \mathbf{v}_1 \in [V_1^0]^2 \\ b_1(\mathbf{u}_1, q_1) = 0 & \forall q_1 \in L^2(\Omega_1) \\ \mathbf{u}_1 = \mathbf{v}_1 & \text{on } \Gamma \\ & \\ a_2(\mathbf{u}_2, \mathbf{v}_2) + b_2(\mathbf{v}_2, p_2) = (\mathbf{f}, \mathbf{v}_2)_{L^2(\Omega_2)} & \forall \mathbf{v}_2 \in [V_2^0]^2 \\ b_2(\mathbf{u}_2, q_2) = 0 & \forall q_2 \in L^2(\Omega_2) \\ a_2(\mathbf{u}_2, E_2\mu) + b_2(E_2\mu, p_2) = (\mathbf{f}, E_2\mu)_{L^2(\Omega_2)} & \\ \qquad +(\mathbf{f}, E_1\mu)_{L^2(\Omega_1)} - a_1(\mathbf{u}_1, E_1\mu) - b_1(E_1\mu, p_1) & \forall \mu \in [\Lambda]^2 \\ & \\ \displaystyle\int_{\Omega_1} p_1 \mathrm{d}\Omega_1 + \int_{\Omega_2} p_2 \mathrm{d}\Omega_2 = 0 & \end{cases} \tag{194}$$

where $\forall \mathbf{w}_i,\ \mathbf{v}_i \in [V_i]^2,\ \forall q_i \in L^2(\Omega_i)$:

$$a_i(\mathbf{w}_i, \mathbf{v}_i) = \int_{\Omega_i} \frac{\nu}{2} \sum_{j,l=1}^{2} \left(\frac{\partial (\mathbf{w}_i)_l}{\partial x_j} + \frac{\partial (\mathbf{w}_i)_j}{\partial x_l} \right) \left(\frac{\partial (\mathbf{v}_i)_l}{\partial x_j} + \frac{\partial (\mathbf{v}_i)_j}{\partial x_l} \right), \tag{195}$$

$$b_i(\mathbf{w}_i, q_i) = -\int_{\Omega_i} (\nabla \cdot \mathbf{w}_i) q_i \mathrm{d}\Omega_i, \tag{196}$$

and $E_i : [\Lambda]^2 \to [V_i]^2$ is a Stokes extension operator.

To construct the interface equation for the Stokes problem, we write $\mathbf{u} = \mathbf{u}_0 + \mathbf{v}$ and $p = p_0 + \pi$, where $(\mathbf{u}_0, p_0)$ is the solution of

$$\begin{cases} \alpha \mathbf{u}_{0,i} - \nu \Delta \mathbf{u}_{0,i} + \nabla p_{0,i} = \mathbf{f} & \text{in } \Omega_i \\ \nabla \cdot \mathbf{u}_{0,i} = 0 & \text{in } \Omega_i \\ \mathbf{u}_{0,i} = \mathbf{0} & \text{on } \partial\Omega_i \end{cases} \qquad i = 1, 2 \tag{197}$$

(as usual $\mathbf{u}_{0,i} = \mathbf{u}_0|_{\Omega_i}$, $p_{0,i} = p_0|_{\Omega_i}$), while $(\mathbf{v}, \pi)$ is the solution of

$$\begin{cases} \alpha \mathbf{v}_i - \nu \Delta \mathbf{v}_i + \nabla \pi_i = \mathbf{0} & \text{in } \Omega_i \\ \nabla \cdot \mathbf{v}_i = 0 & \text{in } \Omega_i \\ \mathbf{v}_i = \mathbf{0} & \text{on } \partial\Omega_i \cap \partial\Omega \\ \mathbf{v}_i = \boldsymbol{\lambda} & \text{on } \partial\Omega_i \cap \Gamma \end{cases} \qquad i = 1, 2 \tag{198}$$

(with $\mathbf{v}_i = \mathbf{v}|_{\Omega_i}$, $\pi_i = \pi|_{\Omega_i}$).

In order to achieve the equivalence with problem (193), $\boldsymbol{\lambda}$ has to be chosen in a way that the following interface equation be satisfied:

$$\nu \frac{\partial \mathbf{u}_1}{\partial \mathbf{n}_1} - p_1 \mathbf{n}_1 = -\nu \frac{\partial \mathbf{u}_2}{\partial \mathbf{n}_2} - p_2 \mathbf{n}_2 \qquad \text{on } \Gamma. \tag{199}$$

This can be equivalently written as:

$$\mathcal{S}_S \boldsymbol{\lambda} = \boldsymbol{\chi}_S \tag{200}$$

with

$$\boldsymbol{\chi}_S = -\left(\nu \frac{\partial \mathbf{u}_{0,1}}{\partial \mathbf{n}_1} - p_1 \mathbf{n}_1 + \nu \frac{\partial \mathbf{u}_{0,2}}{\partial \mathbf{n}_2} - p_2 \mathbf{n}_2 \right) \tag{201}$$

and

$$\mathcal{S}_S \boldsymbol{\lambda} = \mathcal{S}_S^1 \boldsymbol{\lambda} + \mathcal{S}_S^2 \boldsymbol{\lambda} = \left(\nu \frac{\partial \mathbf{v}_1}{\partial \mathbf{n}_1} - \pi_1 \mathbf{n}_1 \right) + \left(\nu \frac{\partial \mathbf{v}_2}{\partial \mathbf{n}_2} - \pi_2 \mathbf{n}_2 \right). \tag{202}$$

Equation (200) is the interface equation for the generalized Stokes problem and $\mathcal{S}_S$ is the associate Steklov-Poincaré operator. To solve the interface equation (200) amounts to solve again two generalized Stokes problems on Ω_i in order to find $(\mathbf{u}_0, p_0)$ and compute $\boldsymbol{\chi}_S$. Then, solve the equation on $\boldsymbol{\lambda}$ and, finally, solve two generalized Stokes problems on Ω_i in order to find $(\mathbf{v}, \pi)$.

As for the Poisson problem we can formulate the finite-dimensional counterpart of (200) and obtain the Schur complement matrix.

Some iterative procedure to solve the discrete interface equation on non-overlapping decompositions are the Dirichlet-Neumann scheme ([67], [54], [70], [11]) and the Neumann-Neumann scheme ([9]).

The Dirichlet-Neumann method for the generalized Stokes system on the two-domain decomposition reads: given $\boldsymbol{\lambda}^{(0)}$, $\forall k \geq 0$ solve:

$$\begin{cases} \alpha \mathbf{u}_1^{(k+1)} - \Delta \mathbf{u}_1^{(k+1)} + \nabla p_1^{(k+1)} = \mathbf{f} & \text{in } \Omega_1 \\ \nabla \cdot \mathbf{u}_1 = 0 & \text{in } \Omega_1 \\ \mathbf{u}_1^{(k+1)} = \mathbf{0} & \text{on } \partial\Omega_1 \cap \partial\Omega \\ \mathbf{u}_1^{(k+1)} = \boldsymbol{\lambda}^{(k)} & \text{on } \Gamma \cap \partial\Omega_1, \end{cases} \tag{203}$$

$$\begin{cases} \alpha \mathbf{u}_2^{(k+1)} - \Delta \mathbf{u}_2^{(k+1)} + \nabla p_2^{(k+1)} = \mathbf{f} & \text{in } \Omega_2 \\ \nabla \cdot \mathbf{u}_2 = 0 & \text{in } \Omega_2 \\ \mathbf{u}_2^{(k+1)} = \mathbf{0} & \text{on } \partial\Omega_2 \cap \partial\Omega \\ \nu \dfrac{\partial \mathbf{u}_2^{(k+1)}}{\partial \mathbf{n}_1} - p_2^{(k+1)} \mathbf{n}_2 = -\nu \dfrac{\partial \mathbf{u}_1^{(k+1)}}{\partial \mathbf{n}_1} - p_1^{(k+1)} \mathbf{n}_1 & \text{on } \Gamma \cap \partial\Omega_2, \end{cases} \tag{204}$$

then set $\boldsymbol{\lambda}^{(k+1)} = \theta \mathbf{u}_2^{(k+1)}|_\Gamma + (1-\theta)\boldsymbol{\lambda}^{(k)}$, θ being a positive acceleration parameter in $(0,1)$.

The initial datum $\boldsymbol{\lambda}^{(0)}$ has to satisfy a compatibility condition with respect to the continuity equation $\nabla \cdot \mathbf{u} = 0$. As a consequence of the divergence theorem, for any $\Omega_k \subset \Omega$, it holds

$$\int_{\Omega_k} \nabla \cdot \mathbf{u} = \int_{\partial\Omega_k} \mathbf{u} \cdot \mathbf{n}_k = 0 \tag{205}$$

and, under the assumption that $\mathbf{u}|_{\partial\Omega} = \mathbf{0}$, we have

$$\int_{\Gamma} \mathbf{u} \cdot \mathbf{n}_k = - \int_{\partial\Omega_k \setminus \Gamma} \mathbf{u} \cdot \mathbf{n}_k = 0. \tag{206}$$

It follows that $\boldsymbol{\lambda}^{(0)}$ must satisfy the condition

$$\int_{\Gamma} \boldsymbol{\lambda}^{(0)} \cdot \mathbf{n}_1 = 0, \tag{207}$$

which guarantees that

$$\int_{\Gamma} \boldsymbol{\lambda}^{(k)} \cdot \mathbf{n}_1 = 0, \qquad \forall k \geq 1. \tag{208}$$

(See [72].)

The Dirichlet-Neumann method converges with a rate independent of N, provided that the inf-sup condition (121) is satisfied with a constant β uniformly bounded from below ([67]). It can also be used as an iterative substructuring procedure to solve the global system arising from a spectral element approximation of the Stokes equations.

Projection Decomposition Method can be considered to solve the interface equation (200) (see [59]). As for the Poisson problem, a well-conditioned basis is constructed to approximate the equation on the interface Γ by the projection Galerkin method. Its rate of convergence is independent of the number of basis functions on the interface as well as the polynomial degree which is used inside each subdomain.

4.3. The Navier-Stokes equations

A multidomain formulation of Navier-Stokes equations can be based on the stabilized SEM. Using the notations of Sec. 1.5 and a decomposition of Ω like (97), a system similar to (138) is obtained. However, now the discrete solution $[\mathbf{u}_\mathcal{H}, p_\mathcal{H}]$ is sought in $\mathbf{V}_\mathcal{H} \times Q_\mathcal{H}$, with

$$\mathbf{V}_\mathcal{H} = [V_\mathcal{H} \cap H_0^1(\Omega)]^2, \qquad Q_\mathcal{H} = V_\mathcal{H} \cap L_0^2(\Omega), \tag{209}$$

and any discrete inner product $(\cdot,\cdot)_{N,\Omega}$ is replaced with its spectral element counterpart. Also the stabilization parameters are referred to the spectral element discretization, in particular T_k plays the role of Ω in (141)–(142).

At each time level the matrix of the system is sparse and its condition number depends heavily on the parameters of the spectral element discretization, say the mesh size H and the polynomial degree N on each element. Numerical results (see [34]) show that the condition number of the matrix $A_{\mathcal{H}}$, arising from the stabilized spectral element approximation, is

$$\chi(A_{\mathcal{H}}) \leq CN^6 H^{-4}, \tag{210}$$

for a suitable constant C independent of N and H. Then it is mandatory to solve the system by a preconditioned iterative method. Instead of a global preconditioner, that requires many storage locations to be stored and incompletely factorized, an element by element bilinear preconditioner based on the same stabilized approach used for the primary problem can be used.

We denote by A_k the square matrix of dimension $3(N+1)^2$ which is the block matrix associated to the stabilized spectral element approximation of Navier Stokes problem on the element $T_k \in \mathcal{T}_H$. Besides, $\mathbf{R}_k^T$ is a rectangular matrix of dimensions $3N_h^2 \times 3(N+1)^2$, (N_h is the total number of nodes in the computational domain Ω) which is the extension matrix associated to T_k (see Sec. 1.5). Then

$$A_{\mathcal{H}} = \sum_{k=1}^{Ne} \mathbf{R}_k^T A_k \mathbf{R}_k. \tag{211}$$

Now we denote by A_k^{fe} the block matrix associated to the stabilized bilinear spectral element matrix on T_k (it has the same dimensions of A_k) and we define a block Jacobi preconditioner as follows:

$$P^{-1} = \sum_{k=1}^{Ne} \mathbf{R}_k^T (A_k^{fe})^{-1} \mathbf{R}_k. \tag{212}$$

Numerical results on a decomposition of Ω in $Ne = M_1 \times M_1$ elements of size H ([32]) show that

$$\chi(P^{-1}A_{\mathcal{H}}) \leq CN^2 H^{-2}. \tag{213}$$

Even if this preconditioner is non-optimal, the numerical results point out substantial reduction of the iteration number with respect to an unpreconditioned system, but also to a diagonally preconditioned system. All results presented in the following section are obtained using the BiCGStab iterative method ([24]) with the preconditioner (212).

5. Numerical results

We present here some numerical results obtained by the approximation of incompressible Navier-Stokes equations (113) with the stabilized spectral element method. Different test cases are addressed.

5.1. Driven cavity flow

This test case shows the motion of a flow inside a plane square domain $\Omega = (0,1)^2$ with tangential velocity prescribed on the top boundary $\mathbf{u}_\infty = (1,0)^T$. We define the Reynolds number as $Re = D\|\mathbf{u}_\infty\|/\nu$ where D is the measure of the side of Ω and $\mathbf{u}_\infty$ is the prescribed tangential velocity on the top side. On the vertical sides and on the bottom horizontal side a no-slip boundary condition is imposed.

The problem has been solved by the stabilized spectral elements (SSE), and by the Euler semi-implicit time advancing scheme (134).

We compare our results with those of Ghia et al. ([35]) about the profiles of the velocity components u and v along vertical and horizontal (respectively) lines through geometric center of the cavity (see Figures 7 and 8). The solid lines represent the numerical solution obtained by SSE at different Reynolds numbers, while the symbols represent the numerical data reported in Tables I and II in [35].

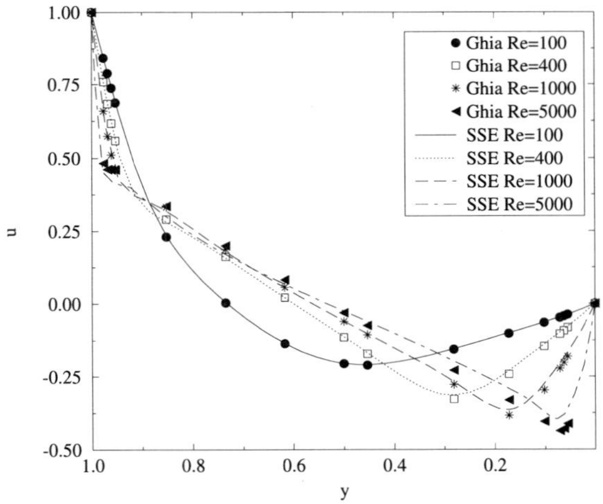

FIGURE 7. Profiles of u velocities along vertical lines through geometric center of the cavity.

In Tab. 3 we compare the space discretization used in [35] and by SSE. N and M stand for the polynomial degree in each spectral element (in each direction) and the number of elements in each direction respectively. Unless otherwise specified the decomposition of Ω is uniform.

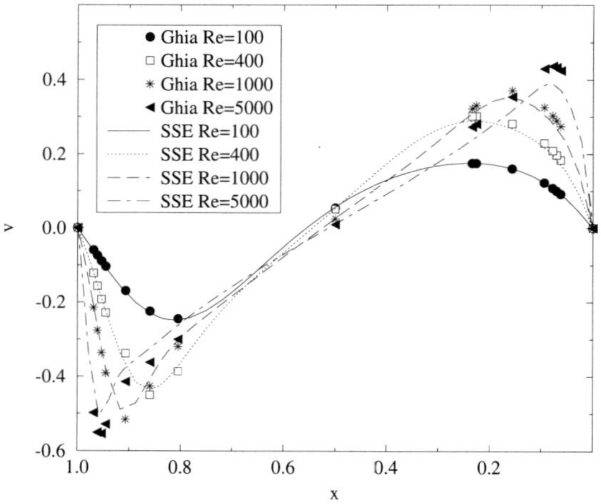

FIGURE 8. Profiles of v velocities along horizontal lines through geometric center of the cavity.

Re	100	400	1000	5000	10000
Ghia discretization	129×129 16641	129×129 16641	129×129 16641	257×257 66049	257×257 66049
SSE discretization	$N = 8$ $M = 6$ (2401)	$N = 8$ $M = 6$ (2401)	$N = 6$ $M = 10$ (3721)	$N = 6$ $M = 12$ (5329)	$N = 6$ $M = 12$ (5329)

TABLE 3. The number of nodes of the space discretization used in the "Driven Cavity" test case. M denotes the number of elements in each space direction.

In Figures 9 and 10 we report the streamline contours for the cavity flow configurations, that we have obtained for *Re*=400, 1000, 5000, 10000. Finally in Fig. 11 we show the pressure for *Re*=5000 and *Re*=10000.

5.2. Creeping flow in a wedge

The computational domain Ω for this test case ([50]), and its spectral element partition are represented in Figure 12. Inside each element, polynomials of degree $N = 6$ have been considered and the Euler semi-implicit scheme with $\Delta t = .01$ has been used. The motion is driven by steady motion of the fluid at the top of the edge; the Reynolds number is $Re = 0.17$ and it is based on the peripheral speed

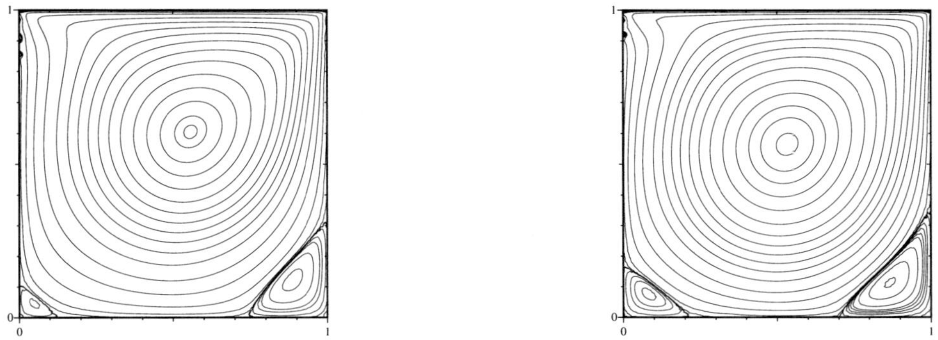

FIGURE 9. Streamlines for Re=400 (left) and Re=1000 (right).

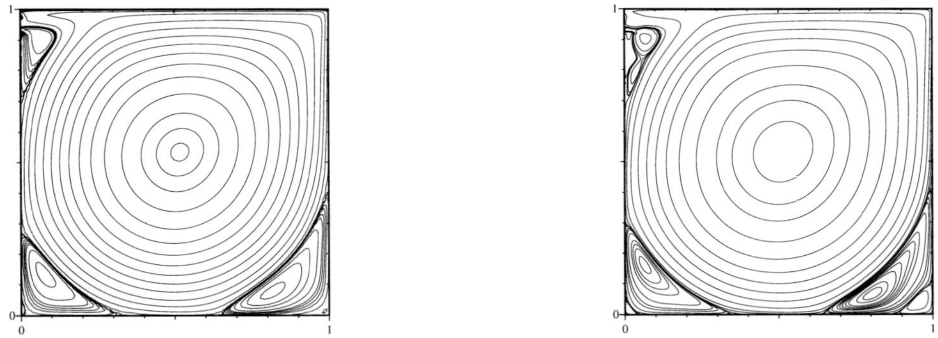

FIGURE 10. Streamlines for Re=5000 (left) and Re=10000 (right).

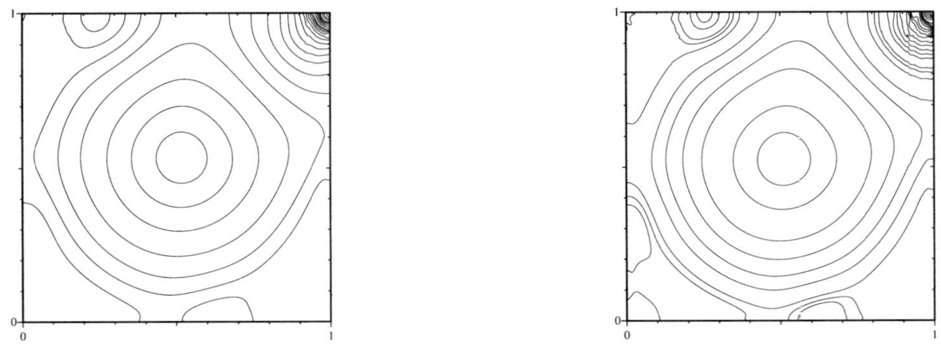

FIGURE 11. The pressure for Re=5000 (left) and Re=10000 (right).

$\mathbf{u}_\infty = (1,0)^T$, $\nu = 36.39$ and the top side length $D = 2\pi$. In Figure 12 (right) the computed stream functions are shown.

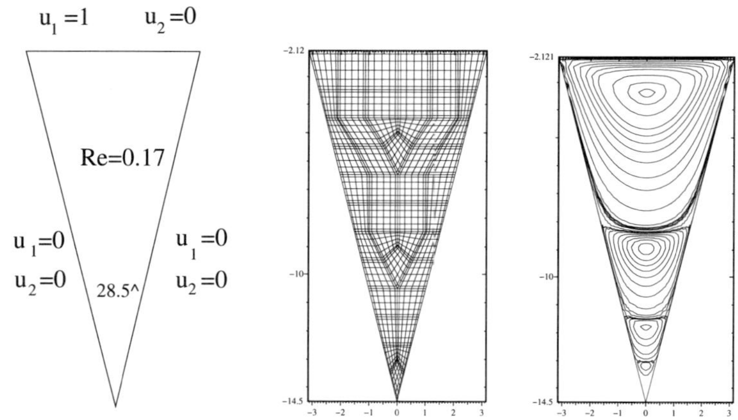

FIGURE 12. The geometry of the problem *creeping flow in a wedge* (left), the domain partition (center) and the stream-functions (right) relative to the *creeping flow in a wedge.*

5.3. Uniform flow past a fence

We consider the motion of a fluid in the domain $\Omega = (0,10) \times (0,5)$ in the presence of a fence of height $D = 1$ and width equal to 0.1. We set the inflow velocity $\mathbf{u}_\infty = (1,0)^T$, a viscosity $\nu = 71.43$ so that $Re = 0.014$. The problem geometry and the boundary conditions used are shown in Fig. 13. The domain has been partitioned in 24 elements with polynomial degree $N = 6$ in each element. The computed stream functions are shown in Fig. 13.

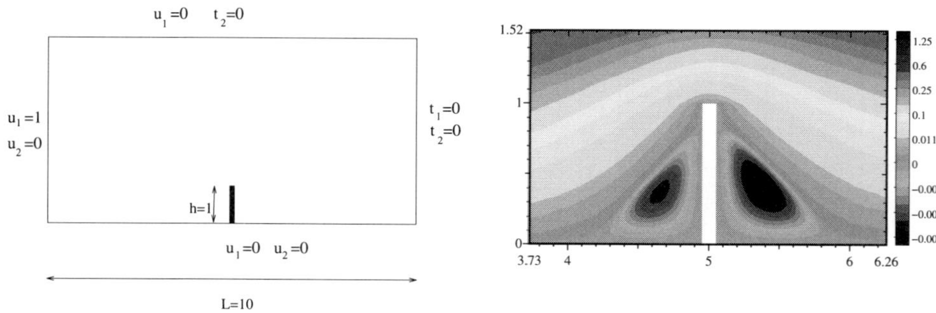

FIGURE 13. The geometry (left) and a zoom on the stream-function (right) relative to the test case: *Uniform flow past a fence.*

5.4. Uniform flow past a circular cylinder

This unsteady problem represents the motion of a flow past a circular obstacle. The computational domain is $\Omega = (-4.5, 15.5) \times (-4.59, 4.5)$ and a circular obstacle, with diameter $D = 1$, is centered at $(0,0)$. The lack of symmetry on the geometry has been adopted on purpose in order to generate periodic motion of the fluid (see [79]). This motion is known in the literature as *Von Karman vortex street.*

The computational domain has been discretized in 64 spectral elements with polynomial degree $N = 6$ in each direction. The total number of nodes is 1700. The boundary conditions are assigned as they can be observed in Fig. 14, where t_1 and t_2 represent the two components of the normal component of the stress tensor $\tilde{T}\mathbf{n} = -p\mathbf{n}+(\mathbf{n}\cdot\nabla)\mathbf{u}$. The semi-implicit Euler scheme has been used with $\Delta t = .05$. After an initially phase of transition, the motion becomes periodic in time with a period $T = 6$, corresponding to a Strouhal number $St = D/(\|\mathbf{u}_\infty\|T) = 0.167$.

In Fig. 14 the discretization of the computational domain is shown while in Figures 15–17 we show the vorticity for Re=200 at six different time steps of a period.

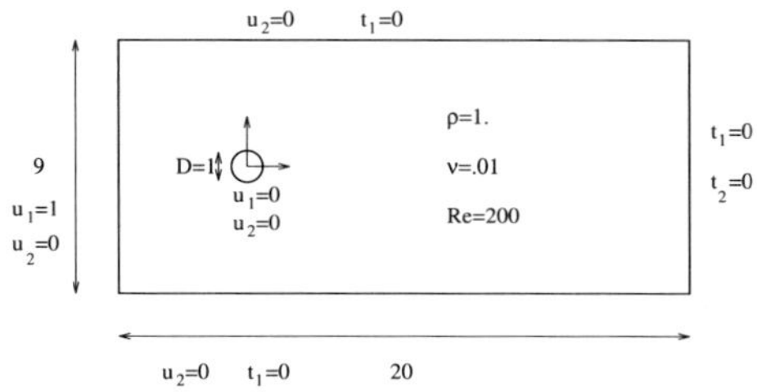

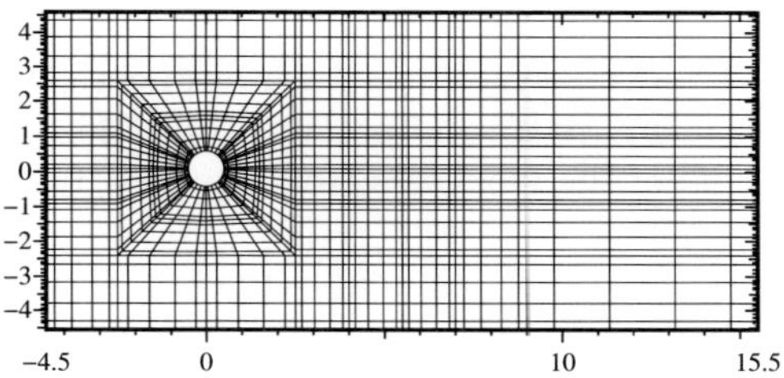

FIGURE 14. The geometry (left) and the partition of the computational domain (right) for the test case of *uniform flow past a circular cylinder.*

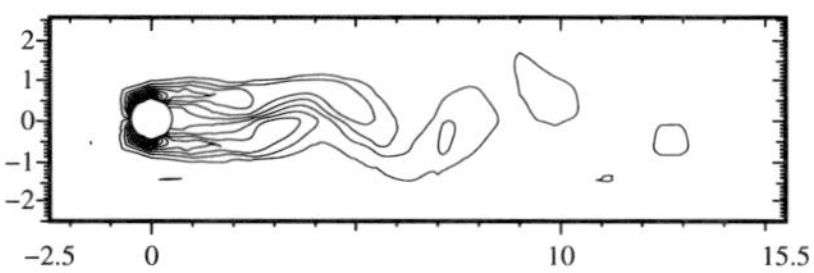

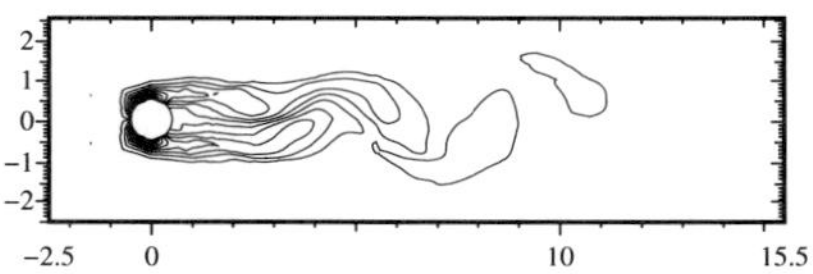

FIGURE 15. The vorticity at $t = 111$ (left) and $t = 112$(right).

5.5. Two-dimensional model of pulsatile Newtonian flow in the human carotid bifurcation

This model problem is considered in biomechanical literature as a simplification of the more complex 3-D problem.

The computational domain and its spectral elements decomposition are shown in Fig. 18. The basic shape of the model agrees with the model of Bharadvaj ([7]) and the geometry parameters are based upon the data described by Ku et al. ([65],

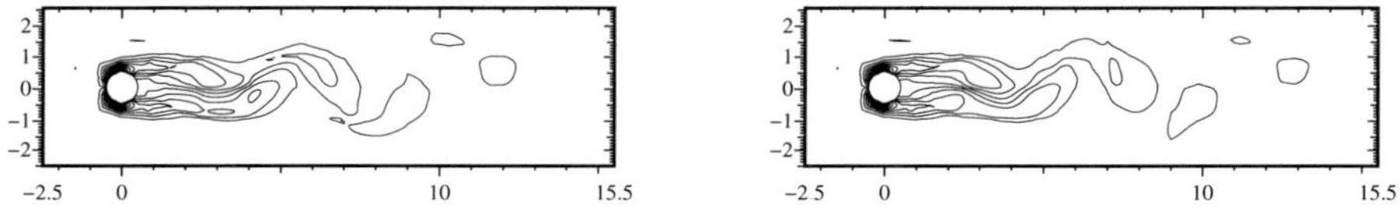

FIGURE 16. The vorticity at $t = 113$ (left) and $t = 114$(right).

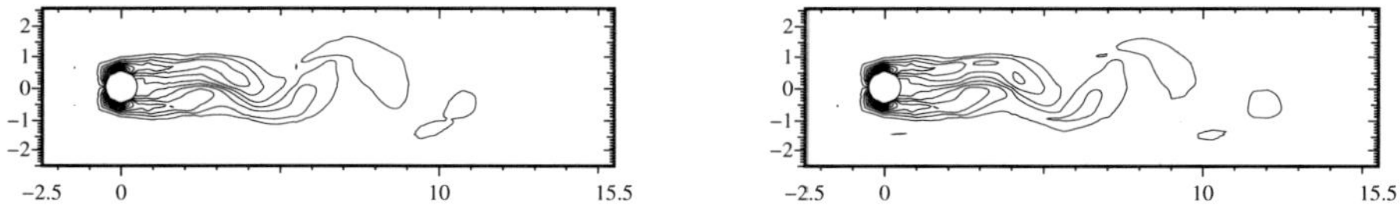

FIGURE 17. The vorticity at $t = 115$ (left) and $t = 116$ (right).

[45]). Using the common carotid diameter $D = 0.62$cm as characteristic length and a reference blood viscosity $\nu = 0.035$, the maximum Reynolds number inside a period of the motion is $Re_{\max} \simeq 800$. The assumed pulse frequency is 80 strokes per minute.
At the inflow boundary (the left vertical side) a fully developed time-dependent velocity profile $\mathbf{g} = \mathbf{g}(x_2, t)$ is prescribed (according to [45], see Fig. 18); at the rigid walls the no slip condition $\mathbf{u} = \mathbf{0}$ is applied, while at the outflow boundary a no-friction condition is imposed (i.e. $\tilde{T}\mathbf{n} = -p\mathbf{n} + (\mathbf{n} \cdot \nabla)\mathbf{u} = \mathbf{0}$).

The discretization is based on 25 spectral elements with polynomial degree $N = 4$ and the global number of nodes is 461. The velocity field and the stream functions in the branching plane during the pulse cycle ($T = 0.8\overline{3}$) are shown in Figures 19–22, in accordance with the results of ([65]).

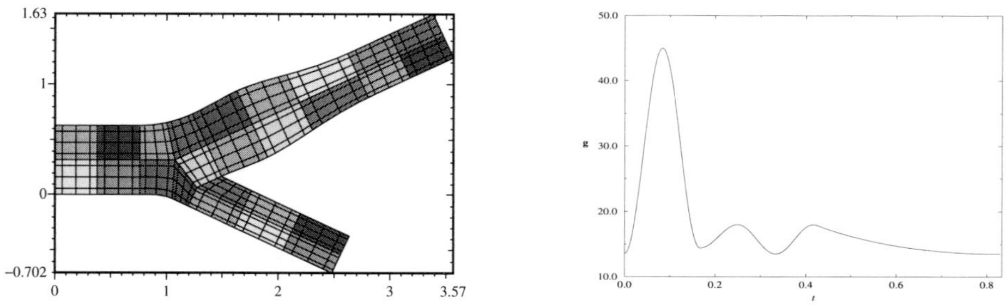

FIGURE 18. The spectral element mesh on the branching plane (left) and the time-dependent velocity profile $\mathbf{g}(x_2, t)$.

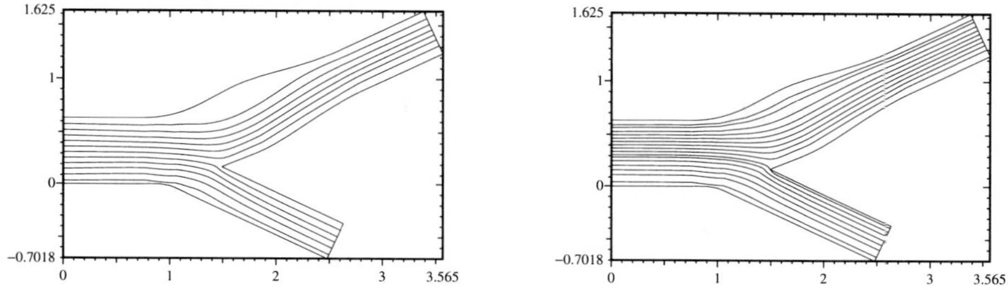

FIGURE 19. The velocity field and the stream functions in the branching plane during the pulse cycle at $t = 0.0$ (left) and $t = 0.05$ (right).

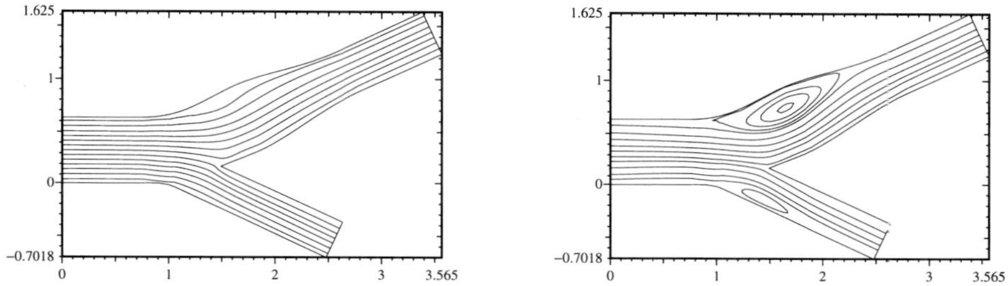

FIGURE 20. The velocity field and the stream functions in the branching plane during the pulse cycle at $t = 0.09$ (left) and $t = 0.14$ (right).

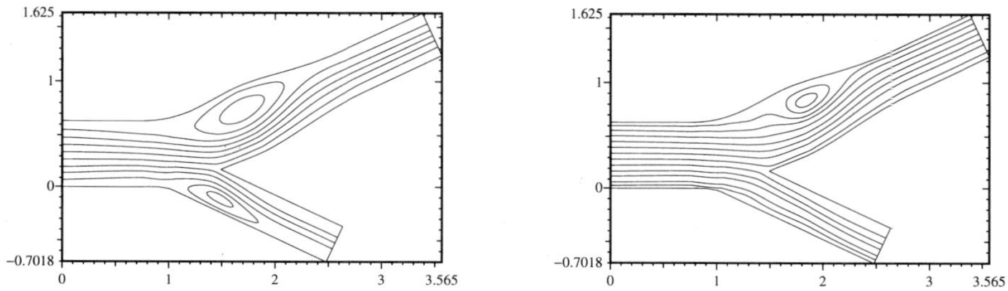

FIGURE 21. The velocity field and the stream functions in the branching plane during the pulse cycle at $t = 0.17$ (left) and $t = 0.26$ (right).

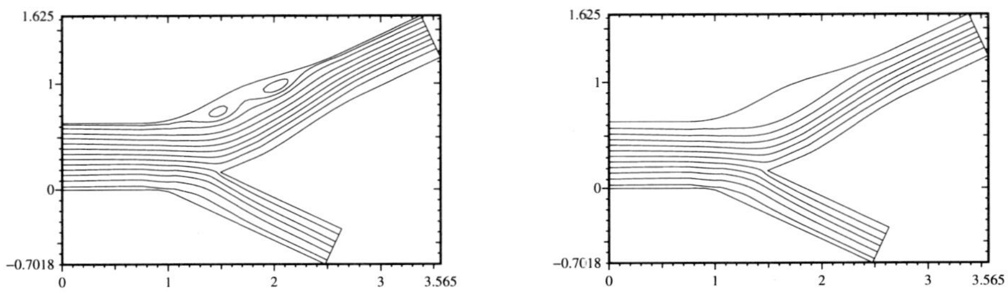

FIGURE 22. The velocity field and the stream functions in the branching plane during the pulse cycle at $t = 0.36$ (left) and $t = 0.8$ (right).

References

[1] V.I. Agoshkov and V.I. Lebedev. The Poincaré-Steklov's operators and the domain decomposition algorithms in variational problems. *Computational Processes and Systems*, 2:173–227, 1985. NAUKA, Moscow, (in Russian).

[2] I. Babuška. The finite element method with Lagrangian multipliers. *Numer. Math.*, 20:179–192, 1973.

[3] I. Babuška and M.R. Dorr. Error estimates for the combined h and p versions of the finite element method. *Numer.Math.*, 37:257–277, 1981.

[4] I. Babuška and M. Suri. The $h-p$ version of the finite element method with quasi-uniform meshes. *Mathematical Modelling and Numerical Analysis*, 21:199–238, 1987.

[5] C. Bernardi and Y. Maday. *Approximations Spectrales de Problèmes aux Limites Elliptiques.* Springer Verlag, Paris, 1992.

[6] C. Bernardi, Y. Maday, and A.T. Patera. A new nonconforming approach to domain decomposition: the mortar element method. In H. Brezis and J.L. Lions, editors, *Nonlinear Partial Differential Equations and Their Applications.* Collège de France Seminar, 1989.

[7] B.K. Bharadvaj, R.F. Mabon, and D.P. Giddens. Steady flow in a model of the human carotid bifurcation. Part I – Flow visualisation. *J. Biomechanics*, 15:349–362, 1982.

[8] P. Bjorstad and O.B. Widlund. Iterative methods for the solution of elliptic problems on regions partitioned into substructures. *SIAM J. Numer. Anal.*, 23:1097–1120, 1986.

[9] J.F. Bourgat, R. Glowinski, P. Le Tallec, and M.Vidrascu. Variational formulation and algorithm for trace operator in domain decomposition calculations. In J.Perieaux T.F.Chan, R.Glowinski and O.B.Widlund, editors, *Domain Decomposition Methods for Partial Differential Equations*, Philadelphia, 1989. SIAM.

[10] J.P Boyd. *Chebyshev and Fourier Spectral Methods.* Springer-Verlag, Berlin, 1989.

[11] J. Bramble, J. Pasciak, and A. Schatz. An iterative method for elliptic problems on regions partitioned into substructures. *Math. Comp.*, 46:361–369, 1986.

[12] F. Brezzi. On the existence, uniqueness and approximation of saddle-point problems arising from Lagrange multipliers. *R.A.I.R.O Anal. Numér.*, 8 (R2):129–151, 1974.

[13] F. Brezzi and J. Pitkaranta. On the stabilization of finite element approximations of the Stokes equations. In *Efficient Solutions of Elliptic Systems, Note on Numerical Fluid Mechanics*, volume 10, pages 11–19. W. Hackbusch, 1984.

[14] A.N. Brooks and T.J.R. Hughes. Streamline Upwind/Petrov-Galerkin formulations for convection dominated flows with particular emphasis on the incompressible Navier-Stokes equations. *Comput. Meth. Appl. Mech. Engrg.*, 32:199–259, 1982.

[15] C. Canuto. Stabilization of spectral methods by finite element bubble functions. *Comput. Methods Appl. Mech. Engrg.*, 116:13–26, 1994.

[16] C. Canuto, M. Y. Hussaini, A. Quarteroni, and T. A. Zang. *Spectral Methods in Fluid Dynamics.* Springer Verlag, Berlin, 1988.

[17] C. Canuto and V. Van Kemenade. Bubble-stabilized spectral methods for the incompressible Navier-Stokes equations. *Comput. Methods Appl. Mech. Engrg.*, 135:35–61, 1996.

[18] C. Canuto and P. Pietra. Boundary and interface conditions within a finite element preconditioner for spectral methods. *J. Comput. Phys.*, 91:310–343, 1990.

[19] C. Canuto and A. Quarteroni. Approximation results for orthogonal polynomials in Sobolev spaces. *Math. Comput.*, 38:67–86, 1982.

[20] C. Canuto and A. Quarteroni. Preconditioned minimal residual methods for Chebyshev spectral calculations. *J. Comput. Phys.*, 60:315–337, 1985.

[21] M.A. Casarin. Quasi-optimal Schwarz methods for the conforming spectral element discretization. *SIAM J. Numer. Anal.*, 34(6):2482–2502, 1997.

[22] T.F. Chan and T.P. Mathew. Domain decomposition algorithms. In *Acta Numerica*, pages 61–143. Cambridge Univ. Press, 1994.

[23] J.W. Cooley and J.W. Tukey. An Algorithm for the Machine Calculation of Complex Fourier Series. *Math. Comput.*, 19:297–301, 1965.

[24] H.A. Van der Vorst. Bi-CGStab: A faster smoothly converging variant of Bi-CG for the solution of nonsymmetric linear systems. *SIAM J. Sci. Stat. Comput.*, 13(2):631–644, 1992.

[25] M.O. Deville and E.H. Mund. Chebyshev pseudospectral solution of second-order elliptic equations with finite element preconditioning. *J. Comput. Phys.*, 60:517–533, 1985.

[26] M.O. Deville and E.H. Mund. Finite element preconditioning for pseudospectral solutions of elliptic problems. *SIAM J. Sci. Stat. Comput.*, 11:311–342, 1990.

[27] M. Dryja and O. Widlund. Some domain decomposition algorithms for elliptic problems. In L.Hayes and D. Kincaid, editors, *Iterative methods for large linear systems*, pages 273–291. Academic, 1989.

[28] M. Dryja and O. Widlund. Additive Schwarz methods for elliptic finite element problems in three dimensions. In T.F.Chan, D.E. Keyes, G.A. Meurant, J.S. Scroggs, and R.G. Voigt, editors, *Fifth Conf. on Domain Decomposition Methods for Partial Differential Equations*, Philadelphia, 1992. SIAM.

[29] M. Dubiner. Spectral methods on triangles and other domains. *J. Sci. Comput.*, 4:345–390, 1991.

[30] L.P. Franca and S.L. Frey. Stabilized finite element methods: II. The Incompressible Navier-Stokes Equations. *Comput. Meth. Appl. Mech. Engrg.*, 99:209–233, 1992.

[31] D. Funaro, A. Quarteroni, and P. Zanolli. An iterative procedure with interface relaxation for domain decomposition methods. *SIAM J.Numer.Anal.*, 25:1213–1236, 1988.

[32] P. Gervasio. *Risoluzione di equazioni alle derivate parziali con metodi spettrali in regioni partizionate in sottodomini.* PhD thesis, Università degli Studi di Milano, 1995.

[33] P. Gervasio, E.I. Ovtchinnikov, and A. Quarteroni. The spectral projection decomposition method for elliptic equations in two dimensions. *SIAM J. Numer. Anal.*, 34(4):1616–1639, 1997.

[34] P. Gervasio and F. Saleri. Stabilized spectral element approximation for the Navier-Stokes equations. *Numerical Methods for Partial Differential Equations*, 14:115–141, 1998.

[35] U. Ghia, K.N. Ghia, and C.T. Shin. High-Re Solutions for Incompressible Flow Using the Navier-Stokes Equations and a Multigrid Method. *J. of Comput. Phys.*, 48:387–411, 1982.

[36] D. Gottlieb and S.A. Orszag. *Numerical Analysis of Spectral Methods: Theory and Applications.* SIAM-CBMS, Philadelphia, 1977.

[37] M.D. Gunzburger. *Finite Element Methods for Viscous Incompressible Flows. A Guide to Theory, Practice and Algorithms.* Academic Press, San Diego, 1989.

[38] P. Haldenwang, G. Labrosse, S. Abboudi, and M. Deville. An element-by-element solution algorithm for problems of structural and solid mechanics. *Comput. Meth. Appl. Mech. Engrg.*, 36:241–254, 1983.

[39] T.J.R. Hughes, L.P. Franca, and M. Balestra. A new finite element formulation for computational fluid dynamics: VII. The Stokes problem with various well-posed boundary conditions: Symmetric formulations that converge for all velocity/pressure spaces. *Comput. Meth. Appl. Mech. Engrg.*, 59:85–99, 1986.

[40] Jr. J. Douglas and J. Wang. An absolutely stabilized finite element method for the Stokes problem. *Math. Comput.*, 52:495–508, 1989.

[41] J. Van Kan. A second-order accurate pressure-correction scheme for viscous incompressible flow. *SIAM J. Sci. Stat. Comput.*, 7(3):870–891, 1986.

[42] G.E. Karniadakis, M. Israeli, and S.A. Orszag. High order splitting methods for the incompressible Navier-Stokes equations. *J. Comput. Phys.*, 97:414–443, 1991.

[43] K.Z. Korczak and A.T. Patera. An isoparametric spectral element method for solution of the Navier-Stokes equations in complex geometries. *J. Comput. Phys.*, 62:361, 1986.

[44] H.O. Kreiss and J. Oliger. Stability of the Fourier method. *SIAM J. Numer. Anal.*, 16:421–433, 1979.

[45] D.N. Ku, D.P. Giddens, C.Z. Zarins, and S. Glagov. Pulsatile flow and atherosclerosis in the human carotid bifurcation. *Arteriosclerosis*, 5:293–302, 1985.

[46] V.I. Lebedev and V.I. Agoshkov. Generalized Schwarz algorithms with variable parameters. Technical report, Report n. 19, Dept. Num. Math., USSR Accademy of Science, 1981. Moscow, (in Russian).

[47] J. L. Lions and E. Magenes. *Nonhomogeneous Boundary Value Problems and Applications.* Springer Verlag, Berlin, 1972.

[48] P.L. Lions. On the Schwarz alternating method III: a variant for nonoverlapping subdomains. In J. Perieaux T.F. Chan, R. Glowinski and O.B. Widlund, editors, *Domain Decomposition Methods for Partial Differential Equations*, pages 202–231, Philadelphia, 1990. SIAM.

[49] Y. Maday, C. Mavripilis, and A.T. Patera. Nonconforming mortar element methods: application to spectral discretizations. In J.Perieaux T.F.Chan, R.Glowinski and O.B.Widlund, editors, *Second International Conference on Domain Decomposition Methods for Partial Differential Equations*, Philadelphia, 1988. SIAM.

[50] Y. Maday, D. Meiron, A.T. Patera, and E.H. Rønquist. Analysis of iterative methods for the steady and unsteady Stokes problem: application to spectral element discretizations. *SIAM J. Sci. Comput.*, 14:310–337, 1993.

[51] Y. Maday and A.T. Patera. Spectral element methods for the incompressible Navier-Stokes equations. In *State-of-the-Art Surveys on Computational Mechanics.* A.K. Noor and J. T. Oden, 1989.

[52] Y. Maday, A.T. Patera, and E.H. Rønquist. An operator-integration-factor splitting method for time-dependent problems: application to incompressible fluid flow. *J. Sci. Comput.*, 5:263–292, 1990.

[53] P.S. Marcus. Simulation of Taylor-Couette flow. Part 1. Numerical methods and comparison with experiments. *J. Fluid Mech.*, 146:45–64, 1984.

[54] L.D. Marini and A. Quarteroni. A relaxation procedure for domain decomposition methods using finite elements. *Numer.Math.*, 55:575–598, 1989.

[55] S.G. Mikhlin and Kh. L. Smolitskii. Approximate Methods for Solving Differential and Integral Equations. *Nauka*, 1965. (In Russian).

[56] Y. Morchoisne. Inhomogeneous flow calculation by spectral methods: monodomain and multidomain techniques. In *Spectral Methods for P.D.E.* R.G. Voight, D. Gottlieb and M.Y. Hussaini, Philadelphia, 1984.

[57] S.A. Orszag and L.C. Kells. Transition to turbolence in plane Pouseuille flow and plane Couette flow. *J. Fluid Mech.*, 96:159–205, 1980.

[58] E. Ovtchinnikov. The construction of a well-conditioned basis for the projection decomposition method. *CALCOLO*, 30(3):255–271, 1993.

[59] E.I. Ovtchinnikov. Projection decomposition method with a well- conditioned basis for the generalised Stokes problem. In *Numerical Methods for Fluid Dynamics V*, pages 515–521, Oxford, 1995. Clarendon Press.

[60] S.S. Pahl. *Schwarz type domain decomposition methods for spectral element discretizations.* PhD thesis, University of the Witwatersand, 1993.

[61] F. Pasquarelli and A. Quarteroni. Effective spectral approximations to convection-diffusion equations. *Comput. Meth. Appl. Mech. Engrg.*, 116:39–51, 1994.

[62] A.T. Patera. A spectral element method for fluid dynamics: laminar flow in a channel expansion. *J. Comput. Phys.*, 54:468–488, 1984.

[63] L. Pavarino. *Domain decomposition methods for the p-version finite element method for elliptic problems.* PhD thesis, Courant Institute of Mathematical Sciences, Department of Mathematics, 1992.

[64] L. Pavarino. Some Schwarz algorithms for the p-version finite element method. In A. Quarteroni, J. Perieaux, Y.A. Kuznetsov, and O.B. Widlund, editors, *Sixth International Symposium on Domain Decomposition Methods for Partial Differential Equations*, volume 157, pages 113–119. Contemporary Mathematics, 1994.

[65] K. Perktold, M. Resch, and H. Florian. Pulsatile non-newtonian flow characteristic in a three-dimensional human carotid bifurcation model. *Journal of Biomechanical Engineering, Transactions of the ASME*, 113:464–475, 1991.

[66] A. Pinelli, A. Vacca, and A. Quarteroni. A spectral multidomain method for the numerical simulation of turbulent flows. *J. Comput. Phys.*, 136(2):546–558, 1997.

[67] A. Quarteroni. Domain decomposition algorithms for the Stokes equations. In *Domain Decomposition Methods*, pages 431–442, Philadelphia, 1989. SIAM.

[68] A. Quarteroni and G. Sacchi Landriani. Domain decomposition preconditioners for the spectral collocation methods. *J. Sci. Comput.*, 3:45–75, (1988).

[69] A. Quarteroni and G. Sacchi Landriani. Parallel algorithms for the capacitance matrix method in domain decompositions. *Calcolo*, 25(1-2):75–102, (1988).

[70] A. Quarteroni and A. Valli. Theory and application of Steklov-Poincaré operators for boundary-value problems. In R. Spigler, editor, *Applied and Industrial Mathematics*, pages 179–203, Kluwer Academic Publisher, Dordest, 1990.

[71] A. Quarteroni and A. Valli. *Numerical Approximation of Partial Differential Equations.* Springer Verlag, Heidelberg, 1994.

[72] A. Quarteroni and A. Valli. *Domain Decomposition Methods for Partial Differential Equations.* Oxford University Press, Oxford, 1998.

[73] A. Quarteroni and E. Zampieri. Finite element preconditioning for Legendre spectral collocation approximation to elliptic equations and systems. *SIAM J. Numer. Anal.*, 29:917–936, 1992.

[74] R. Rannacher. On Chorin's projection method for the incompressible Navier- Stokes equations. In R. Rautmann J.G. Heywood, K. Masuda and S.A. Slonnikov, editors, *The Navier-Stokes Equations II: Theory and Numerical Methods*, pages 167–183. Springer-Verlag, Berlin, 1992.

[75] D. Schötzau, K. Gerdes, and C. Schwab. Galerkin least squares hp-FEM for the Stokes problem. *C.R.A.S., Paris*, 326(Serie I):249–254, 1998.

[76] C. Schwab and M. Suri. Mixed hp-FEM for Stokes and non-Newtonian flow. to appear in Comp. Meth. Appl. Mech. Engrg., 1998.

[77] J. Shen. On error estimates of the projection methods for the Navier Stokes equations: second order schemes. *Math. Comp.*, 65(215):1039–1065, 1997.

[78] S.J. Sherwin and G.E. Karniadakis. A triangular spectral element method; applications to the incompressible Navier-Stokes equations. *Comp. Meth. Appl. Mech. Engrg.*, 123:189–229, 1995.

[79] J.C. Simo and F. Armero. Unconditional stability and long-term behaviour of transient algorithms for the incompressible Navier-Stokes and Euler equations. *Comp. Meth. Appl. Mech. Engrg.*, 111:111–154, 1992.

[80] B.F. Smith. *Domain decomposition algorithms for the partial differential equations of linear elasticity.* PhD thesis, Courant Institute of Mathematical Science, 1990.

[81] B.F. Smith. A domain decomposition algorithm for elliptic problems in three dimensions. *SIAM J. Sci. Comput.*, 13(1):219–234, 1992.

[82] E. Zampieri. A multidomain spectral collocation solver for the elasticity problem. In D. E. Keyes, T.F.Chan, G.Meurant, J.S. Scroggs, and R.G. Voigt, editors, *Fifth International Symposium on Domain Decomposition Methods for Partial Differential Equations*, pages 614–623, Philadelphia, 1992. SIAM.

P. Gervasio
Department of Mathematics
University of Brescia
Brescia (Italy)

A. Quarteroni
Department of Mathematics
Politecnico of Milano
Milano (Italy)
and
EPFL, Lausanne (CH)

F. Saleri
Department of Mathematics
University of Milano
Milano (Italy)

Simple Proofs of Bifurcation Theorems*

John G. Heywood and Wayne Nagata

Abstract. We give short, elementary, constructive proofs of the basic theorems concerning the bifurcation of equilibrium and periodic solutions, from a trivial solution of an ordinary differential equation, as a parameter in the equation passes through a critical value. We begin by considering a finite dimensional equation, formulated to be analogous to an abstract Navier-Stokes equation in Hilbert space. Later in the paper we consider generalizations, including generalizations to partial differential equations, and especially to the Navier-Stokes equations.

1. Introduction

Two of the prototypical problems of bifurcation theory concern the parameterized family of differential equations

$$\frac{du}{dt} = Au + \nu Bu + N(u),$$

for $u(t) \in \mathbf{R}^n$, in which A and B are real $n \times n$ matrices, ν is a real parameter, and $N(u)$ is a nonlinear function which vanishes when u is zero. Under the assumption that the null space of A is one-dimensional, spanned by an eigenvector w_1, one would like to find nontrivial equilibrium solutions ($u(t) \equiv w$) for ν near 0, in fact, a one-parameter family of such solutions $w(\varepsilon)$, $\nu(\varepsilon)$ such that the curve $w(\varepsilon)$ is tangent to w_1 at the origin in $\mathbf{R}^n$. Analogously, under the assumption that A has a pair of simple conjugate imaginary eigenvalues, and hence a two-dimensional eigenspace of elliptical orbits, one would like to find a parameterized family of nontrivial periodic solutions $u(t,\varepsilon)$, $\nu(\varepsilon)$, whose trajectories generate a surface tangent to this eigenspace at the origin.

A natural approach to these problems, following Poincaré [27], is to seek bifurcating solutions as perturbation series. Equilibrium solutions can be sought in the form $w(\varepsilon) = \varepsilon w_1 + \varepsilon^2 w_2 + \cdots$, $\nu(\varepsilon) = \varepsilon \nu_1 + \varepsilon^2 \nu_2 + \cdots$, and periodic solutions analogously, in terms of a series of periodic functions. While such series have been frequently used in formal calculations, direct proofs of their convergence are seldom given. Those that have been given (*e.g.*, [6, 11, 22, 29]) have generally seemed complicated, and inappropriate as the basis for a general theory.

* This work was supported by NSERC, Canada.

Another approach, closely related to the perturbation series, is to formulate an iteration scheme for bifurcating solutions, taking the leading term of the perturbation series as the main part. This is the method of the present paper. Several variants of it have already been given in papers of Rabinowitz [28], Keller [20], Keller and Langford [21], and Langford [25]. The theory developed in the papers [21] and [25] is more general than ours, but also more complicated. Perhaps because of the complexity of these papers, the advantages of the method have seemed to be slow in gaining recognition.

Usually, other, indirect methods are used to prove the existence of bifurcating solutions. Important methods in the current literature use degree theory (*e.g.,* [16, 17, 24, 30, 31, 32, 33]), center manifold theory (*e.g.,* [2, 3, 9, 26]), or the implicit function theorem (*e.g.,* [4, 5, 15]), especially combined with the decomposition of Liapunov and Schmidt (*e.g.,* [3, 7, 8, 12, 13, 18, 19, 22, 23, 30, 31]). While these methods are well-established and have been widely used in applications, we have felt that it is worthwhile to seek simpler and more direct proofs.

In this paper we offer another, simplified, variant of the iteration method, and develop it in a sufficiently general context to treat the bifurcation of both equilibrium and periodic solutions. Our approach is entirely elementary, independent of the implicit function theorem, the Liapunov-Schmidt decomposition, and the center manifold theorem. One corollary is an explicit estimate for the interval of convergence of the perturbation series, which seems not readily available by less direct methods. While we do not go into the treatment of higher dimensional kernels and numerical solutions, these topics are treated by related methods in [21] and [25]. Also, while not included here, we have a closely associated method of treating stablilty, which we plan to present elsewhere.

To emphasize the main ideas, and to make them as widely accessible as possible, even at the undergraduate level, we present them first in the context of a finite system of ordinary differential equations, restricting ourselves initially to the case

$$\frac{du}{dt} = Au + \nu Bu + N(u,u), \tag{1}$$

where $N : \mathbf{R}^n \times \mathbf{R}^n \to \mathbf{R}^n$ is a bilinear map. In subsequent steps, we generalize the argument to treat other nonlinearities, and to treat partial differential equations, particularly the Navier-Stokes equations.

Our proof of the existence of bifurcating equilibrium solutions is given in §2. We treat the bifurcation of periodic solutions, usually referred to as Hopf bifurcation, using the same method, in §3. In §4 we give generalizations to other nonlinearities and to partial differential equations, and briefly discuss the stability of the bifurcating solutions. To understand the Hopf bifurcation theorem, some preliminary information is needed, in common with other approaches, about periodic solutions of linear differential equations. Because we do not know of a concise presentation of this material in the form that is needed here, and in order to make the Hopf bifurcation theorem accessible at the level of texts like [1], [10] or [15], we offer this preliminary material in an appendix.

2. Bifurcation of equilibrium solutions

Our hypotheses in this section are that the $n \times n$ matrix A has a one-dimensional null space $\mathcal{N}(A) = \text{span}\{w_1\}$, and that Bw_1 does not belong to its range $\mathcal{R}(A)$. The first of these hypotheses implies that zero is an eigenvalue of A of geometric multiplicity one. However, zero is permitted to be a multiple root of the characteristic polynomial of A. In the case that zero is a simple root, our hypotheses are equivalent to another condition which has been taken as a hypothesis in may works on bifurcation theory, namely that a simple real eigenvalue of $A+\nu B$ passes through the origin with nonzero speed, as ν increases through 0.

Our objective under the hypotheses of this section is to find small nonzero equilibrium solutions of (1), for ν near 0. This will be accomplished by taking $\varepsilon \in \mathbf{R}$ to be a small parameter, and seeking nontrivial solutions $u(t) \equiv w \in \mathbf{R}^n$, $\nu \in \mathbf{R}$, of

$$Aw + \nu Bw + N(w,w) = 0, \tag{2}$$

in the form

$$\nu = \lambda(\varepsilon), \qquad w = \varepsilon w_1 + v(\varepsilon), \tag{3}$$

with $|\lambda(\varepsilon)| = \mathcal{O}(\varepsilon)$, $|v(\varepsilon)| = \mathcal{O}(\varepsilon^2)$, and $v(\varepsilon) \in \mathcal{V}$, where $\mathcal{V}$ is a subspace of $\mathbf{R}^n$ complementary to the null space $\mathcal{N}(A)$ of A (for example, one may take $\mathcal{V} = \{w_1\}^\perp$).

Substituting (3) into (2), we get

$$-Av = \varepsilon\lambda Bw_1 + \lambda Bv + N(\varepsilon w_1 + v, \varepsilon w_1 + v). \tag{4}$$

For any sufficiently small fixed value of $\varepsilon \neq 0$, we will obtain a solution λ, v of (4) as the limit of solutions λ^k, v^k of the recursive equation

$$-Av^{k+1} = \varepsilon\lambda^{k+1}Bw_1 + \lambda^k Bv^k + N(\varepsilon w_1 + v^k, \varepsilon w_1 + v^k), \tag{5}$$

taking

$$\lambda^0 = 0, \qquad v^0 = 0.$$

In order to solve (5) for v^{k+1}, the right side must be in $\mathcal{R}(A)$. Because $\mathcal{R}(A)$ is $(n-1)$-dimensional, and $Bw_1 \notin \mathcal{R}(A)$, we can choose λ^{k+1} so as to bring the right side of (5) into $\mathcal{R}(A)$.

Indeed, let $\mathcal{N}(A^T) = \text{span}\{\psi\}$, and let $[\phi]$ be defined for any $\phi \in \mathbf{R}^n$ by

$$[\phi] = (\phi, \psi),$$

where $(u,v) = u^T v$, for $u, v \in \mathbf{R}^n$. Since $\mathcal{R}(A) = \{\psi\}^\perp$, $[\phi]$ measures the distance of ϕ from $\mathcal{R}(A)$. We need to choose λ^{k+1} so that the distance of the right side of

(5) from $\mathcal{R}(A)$ is zero, i.e.,

$$\lambda^{k+1} = -\frac{[\lambda^k Bv^k + N(\varepsilon w_1 + v^k, \varepsilon w_1 + v^k)]}{\varepsilon[Bw_1]}. \tag{6}$$

If L_0 is the restriction of $-A$ to $\mathcal{V}$, then $L_0 : \mathcal{V} \to \mathcal{R}(A)$ is 1–1 and onto, and

$$v^{k+1} = L_0^{-1}\{\varepsilon\lambda^{k+1}Bw_1 + \lambda^k Bv^k + N(\varepsilon w_1 + v^k, \varepsilon w_1 + v^k)\}. \tag{7}$$

We will show that the mapping $(\lambda^k, v^k) \mapsto (\lambda^{k+1}, v^{k+1})$ is a contraction, provided ε is small enough. We choose strictly positive numbers α, β, η such that

$$\alpha \geq |L_0^{-1}| = \sup_{u\perp\psi, u\neq 0} \frac{|L_0^{-1}u|}{|u|}, \qquad \beta \geq |B| = \sup_{u\neq 0} \frac{|Bu|}{|u|}, \qquad \eta \geq \sup_{u,v\neq 0} \frac{|N(u,v)|}{|u||v|},$$

and let

$$|w_1| = W, \qquad |\psi| = \Psi, \qquad |[Bw_1]| = \sigma.$$

Note, since $\lambda^0 = 0$ and $v^0 = 0$, that

$$\lambda^1 = -\varepsilon[N(w_1,w_1)]/[Bw_1], \qquad v^1 = L_0^{-1}\{\varepsilon\lambda^1 Bw_1 + \varepsilon^2 N(w_1,w_1)\}, \tag{8}$$

and therefore

$$|\lambda^1| \leq |\varepsilon|\Lambda, \qquad |v^1| \leq |\varepsilon|^2 V, \tag{9}$$

where

$$\Lambda = \eta\Psi W^2/\sigma, \qquad V = \alpha(\beta\Lambda W + \eta W^2).$$

Now, from (6) and (7), we have

$$\begin{aligned}\lambda^{k+1} - \lambda^k \;=\; & -[(\lambda^k - \lambda^{k-1})Bv^k + \lambda^{k-1}B(v^k - v^{k-1}) \\ & + \varepsilon N(w_1, v^k - v^{k-1}) + \varepsilon N(v^k - v^{k-1}, w_1) \\ & + N(v^k - v^{k-1}, v^k) + N(v^{k-1}, v^k - v^{k-1})]/\varepsilon[Bw_1],\end{aligned}$$

$$\begin{aligned}v^{k+1} - v^k \;=\; & L_0^{-1}\{\varepsilon(\lambda^{k+1} - \lambda^k)Bw_1 + (\lambda^k - \lambda^{k-1})Bv^k + \lambda^{k-1}B(v^k - v^{k-1}) \\ & + \varepsilon N(w_1, v^k - v^{k-1}) + \varepsilon N(v^k - v^{k-1}, w_1) \\ & + N(v^k - v^{k-1}, v^k) + N(v^{k-1}, v^k - v^{k-1})\}.\end{aligned}$$

Hence, we obtain the estimates

$$\begin{aligned}|\lambda^{k+1} - \lambda^k| \;\leq\; & (\Psi/|\varepsilon|\sigma)\{\beta|\lambda^k - \lambda^{k-1}||v^k| + \beta|\lambda^{k-1}||v^k - v^{k-1}| \\ & + 2|\varepsilon|\eta W|v^k - v^{k-1}| + \eta(|v^k| + |v^{k-1}|)|v^k - v^{k-1}|\}, \end{aligned} \tag{10}$$

$$\begin{aligned}|v^{k+1} - v^k| \;\leq\; & \alpha\{|\varepsilon|\beta W|\lambda^{k+1} - \lambda^k| + \beta|\lambda^k - \lambda^{k-1}||v^k| + \beta|\lambda^{k-1}||v^k - v^{k-1}| \\ & + 2|\varepsilon|\eta W|v^k - v^{k-1}| + \eta(|v^k| + |v^{k-1}|)|v^k - v^{k-1}|\}.\end{aligned} \tag{11}$$

Theorem 2.1. *There exists $\varepsilon_0 > 0$ such that, for $0 < |\varepsilon| < \varepsilon_0$ and integers $k \geq 0$, there holds*

$$|\lambda^{k+1} - \lambda^k| \leq \frac{|\varepsilon|\Lambda}{2^k}, \qquad |v^{k+1} - v^k| \leq \frac{|\varepsilon|^2 V}{2^k}, \tag{12}$$

and therefore

$$|\lambda^k| \leq 2|\varepsilon|\Lambda, \qquad |v^k| \leq 2|\varepsilon|^2 V.$$

It follows that $\lambda^k = (\lambda^k - \lambda^{k-1}) + (\lambda^{k-1} - \lambda^{k-2}) + \cdots + (\lambda^1 - \lambda^0)$ and $v^k = (v^k - v^{k-1}) + (v^{k-1} - v^{k-2}) + \cdots + (v^1 - v^0)$ converge to a unique solution λ, v of (4) satisfying

$$|\lambda| \leq 2|\varepsilon|\Lambda, \qquad |v| \leq 2|\varepsilon|^2 V. \tag{13}$$

Proof. The proof is by induction. For $k = 0$, (12) and (13) follow from (9). For $k \geq 1$, using (10), (11) and the induction hypotheses, we get

$$\begin{aligned} |\lambda^{k+1} - \lambda^k|/|\varepsilon| &\leq (\Psi/|\varepsilon|^2\sigma)\{2|\varepsilon|^2\beta V|\lambda^k - \lambda^{k-1}| + 2|\varepsilon|\beta\Lambda|v^k - v^{k-1}| \\ &\quad + 2|\varepsilon|\eta W|v^k - v^{k-1}| + 4|\varepsilon|^2\eta V|v^k - v^{k-1}|\} \\ &\leq |\varepsilon|(\Psi/\sigma)(8\beta\Lambda V + 4\eta WV + 8|\varepsilon|\eta V^2)/2^k, \end{aligned} \tag{14}$$

$$\begin{aligned} |v^{k+1} - v^k|/|\varepsilon|^2 &\leq \alpha\{(\beta W/|\varepsilon|)|\lambda^{k+1} - \lambda^k| + 2\beta V|\lambda^k - \lambda^{k-1}| \\ &\quad + 2(\beta\Lambda/|\varepsilon|)|v^k - v^{k-1}| + 2(\eta W/|\varepsilon|)|v^k - v^{k-1}| \\ &\quad + 4\eta V|v^k - v^{k-1}|\} \\ &\leq \alpha\{\beta\Lambda W + |\varepsilon|(8\beta\Lambda V + 4\eta WV + 8|\varepsilon|\eta V^2)\}/2^k. \end{aligned} \tag{15}$$

Now, if we put

$$\varepsilon_0 = \min\{1, \sigma\Lambda/\Psi K, \eta W^2/K\},$$

where

$$K = 8\beta\Lambda V + 4\eta WV + 8\eta V^2,$$

then we get (12), and the existence of a solution satisfying (13) is obvious.

To show uniqueness, suppose λ_1, v_1 and λ_2, v_2 are both solutions of (4) satisfying (13). This implies

$$|\lambda_1 - \lambda_2| \leq 4|\varepsilon|\Lambda, \qquad |v_1 - v_2| \leq 4|\varepsilon|^2 V. \tag{16}$$

On the other hand, using the same arguments that were used to obtain the estimates (14) and (15), we obtain

$$\begin{aligned} |\lambda_1 - \lambda_2|/|\varepsilon| &\leq (\Psi/|\varepsilon|^2\sigma)\{2|\varepsilon|^2\beta V|\lambda_1 - \lambda_2| + 2|\varepsilon|\beta\Lambda|v_1 - v_2| \\ &\quad + 2|\varepsilon|\eta W|v_1 - v_2| + 4|\varepsilon|^2\eta V|v_1 - v_2|\}, \\ |v_1 - v_2|/|\varepsilon|^2 &\leq \alpha\{(\beta W/|\varepsilon|)|\lambda_1 - \lambda_2| + 2\beta V|\lambda_1 - \lambda_2| \\ &\quad + 2(\beta\Lambda/|\varepsilon|)|v_1 - v_2| + 2(\eta W/|\varepsilon|)|v_1 - v_2| + 4\eta V|v_1 - v_2|\}. \end{aligned}$$

Then using (16) in the right sides of these expressions gives

$$|\lambda_1 - \lambda_2|/|\varepsilon| \le |\varepsilon|(\Psi/\sigma)(8\beta\Lambda V + 4\eta W V + 8|\varepsilon|\eta V^2),$$
$$|v_1 - v_2|/|\varepsilon|^2 \le \alpha\{\beta\Lambda W + |\varepsilon|(8\beta\Lambda V + 4\eta W V + 8|\varepsilon|\eta V^2)\},$$

which, for $0 < |\varepsilon| < \varepsilon_0$, imply

$$|\lambda_1 - \lambda_2| \le 2|\varepsilon|\Lambda, \qquad |v_1 - v_2| \le 2|\varepsilon|^2 V.$$

Continuing inductively, we obtain

$$|\lambda_1 - \lambda_2| \le 2^{-k}|\varepsilon|\Lambda, \qquad |v_1 - v_2| \le 2^{-k}|\varepsilon|^2 V,$$

for every integer $k \ge -2$. This implies that $\lambda_1 = \lambda_2$ and $v_1 = v_2$. □

Corollary 2.2. *The functions $\lambda(\varepsilon)$ and $v(\varepsilon)$ can be expressed as convergent power series*

$$\lambda(\varepsilon) = \varepsilon\nu_1 + \varepsilon^2\nu_2 + \dots, \qquad v(\varepsilon) = \varepsilon^2 w_2 + \varepsilon^3 w_3 + \dots. \tag{17}$$

Proof. In the proof of Theorem 2.1, $\lambda^k(\varepsilon)$ and $v^k(\varepsilon)$ are polynomials in ε. Indeed, this is true for $\lambda^1(\varepsilon)$ and $v^1(\varepsilon)$ by (8), and then the recursive equations (6) and (7) continue to return polynomials in ε, when polynomials are substituted into their right sides. For complex ε, the estimates (10)–(11), and hence (12) remain valid. So the functions $\lambda^k(\varepsilon)$, $v^k(\varepsilon)$ are analytic functions of ε, uniformly convergent on compact subsets of the disk $|\varepsilon| < \varepsilon_0$. Thus their limits $\lambda(\varepsilon)$, $v(\varepsilon)$ are analytic. □

The corollary can be applied in the usual way to determine the coefficients of the perturbation series, by substituting (17) into (4), and collecting terms with like powers of ε. The ν_j's are then chosen so that one can solve uniquely for the w_j's in $\mathcal{V}$.

We remark that a simple induction argument shows that the coefficient of ε^j is the same in all iterates $\lambda^k(\varepsilon)$, $k \ge j$ and $v^k(\varepsilon)$, $k \ge j-1$. Thus, sometimes, a simpler way to find the coefficients in the power series for $\lambda(\varepsilon)$ and $v(\varepsilon)$ is to iterate the recursive equations (6) and (7) until one obtains λ^j and v^{j-1}, and then read off the coefficients of ε^j.

3. Bifurcation of periodic solutions

In this section we assume that the matrix A in (1) has a pair of simple conjugate imaginary eigenvalues. By rescaling the independent variable t in (1), we may assume that these eigenvalues are $\pm i$. We assume further that $\pm i$ are simple roots of the characteristic polynomial of A, and that there are no other roots of the form $\pm im$, where m is an integer. Our only assumption concerning B is that the condition (23), formulated below, must be satisfied. Under these hypotheses we prove the existence of bifurcating periodic solutions of (1) by modifying the proof that we gave in the last section.

Clearly, under the hypotheses of this section, the linear equation $du/dt = Au$ has periodic solutions with period 2π. Thus, it is natural to seek nearby solutions of the nonlinear equation (1) with period in the form $2\pi/(1+\tau)$, where τ is a small real number to be determined. Setting

$$u(t) = w(s), \qquad s = (1+\tau)t$$

in (1) transforms the problem of finding $2\pi/(1+\tau)$-periodic solutions $u(t)$ of (1) into that of finding 2π-periodic solutions $w(s)$ of

$$(1+\tau)\frac{dw}{ds} = Aw + \nu Bw + N(w,w). \tag{18}$$

Thus, we seek solutions of a known period, but have introduced an additional unknown parameter τ.

Let $\mathcal{X}_0$ denote the space of continuous 2π-periodic functions $w : \mathbf{R} \to \mathbf{R}^n$, and let its subspace of continuously differentiable functions be denoted by $\mathcal{X}_1$. The spaces $\mathcal{X}_0$, $\mathcal{X}_1$ are Banach spaces with the norms

$$\|w\|_0 = \max_s |w(s)|, \qquad \|w\|_1 = \|w\|_0 + \|dw/ds\|_0,$$

respectively. Clearly, $\|w\|_0 \le \|w\|_1$ for all $w \in \mathcal{X}_1$. We define an inner product on $\mathcal{X}_0$ by setting, for any w, ψ in $\mathcal{X}_0$,

$$(w,\psi) = \frac{1}{2\pi}\int_0^{2\pi} w(s)^T \psi(s)ds, \tag{19}$$

and note that $(w,\psi) \le \|w\|_0\|\psi\|_0$.

Let $c, d \in \mathbf{C}^n$ be complex vectors satisfying

$$Ac = ic, \qquad A^T d = -id, \tag{20}$$

and define real functions w_1, w_2, ψ_1, $\psi_2 \in \mathcal{X}_1$ by

$$w_1(s) = \text{Re}\,(e^{is}c),\ w_2(s) = \text{Im}\,(e^{is}c),\ \psi_1(s) = \text{Re}\,(e^{is}d),\ \psi_2(s) = \text{Im}\,(e^{is}d). \tag{21}$$

Then

$$\frac{dw_j}{ds} - Aw_j = 0, \qquad \frac{d\psi_j}{ds} + A^T\psi_j = 0,$$

for $j = 1, 2$, and we note that the scalar multiples of w_1, or of w_2, generate elliptical orbits that fill the two-dimensional eigenspace span$\{\text{Re}\,c, \text{Im}\,c\}$ of A. The following facts regarding the operator $d/ds - A$ can be found in [7, §VII.2] and [8, §IV.1], but for the reader's convenience we have collected and somewhat simplified their proofs in the appendix.

Proposition 3.1. *The mapping $d/ds - A : \mathcal{X}_1 \to \mathcal{X}_0$ is a bounded linear operator with null space $\mathcal{N} = span\{w_1, w_2\}$, and range $\mathcal{R} = \{f \in \mathcal{X}_0 : (f, \psi_j) = 0, j = 1, 2\}$. The vectors c, d satisfying* (20) *may be chosen so that $\bar{d}^T c = 2$, and hence*

$$(w_j, \psi_k) = \delta_{jk} \text{ for } j,k = 1,2, \qquad \left(\frac{dw_1}{ds}, \psi_1\right) = 0, \qquad \left(\frac{dw_1}{ds}, \psi_2\right) = -1. \tag{22}$$

The restriction L_0 of $d/ds - A$ to $\mathcal{V} = \{\phi \in \mathcal{X}_1 : (\phi, w_j) = 0,\, j = 1,2\}$ is an isomorphism from $\mathcal{V}$ onto $\mathcal{R}$. In particular,

$$\|L_0^{-1}\| = \sup_{f\in\mathcal{R}, f\neq 0} \frac{\|L_0^{-1}f\|_1}{\|f\|_0} < \infty.$$

We need an assumption concerning B analogous to that used in the last section. Provided c, d are chosen as in Proposition 3.1, so that (22) holds, our assumption about B takes the simple form

$$(Bw_1, \psi_1) \neq 0. \tag{23}$$

We note that this condition is independent of the choice of c, d satisfying $\bar{d}^T c = 2$.

Now, by analogy with the method of the previous section, we let $\varepsilon \in \mathbf{R}$ be a small parameter, and seek nontrivial solutions of (18) in the form

$$\nu = \lambda(\varepsilon), \qquad \tau = \rho(\varepsilon), \qquad w = \varepsilon w_1 + v(\varepsilon), \tag{24}$$

with $|\lambda(\varepsilon)| = \mathcal{O}(\varepsilon^2)$, $|\rho(\varepsilon)| = \mathcal{O}(\varepsilon^2)$, $\|v(\varepsilon)\|_1 = \mathcal{O}(\varepsilon^2)$, and $v(\varepsilon) \in \mathcal{V}$. Substituting (24) into (18), we get

$$\frac{dv}{ds} - Av = \varepsilon\lambda Bw_1 - \varepsilon\rho\frac{dw_1}{ds} + \lambda Bv - \rho\frac{dv}{ds} + N(\varepsilon w_1 + v, \varepsilon w_1 + v). \tag{25}$$

We will obtain a solution λ, ρ, v of (25), for fixed small values of $\varepsilon \neq 0$, as the limit of solutions λ^k, ρ^k, v^k of the recursive equation

$$\begin{aligned}\left(\frac{d}{ds} - A\right) v^{k+1} \\ = \varepsilon\lambda^{k+1} Bw_1 - \varepsilon\rho^{k+1}\frac{dw_1}{ds} + \lambda^k Bv^k - \rho^k\frac{dv^k}{ds} + N(\varepsilon w_1 + v^k, \varepsilon w_1 + v^k),\end{aligned} \tag{26}$$

taking

$$\lambda^0 = 0, \qquad \rho^0 = 0, \qquad v^0 = 0.$$

In order to solve (26) for v^{k+1}, we need to choose λ^{k+1} and ρ^{k+1} so as to bring the right side into $\mathcal{R}$. If we define $[\phi]_j$ for $\phi \in \mathcal{X}_0$ by setting

$$[\phi]_j = (\phi, \psi_j), \qquad j = 1, 2,$$

it can be easily seen, using (22) and (23), that this is accomplished by setting

$$\lambda^{k+1} = -\frac{[\lambda^k Bv^k - \rho^k dv^k/ds + N(\varepsilon w_1 + v^k, \varepsilon w_1 + v^k)]_1}{\varepsilon[Bw_1]_1}, \tag{27}$$

$$\rho^{k+1} = -\lambda^{k+1}[Bw_1]_2 - \frac{1}{\varepsilon}\left[\lambda^k Bv^k - \rho^k\frac{dv^k}{ds} + N(\varepsilon w_1 + v^k, \varepsilon w_1 + v^k)\right]_2. \tag{28}$$

Then

$$v^{k+1} = L_0^{-1}\left\{\varepsilon\lambda^{k+1}Bw_1 - \varepsilon\rho^{k+1}\frac{dw_1}{ds} + \lambda^k Bv^k - \rho^k\frac{dv^k}{ds} + N(\varepsilon w_1 + v^k, \varepsilon w_1 + v^k)\right\}. \tag{29}$$

We will show that the mapping $(\lambda^k, \rho^k, v^k) \mapsto (\lambda^{k+1}, \rho^{k+1}, v^{k+1})$ is a contraction for all sufficiently small nonzero ε. We choose a strictly positive number α such that

$$\alpha \geq \|L_0^{-1}\|,$$

and choose β and η as in the previous section. We also let

$$\|w_1\|_1 = W, \qquad \|\psi_j\|_0 = \Psi_j, \qquad |[Bw_1]_j| = \sigma_j, \qquad j = 1, 2.$$

Using the specific forms of w_1, ψ_1, ψ_2, and the bilinearity of N, we find that $N(w_1, w_1) \in \mathcal{R}$, and thus we have very simply

$$\lambda^1 = 0, \qquad \rho^1 = 0, \qquad v^1 = \varepsilon^2 L_0^{-1} N(w_1, w_1). \tag{30}$$

Analogously to the derivation of the estimates (10)–(11) from (6)–(7), we now obtain, from (27)–(29), the estimates

$$\begin{aligned} |\lambda^{k+1} - \lambda^k| \leq (\Psi_1/|\varepsilon|\sigma_1)\{&\beta|\lambda^k - \lambda^{k-1}|\,\|v^k\|_1 + \beta|\lambda^{k-1}|\,\|v^k - v^{k-1}\|_1 \\ &+ |\rho^k - \rho^{k-1}|\,\|v^k\|_1 + |\rho^{k-1}|\,\|v^k - v^{k-1}\|_1 \\ &+ 2|\varepsilon|\eta W\|v^k - v^{k-1}\|_1 + \eta(\|v^k\|_1 + \|v^{k-1}\|_1)\|v^k - v^{k-1}|\,_1\}, \end{aligned} \tag{31}$$

$$\begin{aligned} |\rho^{k+1} - \rho^k| \leq \sigma_2|\lambda^{k+1} - \lambda^k| + (\Psi_2/|\varepsilon|)\{&\beta|\lambda^k - \lambda^{k-1}|\,\|v^k\|_1 + \beta|\lambda^{k-1}|\,\|v^k - v^{k-1}\|_1 \\ &+ |\rho^k - \rho^{k-1}|\,\|v^k\|_1 + |\rho^{k-1}|\,\|v^k - v^{k-1}\|_1 \\ &+ 2|\varepsilon|\eta W\|v^k - v^{k-1}\|_1 + \eta(\|v^k\|_1 + \|v^{k-1}\|_1)\|v^k - v^{k-1}\|_1\}, \end{aligned} \tag{32}$$

$$\begin{aligned} \|v^{k+1} - v^k\|_1 \leq \alpha\{&|\varepsilon|\beta W|\lambda^{k+1} - \lambda^k| + |\varepsilon|W|\rho^{k+1} - \rho^k| + \beta|\lambda^k - \lambda^{k-1}|\,\|v^k\|_1 \\ &+ \beta|\lambda^{k-1}|\,\|v^k - v^{k-1}\|_1 + |\rho^k - \rho^{k-1}|\,\|v^k\|_1 + |\rho^{k-1}|\,\|v^k - v^{k-1}\|_1 \\ &+ 2|\varepsilon|\eta W\|v^k - v^{k-1}\|_1 + \eta(\|v^k\|_1 + \|v^{k-1}\|_1)\|v^k - v^{k-1}\|_1\}. \end{aligned} \tag{33}$$

Theorem 3.2. *There exist $\Lambda, P, V, \varepsilon_0 > 0$ such that, for $0 < |\varepsilon| < \varepsilon_0$ and integers $k \geq 0$, there holds*

$$|\lambda^{k+1} - \lambda^k| \leq \frac{|\varepsilon|^2\Lambda}{2^k}, \qquad |\rho^{k+1} - \rho^k| \leq \frac{|\varepsilon|^2 P}{2^k}, \qquad \|v^{k+1} - v^k\|_1 \leq \frac{|\varepsilon|^2 V}{2^k}, \tag{34}$$

and therefore

$$|\lambda^k| \leq 2|\varepsilon|^2\Lambda, \qquad |\rho^k| \leq 2|\varepsilon|^2 P, \qquad \|v^k\|_1 \leq 2|\varepsilon|^2 V.$$

It follows that λ^k, ρ^k, v^k converge to a unique solution λ, ρ, v of (25) *satisfying*

$$|\lambda| \leq 2|\varepsilon|^2\Lambda, \qquad |\rho| \leq 2|\varepsilon|^2 P, \qquad \|v\|_1 \leq 2|\varepsilon|^2 V. \tag{35}$$

Furthermore, λ and ρ are even in ε.

Proof. For $k = 0$, (34) and (35) follow from (30), for any positive Λ, P and for

$$V = \alpha\eta W^2.$$

For $k \geq 1$, using (31)–(33) and the induction hypotheses, we get

$$|\lambda^{k+1} - \lambda^k|/|\varepsilon|^2 \leq (\Psi_1/\sigma_1)\{4\eta WV + 8|\varepsilon|(\beta\Lambda + P + \eta V)V\}/2^k,$$

$$|\rho^{k+1} - \rho^k|/|\varepsilon|^2 \leq \sigma_2|\lambda^{k+1} - \lambda^k|/|\varepsilon|^2 + \Psi_2\{4\eta WV + 8|\varepsilon|(\beta\Lambda + P + \eta V)V\}/2^k,$$

$$\begin{aligned}\|v^{k+1} - v^k\|_1/|\varepsilon|^2 \;\leq\; & \alpha\beta W|\lambda^{k+1} - \lambda^k|/|\varepsilon| + \alpha W|\rho^{k+1} - \rho^k|/|\varepsilon| \\ & + |\varepsilon|\alpha\{4\eta WV + 8||\varepsilon||(\beta\Lambda + P + \eta V)V\}/2^k.\end{aligned}$$

If we now put

$$\Lambda = 5\Psi_1\eta WV/\sigma_1, \qquad P = \sigma_2\Lambda + 5\Psi_2\eta WV, \qquad K = 8(\beta\Lambda + P + \eta V)V,$$

and let

$$|\varepsilon|_0 = \min\{\eta WV/K,\, V/(\alpha\beta W\Lambda + \alpha WP + 5\alpha\eta WV)\},$$

we get (34).

The proof of the uniqueness of the solution is a straightforward modification of the proof used in Theorem 2.1.

To show that $\lambda(\varepsilon)$ and $\rho(\varepsilon)$ are even functions of ε, we suppose that $\lambda(\varepsilon)$, $\rho(\varepsilon)$, $v(\varepsilon)$ is a solution of (25). Replacing ε by $-\varepsilon$ in (25), we observe that $\tilde{\lambda} = \lambda(-\varepsilon)$, $\tilde{\rho} = \rho(-\varepsilon)$, $\tilde{v} = v(-\varepsilon)$ is a solution of

$$\frac{d\tilde{v}}{ds} - A\tilde{v} = -\varepsilon\tilde{\lambda}Bw_1 + \varepsilon\tilde{\rho}\frac{dw_1}{ds} + \lambda B\tilde{v} - \tilde{\rho}\frac{d\tilde{v}}{ds} + N(-\varepsilon w_1 + \tilde{v}, -\varepsilon w_1 + \tilde{v}). \tag{36}$$

On the other hand, for any function $v \in \mathcal{X}_0$, let

$$v^\pi(s) = v(s + \pi)$$

denote the function phase shifted by half a period. Shifting the phase in (25) by half a period, and observing that $w_1^\pi = -w_1$, we have that $\tilde{\lambda} = \lambda(\varepsilon)$, $\tilde{\rho} = \rho(\varepsilon)$, $\tilde{v} = v^\pi(\varepsilon)$ is also a solution of (36). Thus by the uniqueness of solutions of (36), we have

$$\lambda(-\varepsilon) = \lambda(\varepsilon), \qquad \rho(-\varepsilon) = \rho(\varepsilon), \qquad v(-\varepsilon) = v^\pi(\varepsilon).$$

In particular, λ and ρ are even in ε. □

In the same manner as in the previous section, we obtain

Corollary 3.3. *The functions $\lambda(\varepsilon)$, $\rho(\varepsilon)$ and $v(\varepsilon)$ can be expressed as convergent power series*

$$\begin{aligned}\lambda(\varepsilon) &= \varepsilon^2\nu_2 + \varepsilon^4\nu_4 + \dots, \\ \rho(\varepsilon) &= \varepsilon^2\tau_2 + \varepsilon^4\tau_4 + \dots, \\ v(\varepsilon) &= \varepsilon^2 w_2 + \varepsilon^3 w_3 + \varepsilon^4 w_4 + \dots.\end{aligned} \tag{37}$$

4. Generalizations

The iteration method can also be used to treat more general nonlinearities. Suppose that we have a parameterized family of differential equations

$$\frac{dv}{dt} = F(v; \mu), \tag{38}$$

where F is a mapping from an open region in $\mathbf{R}^n \times \mathbf{R}$ into $\mathbf{R}^n$, whose partial derivatives D_1F, D_2F, $D_{11}F$, $D_{12}F$ exist and are continuous. Assume that there is a continuously differentiable curve $\bar{v}(\mu)$, from an open interval I in $\mathbf{R}$ into $\mathbf{R}^n$, of 'trivial' equilibrium solutions satisfying $F(\bar{v}(\mu); \mu) = 0$ for all $\mu \in I$. If there is a critical parameter value μ_0 such that the linearization $D_1F(\bar{v}(\mu_0); \mu_0)$ about the trivial solution has a one-dimensional null space, or has a pair of simple imaginary eigenvalues, we put

$$v = \bar{v}(\mu) + u, \qquad \mu = \mu_0 + \nu$$

in (38), and expand the right side using Taylor's theorem to obtain

$$\frac{du}{dt} = Au + \nu Bu + N(u; \nu), \tag{39}$$

where $A = D_1F(\bar{v}(\mu_0); \mu_0)$ and $B = D_{11}F(\bar{v}(\mu_0); \mu_0)\bar{v}\,'(\mu_0) + D_{12}F(\bar{v}(\mu_0); \mu_0)$ are $n \times n$ matrices, and N is a mapping whose partial derivatives D_1N, D_2N, $D_{12}N$ exist and are continuous in a neighbourhood of $(0, 0) \in \mathbf{R}^n \times \mathbf{R}$. The mapping N has the form

$$N(u; \nu) = C(\nu)u + R(u; \nu),$$

where C is a continuously differentiable mapping from $\mathbf{R}$ into the $n \times n$ matrices satisfying $C(0) = O$ and $C\,'(0) = O$, and R satisfies $R(0; \nu) = 0$, $D_1R(0; \nu) = 0$ for all ν belonging to an open interval containing 0. The more general nature of the nonlinear term $N(u; \nu)$ on the right side of (39) requires only straightforward modifications to the iterative schemes used above for (1).

Indeed, to obtain bifurcating equilibrium solutions of (39) in the case where A has a one-dimensional null space $\mathcal{N}(A) = \text{span}\{w_1\}$, and Bw_1 does not belong to the range $\mathcal{R}(A)$, we seek nontrivial solutions of

$$Aw + \nu Bw + N(w; \nu) = 0,$$

in the form

$$\nu = \lambda(\varepsilon), \qquad w = \varepsilon w_1 + v(\varepsilon),$$

with $|\lambda(\varepsilon)| = o(1)$ and $|v(\varepsilon)| = o(\varepsilon)$ as $\varepsilon \to 0$, and $v(\varepsilon) \in \mathcal{V}$. We then solve

$$-Av = \varepsilon\lambda Bw_1 + \lambda Bv + N(\varepsilon w_1 + v; \lambda) \tag{40}$$

by using the recursive equation

$$-Av^{k+1} = \varepsilon\lambda^{k+1}Bw_1 + \lambda^k Bv^k + N(\varepsilon w_1 + v^k; \lambda^k), \tag{41}$$

taking $\lambda^0 = 0$, $v^0 = 0$. To show that $(\lambda^k, v^k) \mapsto (\lambda^{k+1}, v^{k+1})$ is a contraction, let α, β, W, Ψ, σ be as defined in §2, choose any $\Lambda > 0$ (say, $\Lambda = 1$), put $V = \alpha(\beta\Lambda W + \sigma\Lambda/\Psi)$, and choose a strictly positive γ satisfying

$$\gamma \leq \sigma\Lambda\Psi/(8\beta\Lambda W + 8\Lambda W + 4V). \tag{42}$$

Since C is continuously differentiable and $C(0) = O$, $C'(0) = O$, we may choose a strictly positive $\delta_0 \leq \min(\gamma, W/2V)$ such that

$$|C(\nu)| \leq \gamma,\ |C'(\nu)| \leq \gamma, \quad \text{for all } |\nu| \leq 2\delta_0\Lambda. \tag{43}$$

Theorem 4.1. *Given any $\delta > 0$, there exists $\varepsilon_0 > 0$ such that, for $0 < |\varepsilon| < \varepsilon_0$ and integers $k \geq 0$, there holds*

$$|\lambda^{k+1} - \lambda^k| \leq \frac{\delta\Lambda}{2^k}, \qquad |v^{k+1} - v^k| \leq \frac{|\varepsilon|\delta V}{2^k}, \tag{44}$$

and therefore

$$|\lambda^k| \leq 2\delta\Lambda, \qquad |v^k| \leq 2\delta|\varepsilon|V.$$

It follows that λ^k and v^k converge to a unique solution λ, v of (40) satisfying

$$|\lambda| \leq 2\delta\Lambda, \qquad |v| \leq 2\delta|\varepsilon|V. \tag{45}$$

Proof. Without loss of generality, we may assume that $\delta \leq \delta_0$. For $k = 0$, our recursive procedure gives

$$\begin{aligned} \lambda^1 &= -[R(\varepsilon w_1; 0)]/\varepsilon[Bw_1], \\ v^1 &= L_0^{-1}\{\varepsilon\lambda^1 Bw_1 + R(\varepsilon w_1; 0)\}. \end{aligned}$$

Because $R(u; 0) = o(|u|)$ as $u \to 0$, we can choose ε_0 sufficiently small so that, if $0 < |\varepsilon| < \varepsilon_0$, then

$$|R(\varepsilon w_1; 0)|/|\varepsilon| \leq \delta\sigma\Lambda/\Psi \tag{46}$$

and hence, by (46) and our choice of V, we have

$$|\lambda^1| \leq \delta\Lambda, \qquad |v^1|/|\varepsilon| \leq \delta V.$$

For $k \geq 1$, we find

$$\begin{aligned} |\lambda^{k+1} - \lambda^k| \leq\ & (\Psi/|\varepsilon|\sigma)\{\beta|\lambda^k - \lambda^{k-1}||v^k| + \beta|\lambda^{k-1}||v^k - v^{k-1}| \\ & + |C(\lambda^k)(\varepsilon w_1 + v^k) - C(\lambda^{k-1})(\varepsilon w_1 + v^{k-1})| \\ & + |R(\varepsilon w_1 + v^k; \lambda^k) - R(\varepsilon w_1 + v^{k-1}; \lambda^{k-1})|\}, \end{aligned} \tag{47}$$

$$\begin{aligned} |v^{k+1} - v^k| \leq\ & \alpha\{|\varepsilon|\beta W|\lambda^{k+1} - \lambda^k| + \beta|\lambda^k - \lambda^{k-1}||v^k| + \beta|\lambda^{k-1}||v^k - v^{k-1}| \\ & + |C(\lambda^k)(\varepsilon w_1 + v^k) - C(\lambda^{k-1})(\varepsilon w_1 + v^{k-1})| \\ & + |R(\varepsilon w_1 + v^k; \lambda^k) - R(\varepsilon w_1 + v^{k-1}; \lambda^{k-1})|\}. \end{aligned} \tag{48}$$

Using the mean value theorem and (43), we obtain

$$|C(\lambda^k)(\varepsilon w_1 + v^k) - C(\lambda^{k-1})(\varepsilon w_1 + v^k)| \leq \gamma(|\varepsilon w_1 + v^k||\lambda^k - \lambda^{k-1}| + |v^k - v^{k-1}|), \tag{49}$$

provided $|\lambda^k|, |\lambda^{k-1}| \leq 2\delta_0\Lambda$. Also, because D_1R, $D_{12}R$ are continuous, and because $D_1R(0;\nu) = 0$, $D_{12}R(0;\nu) = 0$, we can choose ε_0 sufficiently small so that, if $0 < |\varepsilon| < \varepsilon_0$, then

$$|D_1R(u;\nu)| \leq \gamma, \ |D_{12}R(u;\nu)| \leq \gamma, \quad \text{for all } |u| \leq 2|\varepsilon|W, \ |\nu| \leq 2\delta\Lambda. \tag{50}$$

Thus the mean value theorem and (50) give

$$|R(\varepsilon w_1 + v^k; \lambda^k) - R(\varepsilon w_1 + v^{k-1}; \lambda^{k-1})| \leq \gamma(2|\varepsilon|W|\lambda^k - \lambda^{k-1}| + |v^k - v^{k-1}|), \tag{51}$$

provided $|\varepsilon w_1 + v^k|, |\varepsilon w_1 + v^{k-1}| \leq 2|\varepsilon|W$ and $|\lambda^k|, |\lambda^{k-1}| \leq 2\delta\Lambda$. Since the induction hypotheses together with $\delta \leq \delta_0 \leq W/2V$ imply that $|\varepsilon w_1 + v^k|, |\varepsilon w_1 + v^{k-1}| \leq 2|\varepsilon|W$ and $|\lambda^k|, |\lambda^{k-1}| \leq 2\delta\Lambda$, we obtain (44) by substituting (49) and (51) into (47) and (48), and using $\delta \leq \delta_0 \leq \gamma$ and (42).

The proof that the solutions are unique is much the same as in the proof of Theorem 2.1. □

We note that if the nonlinear term N in (39) is C^p, $p \geq 2$, then we may expand N using Taylor's theorem and obtain asymptotic expansions

$$\lambda(\varepsilon) = \varepsilon\nu_1 + \cdots + \varepsilon^{p-1}\nu_{p-1} + o(\varepsilon^{p-1}), \qquad v(\varepsilon) = \varepsilon^2 w_2 + \cdots + \varepsilon^p w_p + o(\varepsilon^p),$$

as $\varepsilon \to 0$, while if N is analytic, then Corollary 2.2 holds.

If A has a pair of simple eigenvalues $\pm i$ and no other eigenvalues of the form $\pm im$, where m is an integer, Hopf bifurcation for (39) is proved in essentially the same way as in §3. We seek real numbers ν and τ, and a 2π-periodic function $w(s)$, satisfying

$$(1+\tau)\frac{dw}{ds} = Aw + \nu Bw + N(w;\nu).$$

Using the same procedure as in §3, we seek solutions in the form

$$\nu = \lambda(\varepsilon), \qquad \tau = \rho(\varepsilon), \qquad w = \varepsilon w_1 + v(\varepsilon),$$

but now with $|\lambda(\varepsilon)| = o(1)$, $|\rho(\varepsilon)| = o(1)$, $\|v(\varepsilon)\|_1 = o(\varepsilon)$, as $\varepsilon \to 0$. This leads to the equation

$$\left(\frac{d}{ds} - A\right) v = \varepsilon\lambda Bw_1 - \varepsilon\rho\frac{dw_1}{ds} + \lambda Bv - \rho\frac{dv}{ds} + N(\varepsilon w_1 + v; \lambda), \tag{52}$$

which we solve by using the recursive relation

$$\left(\frac{d}{ds} - A\right) v^{k+1} = \varepsilon\lambda^{k+1}Bw_1 - \varepsilon\rho^{k+1}\frac{dw_1}{ds} + \lambda^k Bv^k - \rho^k\frac{dv^k}{ds} + N(\varepsilon w_1 + v^k; \lambda^k),$$

taking

$$\lambda^0 = 0, \qquad \rho^0 = 0, \qquad v^0 = 0.$$

Using arguments very similar to those above, we obtain

Theorem 4.2. *There exist positive constants Λ, P, V with the following property: given any $\delta > 0$, there exists $\varepsilon_0 > 0$ such that for $0 < |\varepsilon| < \varepsilon_0$ and integers $k \geq 0$, there holds*

$$|\lambda^{k+1} - \lambda^k| \leq \frac{\delta\Lambda}{2^k}, \qquad |\rho^{k+1} - \rho^k| \leq \frac{\delta P}{2^k}, \qquad \|v^{k+1} - v^k\|_1 \leq \frac{|\varepsilon|\delta V}{2^k},$$

and therefore

$$|\lambda^k| \leq 2\delta\Lambda, \qquad |\rho^k| \leq 2\delta P, \qquad \|v^k\|_1 \leq 2\delta|\varepsilon|V.$$

It follows that λ^k, ρ^k and v^k converge to a unique solution λ, ρ, v of (52) satisfying

$$|\lambda| \leq 2\delta\Lambda, \qquad |\rho| \leq 2\delta P, \qquad \|v\|_1 \leq 2\delta|\varepsilon|V,$$

and that λ and ρ are even in ε.

If the nonlinear term N in (39) has sufficiently many derivatives, we may obtain asymptotic expansions for $\lambda(\varepsilon)$, $\rho(\varepsilon)$ and $v(\varepsilon)$. In particular, if N is C^3, then the coefficients of ε in the expansions for $\lambda(\varepsilon)$ and $\rho(\varepsilon)$ vanish for the same reasons as they do in §3, and we have $\lambda(\varepsilon) = \varepsilon^2\nu_2 + o(\varepsilon^2)$, $\rho(\varepsilon) = \varepsilon^2\tau_2 + o(\varepsilon^2)$ and $v(\varepsilon) = \varepsilon^2 w_2 + \varepsilon^3 w_3 + o(\varepsilon^3)$, as $\varepsilon \to 0$. If N is analytic, then Corollary 3.2 holds.

The iteration method also generalizes to infinite-dimensional bifurcation problems. Suppose that $\mathcal{Y}$ and $\mathcal{Z}$ are real Banach spaces with $\mathcal{Y} \subseteq \mathcal{Z}$, that $F : \mathcal{Y} \times \mathbf{R} \to \mathcal{Z}$ is a mapping whose Fréchet derivatives D_1F, D_2F, $D_{11}F$, $D_{12}F$ exist and are continuous in an open region of $\mathcal{Y} \times \mathbf{R}$, and that there is a C^1 'trivial' solution $\bar{v}(\mu)$ satisfying $F(\bar{v}(\mu);\mu) = 0$ for all μ belonging to some interval. Then we seek solutions of (38) bifurcating from the trivial solution at $\mu = \mu_0$, if $D_1F(\bar{v}(\mu_0);\mu_0)$ has a one-dimensional null space, or a pair of simple imaginary eigenvalues. Expanding the right side of (38), as in the finite-dimensional case, we obtain (39), where A and B are now bounded linear mappings from $\mathcal{Y}$ into $\mathcal{Z}$, and the nonlinear term N is a mapping whose Fréchet derivatives D_1N, D_2N and $D_{12}N$ exist and are continuous in a neighbourhood of $(0,0) \in \mathcal{Y} \times \mathbf{R}$.

In order to obtain bifurcating equilibrium solutions of (39) in the infinite-dimensional case, we assume that $\mathcal{N}(A) = \operatorname{span}\{w_1\}$ for some w_1 in $\mathcal{Y}$, and that there is some ψ in the dual space $\mathcal{Z}'$ such that $\mathcal{R}(A) = \{f \in \mathcal{Z} : \langle f, \psi\rangle = 0\}$, where $\langle\cdot,\cdot\rangle$ denotes the duality pairing between $\mathcal{Z}$ and $\mathcal{Z}'$. We also assume that $Bw_1 \notin \mathcal{R}(A)$. These hypotheses have been verified in many applications to physical problems.

Under these assumptions, one can choose a complementary subspace $\mathcal{V}$ to $\mathcal{N}(A)$ in $\mathcal{Y}$, and it follows that the restriction $L_0 : \mathcal{V} \to \mathcal{R}(A)$, of $-A$ to $\mathcal{V}$, is an isomorphism. Strictly positive numbers α, β can be chosen such that

$$\begin{aligned} \|L_0^{-1}u\|_{\mathcal{Y}} &\leq \alpha\|u\|_{\mathcal{Z}} \quad \text{for all } u \in \mathcal{R}(A), \\ \|Bu\|_{\mathcal{Z}} &\leq \beta\|u\|_{\mathcal{Y}} \quad \text{for all } u \in \mathcal{Y}, \end{aligned}$$

and one can set

$$\|w_1\|_{\mathcal{Y}} = W, \qquad \|\psi\|_{\mathcal{Z}'} = \Psi, \qquad |[Bw_1]| = \sigma,$$

where $[\phi] = \langle \phi, \psi \rangle$. Sequences $\lambda^k \in \mathbf{R}$, $v^k \in \mathcal{V}$ can be defined as before by the recursion relation (41), and the proof of their convergence, for all sufficiently small ε, is the same as in the finite-dimensional case, provided that the expressions $|v^{k+1} - v^k|$, *etc.* in the finite-dimensional case are replaced by their analogs $\|v^{k+1} - v^k\|_{\mathcal{Y}}$, *etc.* in the infinite-dimensional case. In the papers of Keller [20] and of Keller and Langford [21], a version of the iteration method is used to prove the existence of bifurcating equilibrium solutions in infinite-dimensional spaces.

Many interesting examples of bifurcation phenomena concern the system of Navier-Stokes equations

$$\rho\left(\frac{\partial v}{\partial t} + v \cdot \nabla v\right) = -\nabla p + \nu \Delta v, \qquad \nabla \cdot v = 0,$$

for the velocity v and pressure p of a viscous incompressible fluid. In order to eliminate the pressure, one can define $\mathcal{Z}$ to be a space of divergence-free velocity fields which is closed as a subspace of the space $\mathbf{L}^2$ of all square-integrable vector fields. Then, defining P to be the orthogonal projection of $\mathbf{L}^2$ onto $\mathcal{Z}$, and, for convenience, normalizing the density ρ and viscosity ν, the Navier-Stokes equations can be written in the form

$$\frac{dv}{dt} = P\Delta v - P(v \cdot \nabla)v. \tag{53}$$

One situation of interest, the Taylor-Couette problem, concerns the behavior of a fluid that fills the space between two coaxial cylinders. The fluid is observed to adhere to the rigid cylindrical boundaries. Let us suppose that the cylinders are infinite, that the outer cylinder is fixed, and that the inner one rotates with a constant specified speed μ. Then, corresponding to the speed $\mu = 1$, there is a steady solution $\bar{v}$ (unit Couette flow) which is both axially and angularly symmetric (and easily found explicitly). It is easily seen (by symmetry) that both $P(\bar{v}\cdot\nabla)\bar{v} = 0$ and $P\Delta\bar{v} = 0$. Therefore, as μ varies, there is a curve of steady solutions of (53) of the form $\bar{v}(\mu) = \mu\bar{v}$. Putting $v = \mu\bar{v} + u$ in (53), we have

$$\frac{du}{dt} = P\Delta u - \mu P\{(\bar{v} \cdot \nabla)u + (u \cdot \nabla)\bar{v}\} - P(u \cdot \nabla)u. \tag{54}$$

In this case, the nonlinear part is $N(u,u) = -P(u \cdot \nabla)u$, which is in fact bilinear and independent of μ.

In experiments, Couette flow is the only observed flow for small values of μ. But, for μ slightly exceeding some critical value μ_0, steady secondary flows are observed which may be regarded as the superposition of Couette flow with certain angularly symmetric, axially periodic convection cells. These convection cells, commonly referred to as Taylor vortices, convect the more rapidly rotating fluid near the inner cylinder towards the outer cylinder. There is a wealth of further bifurcation phenomena which occurs as the speeds of the rotations of the cylinders are varied, including the bifurcation of time periodic flows.

To study the first appearance of Taylor vortices mathematically, one can specify a length of axial periodicity, and then choose $\mathcal{Z}$ to consist of those divergence free velocity fields w, with this specified axial periodicity, which satisfy $w \cdot n = 0$ at the cylindrical boundaries (here n denotes the unit normal to the boundaries), and are square-integrable over any spatial period of the flow region. The subspace $\mathcal{Y}$ of $\mathcal{Z}$ should then consist of those elements w of $\mathcal{Z}$ whose first and second derivatives are also square-integrable over any spatial period of the domain, and which satisfy $w = 0$ at the cylindrical boundaries. It can be shown that there exists a positive number μ_0 such that the linear operators $A, B : \mathcal{Y} \to \mathcal{Z}$, defined by

$$\begin{aligned} Au &= P\Delta u - \mu_0 P\{(\bar{v} \cdot \nabla)u + (u \cdot \nabla)\bar{v}\}, \\ Bu &= P\{(\bar{v} \cdot \nabla)u + (u \cdot \nabla)\bar{v}\}, \end{aligned}$$

satisfy the hypotheses mentioned above [16, 23, 33]. Setting $\nu = \mu - \mu_0$, the equation (54) is cast in the form (39). The bifurcating equilibrium solutions correspond to the steady Taylor vortices in the fluid.

To treat Hopf bifurcations in infinite-dimensional systems using the iteration method, one must first choose spaces of functions so that a suitable analog of Proposition 3.1 can be proved. In addition, the nature of the nonlinearity influences the choice of function spaces. For problems involving Navier-Stokes or similar equations, examples of suitable spaces of time-periodic functions use Schauder type spaces [19, 29, 31], Sobolev spaces [13, 18], or domains of fractional powers of unbounded operators [5, 9] for the spatial variables. Perhaps the simplest choice of spaces for such problems is that of Iooss [14], which uses Sobolev spaces also with respect to the time variable. Assuming that spaces have been chosen so that the linear and nonlinear parts of the problem can be proven to have suitable properties, then the proof of the infinite-dimensional version of Theorem 4.2 proceeds in essentially the same way as in the finite-dimensional case.

Along with the bifurcation theorems themselves, there is a natural interest in the exchange of stabilities between solutions. In order to determine the nonlinear stability of the bifurcating solutions, it is sufficient, by the principle of linearized stability, to linearize the differential equation about the solutions, and show that the eigenvalues (in the case of equilibrium solutions), or nontrivial Floquet exponents (in the case of periodic solutions), have nonzero real parts. Our method can be used to construct and prove the convergence of an expansion in powers of ε, of the critical eigenvalue or Floquet exponent that determines the stability of the solutions near the bifurcation point. This requires only straightforward modifications of the methods used in §§2–3.

Appendix A. Proof of Proposition 3.1

In this appendix we supply the proofs of several well known facts about the linear operator $d/ds - A$, which were collected together in Proposition 3.1. We keep the notation of §3, in particular using $\mathcal{N}$ and $\mathcal{R}$ to denote the null space and range of $d/ds - A : \mathcal{X}_1 \to \mathcal{X}_0$. We also use $\mathcal{N}_T$ and $\mathcal{R}_T$ to denote the null space and range of $d/ds + A^T : \mathcal{X}_1 \to \mathcal{X}_0$.

Proposition A.1. $\mathcal{R} = \{f \in \mathcal{X}_0 : (f, \psi) = 0 \text{ for all } \psi \in \mathcal{N}_T\}$.

Proof. Clearly, $\mathcal{N}$ consists of all functions which can be written in the form $w(s) = e^{sA}x$, for some x satisfying

$$[e^{2\pi A} - I\,]x = 0. \tag{55}$$

Similarly, $\mathcal{N}_T$ consists of all functions which can be written in the form $\psi(s) = e^{-sA^T}y$, for some y satisfying

$$[e^{-2\pi A^T} - I\,]y = 0. \tag{56}$$

By definition, $f \in \mathcal{R}$ if and only if there exists a function $\phi \in \mathcal{X}_1$ such that $d\phi/ds - A\phi = f$. All solutions of this differential equation are given by the variation of constants formula $\phi(s) = e^{sA}[x + \int_0^s e^{-\sigma A} f(\sigma)\, d\sigma]$, for some $x \in \mathbf{R}^n$. Such functions ϕ belong to $\mathcal{X}_1$ if and only if they are 2π-periodic, *i.e.*, if and only if x satisfies

$$[e^{-2\pi A} - I\,]x = \int_0^{2\pi} e^{-\sigma A} f(\sigma)\, d\sigma,$$

or equivalently, using the identity $(e^{-sA^T})^T = e^{-sA}$,

$$[e^{-2\pi A^T} - I\,]^T x = \int_0^{2\pi} (e^{-\sigma A^T})^T f(\sigma)\, d\sigma. \tag{57}$$

Thus, $f \in \mathcal{R}$ if and only if (57) is solvable for x. By the Fredholm alternative for matrices, (57) is solvable if and only if

$$y^T \int_0^{2\pi} (e^{-\sigma A^T})^T f(\sigma)\, d\sigma = 0, \tag{58}$$

for every y satisfying (56). Since $y^T(e^{-\sigma A^T})^T = (e^{-\sigma A^T} y)^T$, this condition is equivalent to the condition

$$\int_0^{2\pi} \psi(\sigma)^T f(\sigma)\, d\sigma = 0, \tag{59}$$

for every $\psi \in \mathcal{N}_T$. □

Proposition A.2. *Let $\mathcal{V} = \{\phi \in \mathcal{X}_1 : (\phi, w) = 0$ for all $w \in \mathcal{N}\}$, and let L_0 denote the restriction of the operator $d/ds - A$ to $\mathcal{V}$. Then L_0 is an isomorphism from $\mathcal{V}$ onto $\mathcal{R}$. In particular, there exists a constant $c_{\mathcal{V}}$ such that if $d\phi/ds - A\phi = f$, with $\phi \in \mathcal{V}$ and $f \in \mathcal{R}$, then $\|\phi\|_1 = \|L_0^{-1} f\|_1 \le c_{\mathcal{V}} \|f\|_0$.*

Proof. Using Proposition A.1, it is easily seen that $\mathcal{R}$ is a closed subspace of $\mathcal{X}_0$. It is also easy to see that $\mathcal{V}$ is a closed subspace of $\mathcal{X}_1$. Clearly $L_0 : \mathcal{V} \to \mathcal{R}$ is continuous, one-to-one and onto. Therefore, by the open mapping theorem, L_0^{-1} is also continuous. □

Proposition A.3. *If the characteristic polynomial of the real $n \times n$ matrix A has simple roots $\pm i$, and no other roots of the form $\pm ik$, where k is an integer, then the null spaces $\mathcal{N}$ and $\mathcal{N}_T$ are two-dimensional. Moreover, we can choose $\{w_1, w_2\}$ spanning $\mathcal{N}$, and $\{\psi_1, \psi_2\}$ spanning $\mathcal{N}_T$, such that*

$$(w_j, \psi_k) = \delta_{jk}, \qquad (w_j, w_k) = \delta_{jk}, \tag{60}$$

for $j, k = 1, 2$, and

$$\left(\frac{dw_1}{ds}, \psi_1\right) = 0, \qquad \left(\frac{dw_1}{ds}, \psi_2\right) = -1, \tag{61}$$

Proof. To determine the dimension of $\mathcal{N}$, we may assume that A is in real canonical form (see [10, pp.126–136]). Then the 2×2 block corresponding to the simple roots $\pm i$ produces two independent 2π-periodic solutions of $dw/ds - Aw = 0$, and no other block produces any further 2π-periodic solutions. So the dimension of $\mathcal{N}$ is two, and similarly, so is that of $\mathcal{N}_T$.

Next, we claim that complex eigenvectors $c, d \in \mathbf{C}^n$, satisfying $Ac = ic$ and $A^T d = -id$, can be chosen to satisfy the additional conditions

$$\bar{c}^T c = 2, \qquad \bar{d}^T c = 2, \qquad d^T c = 0. \tag{62}$$

The last of these conditions is automatically satisfied by any choice of the eigenvectors c, d since $d^T c = (-id^T)(ic) = (d^T A)(Ac) = d^T(A^2 c) = -d^T c$. Notice further, that if c, d are any eigenvectors satisfying $\bar{d}^T c \neq 0$, they can be normalized by multiplication by suitable complex scalars, so as to satisfy $\bar{c}^T c = 2$, $\bar{d}^T c = 2$.

To obtain eigenvectors satisfying $\bar{d}^T c \neq 0$, set $c = Pe_1$ and $d^T = {e_1}^T \bar{P}^{-1}$, where $e_1 \in \mathbf{C}^n$ denotes the column vector with first component 1 and all other components 0, and where P is a nonsingular complex matrix such that $AP = PJ$, where J is block-diagonal with upper left 1×1 block $[\, i \,]$ (*e.g.*, J can be the Jordan canonical form of A). Then $APe_1 = PJe_1 = iPe_1$, which shows that $c = Pe_1$ satisfies $Ac = ic$. Also, $\bar{P}^{-1} A = \bar{J} \bar{P}^{-1}$, and hence ${e_1}^T \bar{P}^{-1} A = {e_1}^T \bar{J} \bar{P}^{-1} = -i{e_1}^T \bar{P}^{-1}$, which shows that $d^T = {e_1}^T \bar{P}^{-1}$ satisfies $d^T A = -id^T$, and hence $A^T d = -id$. Finally, $\bar{d}^T c = {e_1}^T P^{-1} P e_1 = 1$, which is nonzero.

For eigenvectors c, d satisfying (62), we write $c = c_1 + ic_2$, $d = d_1 + id_2$, where c_j, $d_j \in \mathbf{R}^n$, $j = 1, 2$ and put

$$\begin{aligned} w_1(s) &= \text{Re}\,(e^{is} c) = c_1 \cos s - c_2 \sin s, \\ w_2(s) &= \text{Im}\,(e^{is} c) = c_2 \cos s + c_1 \sin s, \\ \psi_1(s) &= \text{Re}\,(e^{is} d) = d_1 \cos s - d_2 \sin s, \\ \psi_2(s) &= \text{Im}\,(e^{is} d) = d_2 \cos s + d_1 \sin s. \end{aligned} \tag{63}$$

Clearly $\{w_1, w_2\}$ is a basis for $\mathcal{N}$, and $\{\psi_1, \psi_2\}$ is a basis for $\mathcal{N}_T$. The relations (60)–(61) are easily verified by a straightforward computation using (62) and (19). □

Acknowledgement. We thank Professor Gérard Iooss for suggesting several improvements in the paper.

References

[1] V. I. Arnold. *Ordinary Differential Equations.* M.I.T. Press, Cambridge, 1973.

[2] J. Carr. *Applications of Centre Manifold Theory.* Springer-Verlag, New York, 1981.

[3] S. N. Chow and J. K. Hale. *Methods of Bifurcation Theory.* Springer-Verlag, New York, 1982.

[4] M. G. Crandall and P. H. Rabinowitz. Bifurcation from simple eigenvalues. *J. Funct. Anal.* **8**, 321–340 (1971).

[5] ———. The Hopf bifurcation theorem in infinite dimensions. *Arch. Rational Mech. Anal.* **67**, 53–72 (1977).

[6] P. C. Fife and D. D. Joseph. Existence of convective solutions of the Bénard problem which are analytic in their norm. *Arch. Rational Mech. Anal.* **33**, 117–138 (1969).

[7] M. Golubitsky and D. G. Schaeffer. *Singularities and Groups in Bifurcation Theory,* Vol. 1. Springer-Verlag, New York, 1985.

[8] J. Hale. *Ordinary Differential Equations.* Krieger, Malabar, 1980.

[9] D. Henry. *Geometric Theory of Semilinear Parabolic Equations,* Lecture Notes in Mathematics, Vol. 840. Springer-Verlag, Berlin Heidelberg, 1981.

[10] M. W. Hirsch and S. Smale. *Differential Equations, Dynamical Systems, and Linear Algebra.* Academic Press, New York, 1974.

[11] E. Hopf. Abzweigung einer periodischen Lösung von einer stationären Lösung eines Differentialsystems. *Ber. Math.-Phys. Kl. Säch. Akad. Wiss. Leipzig.* XCIV, 1–22 (1942). English transl. in [26].

[12] G. Iooss. Théorie non linéaire de la stabilité des écoulements laminaires dans le cas de "l'échange des stabilités". *Arch. Rational Mech. Anal.* **40** (1971), 166–208.

[13] ———. Existence et stabilité de la solution périodique secondaire intervenant dans les problèmes d'évolution du type Navier-Stokes. *Arch. Rational Mech. Anal.* **47** (1972), 301–329.

[14] ———. Bifurcation and transition to turbulence in hydrodynamics, in: *Bifurcation Theory and Applications* (L. Salvadori, Ed.), pp. 152–201, Lecture Notes in Mathematics, Vol. 1057. Springer-Verlag, Berlin Heidelberg, 1984.

[15] G. Iooss and D. D. Joseph. *Elementary Stability and Bifurcation Theory.* Springer-Verlag, New York, 1990.

[16] V. I. Iudovich. Secondary flows and fluid instability between rotating cylinders. *Prikl. Mat. Meh.* **30**, 688–698 (1966). English transl. *J. Appl. Math.* **30**, 822–833 (1966).

[17] ———. On the origin of convection. *Prikl. Mat. Meh.* **30**, 1000–1005 (1966). English transl. *J. Appl. Math.* **30**, 1193–1199 (1966).

[18] ———. The onset of auto-oscillations in a fluid. *Prikl. Mat. Meh.* **35**, 638–655 (1971). English transl. *J. Appl. Math.* **35**, 587–603 (1971).

[19] D. D. Joseph and D. H. Sattinger. Bifurcating time periodic solutions and their stability. *Arch. Rational Mech. Anal.* **45** 79–109, (1972).

[20] H. B. Keller. Nonlinear bifurcation. *J. Differential Equations* **7**, 417–434 (1970).

[21] H. B. Keller and W. F. Langford. Iterations, perturbations and multiplicities for nonlinear bifurcation problems. *Arch. Rational Mech. Anal.* **48**, 83–108 (1972).

[22] K. Kirchgässner. Bifurcation in nonlinear hydrodynamic stability. *SIAM Review* **17**, 652–683 (1975).

[23] K. Kirchgässner and P. Sorger. Branching analysis for the Taylor problem. *Quart. J. Mech. Appl. Math.* **22**, 183–209 (1969).

[24] M. A. Krasnoselskii. *Topological Methods in the Theory of Nonlinear Integral Equations.* Pergamon, Oxford, 1964.

[25] W. F. Langford. Numerical solution of bifurcation problems for ordinary differential equations. *Numer. Math.* **28**, 171–190 (1977).

[26] J. E. Marsden and M. McCracken. *The Hopf Bifurcation and Its Applications.* Springer-Verlag, New York, 1976.

[27] H. Poincaré. *Les Méthodes Nouvelles de la Mécanique Céleste,* Vol. 3. Gauthier-Villars, Paris, 1892.

[28] P. H. Rabinowitz. Existence and nonuniqueness of rectangular solutions of the Bénard problem. *Arch. Rational Mech. Anal.* **29**, 32–57 (1968).

[29] D. H. Sattinger. Bifurcation of periodic solutions of the Navier-Stokes equations. *Arch. Rational Mech. Anal.* **41**, 66–80 (1971).

[30] ———. Stability of bifurcating solutions by Leray-Schauder degree. *Arch. Rational Mech. Anal.* **43**, 154–166 (1971).

[31] ———. *Topics in Stability and Bifurcation Theory,* Lecture Notes in Mathematics, Vol. 309. Springer-Verlag, Berlin Heidelberg, 1973.

[32] W. Velte. Stabilitätsverhalten und Verzweigung stationärer Lösungen der Navier-Stokesschen Gleichungen. *Arch. Rational Mech. Anal.* **16**, 97–125 (1964).

[33] ———. Stabilität und Verzweigung stationärer Lösungen der Navier-Stokesschen Gleichungen beim Taylor Problem. *Arch. Rational Mech. Anal.* **22**, 1–14 (1964).

John G. Heywood
Wayne Nagata

Department of Mathematics
University of British Columbia
Vancouver, B.C., Canada V6T 1Z2

On The Steady Transport Equation*

John G. Heywood and Mariarosaria Padula

Abstract. We give a simple approach to the theory of the steady transport equation, providing some results that are needed in the theory of steady compressible viscous flow. In particular, we prove the assertions that are collected together in Lemma 3 of our accompanying paper [11]. Our methods are based on a construction of solutions by Galerkin approximation and on duality arguments with the adjoint equation. We conclude with some heuristic observations based on the theory of characteristics, explaining, particularly, the need for successively stronger restrictions on the size of the velocity field, in order to prove successively more regularity of the solution.

1. Introduction

The theory of degenerate elliptic and parabolic equations, and of related hyperbolic problems, has received the attention of many excellent scientists, beginning with its initiation in the nineteen fifties by Fichera [7], [8] and Friedrichs [9], and continued in further pioneering works of Oleinik [19], Lax and Phillips [13], Kohn and Nirenberg [12], and Mizohata [14]. One of its recent applications has been to the so called steady transport equation, which plays a fundamental role related to mass conservation in the existence theory for steady compressible viscous flow; see Padula [20], [21], [22], [23], [24], Beirao da Veiga [1], Valli [26], Farwig [6], Novotny and Padula [18], and Novotny [15], [16]. However, in the case of the steady transport equation, the tools which were developed for the general theory of degenerate second order equations seem overly abstract and difficult. Seeking simpler methods as well as new results needed in the theory of compressible viscous flow, and of non-Newtonian flow, new existence theorems for the steady transport equation have been given by Beirao da Veiga [2], [3], [4], Farwig [6], Galdi [10], Coscia and Galdi [5], and Novotny [17]. Our objective here is to give still another variant of this theory, providing further simplifications, and giving additional results that are needed in our accompanying paper [11] on steady compressible viscous flow.

* This work has been supported by the Natural Sciences and Engineering Research Council of Canada, contracts MURST and the GNFM of Italian CNR, and the Deutsche Forschungsgemeinschaft.

There are several variants of the steady transport equation, without a standard terminology. We will refer to

$$k\sigma + \gamma\nabla \cdot (\sigma v) = H$$

as the steady transport equation because of its central role in [11]. The unknown is a scalar function σ, to be sought in a bounded domain $\Omega \subset R^n$, $n \geq 1$. The vector field v and the right side H are given, as are the constants k and γ, which we assume to be positive. In our application to compressible flow, σ represents a perturbation in the mass density, v the fluid velocity, H an effective pressure function, k the reciprocal of the square of the Mach number, and γ a viscosity coefficient. Our theory of the transport equation is developed simultaneously with that of its adjoint

$$k\sigma - \gamma v \cdot \nabla\sigma = H \ .$$

Some authors refer to this second equation as the transport equation. Much of our theory depends upon duality arguments between these two equations.

To study the regularity of the solutions of these scalar equations, we also consider the vector transport equations satisfied by their gradients $\nabla\sigma$. The theory of these equations for vector unknowns is similar to that for the scalar equations. By an evident induction, this generalization is carried further to a general class of transport equations for multi-indexed solutions, which provides the setting for a general regularity theory. It is included for its own interest in Section 5, although it is not need in [11].

Our existence theorems are based on Galerkin approximation, using eigenfunctions of the Laplacian as the basis. For the boundary conditions and duality arguments we need, it is important to use the eigenfunctions of the Neuman problem in some arguments, and the eigenfunctions of the Dirichlet problem in others. In contrast, the papers [2], [3], [4], [6], and [17], are based on approximation by elliptic regularization, using only one boundary condition or the other, and on some rather complicated mollification arguments. Another approach to the theory by Galerkin approximation has been given by Galdi [10] and by Coscia and Galdi [5], using a special basis that was introduced for the study of the Euler equation by Temam [25]. This provides an even simpler approach to the theory of the transport equation than given here, but the underlying a priori estimates are different than those we require in [11], particularly as we need the estimates for the Laplacian given in Section 4.

Finally, we mention that the constants in the estimates of this paper have only a slight dependence on the regularity of the boundary, coming through the use of Sobolev's inequalities. We assume that $\partial\Omega$ is of class $C^{1,1}$ only to ensure that the eigenfunctions of the Laplacian belong to $W^{2,2}(\Omega)$, so that the Galerkin approximations will also belong to this space and can be freely integrated by parts. It may be that by considering the limit of a sequence of domains, the results of this paper can be extended to all domains for which the required Sobolev inequalities are valid.

Our notation is standard, but we should mention several of our conventions. First, L^2-norms are denoted simply by $\|\cdot\|$, while all other norms are distinguished by subscripts. For example, $\|\cdot\|_p$ denotes the norm for $L^p(\Omega)$, and $\|\cdot\|_{m,p}$ the norm for the Sobolev space $W^{m,p}(\Omega)$, consisting of all functions that possess derivatives in $L^p(\Omega)$ up to order m. Second, we do not distinguish through our notation between scalar and vector valued functions, or function spaces, or norms. But in our usage, σ, ψ and H are always real valued scalar functions, while v, Φ, Ψ, and E are usually R^n-valued functions. In Section 5, Φ, Ψ, and E can also be matrix valued functions. The eigenfunctions of the scalar and vector Laplacians are denoted by a^k and A^k respectively, so that in the vector case the components can be denoted by A_l^k.

2. Existence in $W^{1,2} \cap L^q$ for the scalar transport equation

We consider the steady transport equation

$$k\sigma + \gamma\nabla \cdot (\sigma v) = H \; , \tag{1}$$

and its adjoint

$$k\sigma - \gamma v \cdot \nabla\sigma = H \; , \tag{2}$$

in a bounded subdomain Ω of R^n, $n \geq 1$, with a boundary of class $C^{1,1}$. For either equation, the unknown is a scalar function σ. The vector field v and the scalar right side H are given, as are the constants k and γ, which we assume to be positive. We assume throughout that the velocity field v belongs to $W^{2,p}(\Omega)$, for some $p > n$, and satisfies $v \cdot n|_{\partial\Omega} = 0$. Stronger assumptions on v are assumed as needed in Sections 4 and 5. We also assume throughout that $H \in L^2(\Omega)$, and make stronger assumptions as needed.

For $\sigma \in W^{1,2}(\Omega)$, we have the key identity

$$(\nabla \cdot (\sigma v), \sigma) = -(v \cdot \nabla\sigma, \sigma) = -\int_\Omega v \cdot \nabla\left(\frac{\sigma^2}{2}\right) dx = \tfrac{1}{2}\int_\Omega \sigma^2 \nabla \cdot v \, dx \, . \tag{3}$$

Thus, solutions $\sigma \in W^{1,2}(\Omega)$ of either (1) or (2) satisfy

$$k\|\sigma\|^2 + \tfrac{\gamma}{2}\int_\Omega \sigma^2 \nabla \cdot v \, dx = (H, \sigma) \; , \tag{4}$$

and hence

$$\left(k - \tfrac{\gamma}{2}\|\nabla \cdot v\|_\infty\right)\|\sigma\| \leq \|H\| \; . \tag{5}$$

Provided $k > \frac{\gamma}{2}\|\nabla \cdot v\|_\infty$, this gives an a priori L^2-estimate for solutions $\sigma \in W^{1,2}(\Omega)$, and also their uniqueness, since the equations (1) and (2) are linear.

We will have a need to consider L^2-*solutions* of the transport equation (1), defined as functions $\sigma \in L^2(\Omega)$ such that

$$(\sigma, k\psi - \gamma v \cdot \nabla\psi) = (H, \psi) \; , \qquad \forall\psi \in W_0^{1,2}(\Omega) \; , \tag{6}$$

and also L^2-*solutions* of the adjoint equation (2), defined as functions $\sigma \in L^2(\Omega)$, such that

$$(\sigma, k\psi + \gamma \nabla \cdot (\psi v)) = (H, \psi) \ , \qquad \forall \psi \in W_0^{1,2}(\Omega) \ . \tag{7}$$

The uniqueness of L^2-solutions is an important issue in what follows, especially for the vector transport equation in the next section. This is because, to prove that $\nabla\sigma \in W^{1,2}$, we proceed by identifying $\nabla\sigma$ with the solution $\Phi \in W^{1,2}$ of a vector transport equation, which $\nabla\sigma$ is first shown to satisfy only as an L^2-solution. The uniqueness of L^2-solutions is based on a duality argument with the adjoint equation.

Theorem 1. *If $k > \frac{\gamma}{2}\|\nabla \cdot v\|_\infty$, then, for every $H \in L^2(\Omega)$, the equations (1) and (2) possess L^2-solutions. The solutions which are constructed here satisfy the estimate (5). They also satisfy the identities (6) and (7), in the weak formulations of the equations, for all test functions $\psi \in W^{1,2}(\Omega)$. As a consequence, the solution of (1) which is constructed here satisfies $\int_\Omega \sigma\, dx = 0$, if $\int_\Omega H\, dx = 0$.*

Proof. We consider first the equation (1). Let $\{a^1, a^2, \dots\}$ and $\{\lambda_1, \lambda_2, \dots\}$ denote the eigenfunctions and eigenvalues of the Neumann problem

$$-\Delta a = \lambda a \quad , \quad \left.\frac{\partial a}{\partial n}\right|_{\partial\Omega} = 0 \, , \tag{8}$$

and let $\sigma_n = \sum_{k=1}^{n} c_k a^k$ be the unique solution of the system of linear equations

$$\left(k\sigma_n + \gamma\nabla \cdot (\sigma_n v)\, , a^k\right) = \left(H, a^k\right) \ , \qquad k = 1, 2, \cdots, n \, . \tag{9}$$

Uniqueness is evident here because multiplying (9) by c_k, and summing 1 to n, we obtain the estimate (5) for σ_n, which shows that zero is the only solution in the homogeneous case. Since (9) is equivalent to

$$(\sigma_n, ka_k - \gamma v \cdot \nabla a_k) = (H, a_k) \ , \qquad k = 1, 2, \cdots, n \, , \tag{10}$$

and since the sequence $\|\sigma_n\|$ satisfies the bound (5), a subsequence of $\{\sigma_n\}$, again denoted by $\{\sigma_n\}$, converges weakly in $L^2(\Omega)$ to a limit σ satisfying (5) and

$$(\sigma, ka_k - \gamma v \cdot \nabla a_k) = (H, a_k) \ , \qquad k = 1, 2, \cdots . \tag{11}$$

It follows that σ satisfies (6) for all test functions $\psi \in W^{1,2}(\Omega)$, since the eigenfunctions $\{a_1, a_2, \cdots\}$ of the Neumann problem are complete in $W^{1,2}(\Omega)$. Finally, by setting ψ equal to a constant in (6), it is evident that the mean of σ must be zero if the mean of H is zero.

The existence of an L^2-solution of the adjoint equation (2) is proved similarly, taking $\sigma_n = \sum_{k=1}^{n} c_k a^k$ to be the unique solution of the system of linear equations

$$\left(k\sigma_n - \gamma v \cdot \nabla\sigma_n\, , a^k\right) = \left(H, a^k\right) \ , \qquad k = 1, 2, \cdots, n \, , \tag{12}$$

and proceeding as before. The only difference is that the result concerning the mean of σ does not hold for the adjoint equation. This completes the proof.

Theorem 2. *If $k > \frac{\gamma}{2}\|\nabla\cdot v\|_\infty + \gamma\|Dv\|_\infty$, then, for every $H\in W^{1,2}(\Omega)$, the equations (1) and (2) each possess a unique solution $\sigma\in W^{1,2}(\Omega)$. For either equation, $\sigma\in W_0^{1,2}(\Omega)$ if $H\in W_0^{1,2}(\Omega)$. For either equation, the weaker assumption that $k > \frac{\gamma}{2}\|\nabla\cdot v\|_\infty$ suffices to imply uniqueness within the class of solutions that belong to $W^{1,2}(\Omega)$. The solution of the adjoint equation (2) satisfies the estimate (17) below, which is of the form*

$$\|\nabla\sigma\| \le c\,\|\nabla H\|\,, \tag{13}$$

while the solution of the original transport equation (1) satisfies the estimate (23) below, which is of the form

$$\|\nabla\sigma\| \le c\,\|H\|_{1,2}\,. \tag{14}$$

The constant in (13) is simply $\left(k-\frac{\gamma}{2}\|\nabla\cdot v\|_\infty - \gamma\|Dv\|_\infty\right)^{-1}$.
The more complicated constant in (14) depends also on (and increases monotonically with) $\|\nabla\nabla\cdot v\|_p$, for p as specified at the beginning of this section.

Proof. We consider first the adjoint equation (2), since it is the easier case, and begin by seeking a formal estimate for $\nabla\sigma$, by taking the $\partial/\partial x_j$ derivative of (2), multiplying through by $\partial\sigma/\partial x_j$, and integrating over Ω. Note, similarly to (3), that

$$\begin{aligned}\left(\frac{\partial}{\partial x_j}\left(v_i\frac{\partial\sigma}{\partial x_i}\right),\frac{\partial\sigma}{\partial x_j}\right) &= \left(v_i\frac{\partial}{\partial x_i}\left(\frac{\partial\sigma}{\partial x_j}\right),\frac{\partial\sigma}{\partial x_j}\right)+\left(\frac{\partial\sigma}{\partial x_i}\frac{\partial v_i}{\partial x_j},\frac{\partial\sigma}{\partial x_j}\right)\\ &= -\tfrac{1}{2}\int_\Omega|\nabla\sigma|^2\,\nabla\cdot v\,dx+\left(\frac{\partial\sigma}{\partial x_i}\frac{\partial v_i}{\partial x_j},\frac{\partial\sigma}{\partial x_j}\right).\end{aligned} \tag{15}$$

Let us also set

$$Dv=\frac{1}{2}\left(\nabla v+(\nabla v)^\top\right),\quad |Dv|=\left(\sum_{i,j}(Dv)_{i,j}^2\right)^{1/2},\quad \|Dv\|_\infty=\sup_\Omega|Dv|\,.$$

Then, from (2) we obtain

$$\begin{aligned}k\|\nabla\sigma\|^2 &= (\nabla H,\nabla\sigma)-\tfrac{\gamma}{2}\int_\Omega|\nabla\sigma|^2\,\nabla\cdot v\,dx+\gamma\left(\frac{\partial\sigma}{\partial x_i}\frac{\partial v_i}{\partial x_j},\frac{\partial\sigma}{\partial x_j}\right)\\ &\le \|\nabla H\|\,\|\nabla\sigma\|+\tfrac{\gamma}{2}\|\nabla\cdot v\|_\infty\|\nabla\sigma\|^2+\gamma\|Dv\|_\infty\|\nabla\sigma\|^2,\end{aligned} \tag{16}$$

and hence

$$\left(k-\tfrac{\gamma}{2}\|\nabla\cdot v\|_\infty-\gamma\|Dv\|_\infty\right)\|\nabla\sigma\|\le\|\nabla H\|\,. \tag{17}$$

This estimate can be proven for the Galerkin approximations that were used in constructing the L^2-solution. Indeed, multiplying (12) by $c_k\lambda_k$, and summing over k from 1 to n, we obtain

$$(k\sigma_n-\gamma v\cdot\nabla\sigma_n\,,-\Delta\sigma_n)=(H,-\Delta\sigma_n)\,. \tag{18}$$

Since σ_n satisfies the boundary condition for the homogeneous Neuman problem, (18) can be integrated by parts to obtain analogues of (16) and (17). The estimate

(17) implies that the solution σ, that was found as a weak L^2-limit in Theorem 1, actually belongs to $W^{1,2}(\Omega)$ and satisfies (17). The uniqueness of solutions belonging to $W^{1,2}(\Omega)$ follows from (5), as was already pointed out.

If $H \in W_0^{1,2}(\Omega)$, we obtain a solution $\sigma \in W_0^{1,2}(\Omega)$ by constructing it using the eigenfunctions of the Dirichlet problem, $-\triangle a = \lambda a$, $a|_{\partial\Omega} = 0$. Since these functions are complete in $W_0^{1,2}(\Omega)$, we obtain (7) with $W_0^{1,2}(\Omega)$ as test space. The integration by parts $(H, -\triangle\sigma_n) = (\nabla H, \nabla\sigma_n)$ is justified now by virtue of the boundary condition for H, rather than for σ_n. Hence we again obtain the estimate (17), and incorporate it into the existence theorem as before.

Consider now the transport equation (1). Again, we obtain a formal estimate for $\nabla\sigma$ by taking the $\partial/\partial x_j$ derivative of (1), multiplying through by $\partial\sigma/\partial x_j$, and integrating over Ω. Analogously to (15), we obtain the identity

$$\begin{aligned}\left(\frac{\partial}{\partial x_j}\nabla\cdot(\sigma v), \frac{\partial\sigma}{\partial x_j}\right) &= \left(\frac{\partial}{\partial x_i}\frac{\partial}{\partial x_j}[\sigma v_i], \frac{\partial\sigma}{\partial x_j}\right)\\ &= \left(\frac{\partial}{\partial x_i}\left[\frac{\partial\sigma}{\partial x_j}v_i\right], \frac{\partial\sigma}{\partial x_j}\right) + \left(\frac{\partial}{\partial x_i}\left[\sigma\frac{\partial v_i}{\partial x_j}\right], \frac{\partial\sigma}{\partial x_j}\right)\\ &= \tfrac{1}{2}\int_\Omega |\nabla\sigma|^2\,\nabla\cdot v\,dx + \left(\frac{\partial\sigma}{\partial x_i}\frac{\partial v_i}{\partial x_j}, \frac{\partial\sigma}{\partial x_j}\right) + \left(\sigma\frac{\partial^2 v_i}{\partial x_j\partial x_i}, \frac{\partial\sigma}{\partial x_j}\right),\end{aligned} \tag{19}$$

using (3) in the last line, with σ replaced by $\partial\sigma/\partial x_j$. Thus any smooth solution of (1) satisfies

$$\begin{aligned}&k\|\nabla\sigma\|^2\\ &=(\nabla H,\nabla\sigma) - \tfrac{\gamma}{2}\int_\Omega|\nabla\sigma|^2\nabla\cdot v\,dx - \gamma\left(\frac{\partial\sigma}{\partial x_i}\frac{\partial v_i}{\partial x_j}, \frac{\partial\sigma}{\partial x_j}\right) - \gamma\left(\sigma\frac{\partial^2 v_i}{\partial x_i\partial x_j}, \frac{\partial\sigma}{\partial x_j}\right)\\ &\le \|\nabla H\|\|\nabla\sigma\| + \tfrac{\gamma}{2}\|\nabla\cdot v\|_\infty\|\nabla\sigma\|^2 + \gamma\|Dv\|_\infty\|\nabla\sigma\|^2 + \gamma\|\nabla\nabla\cdot v\|_p\|\sigma\|_q\|\nabla\sigma\|\end{aligned}$$

and hence

$$\left(k - \tfrac{\gamma}{2}\|\nabla\cdot v\|_\infty - \gamma\|Dv\|_\infty\right)\|\nabla\sigma\| \le \|\nabla H\| + \gamma\|\nabla\nabla\cdot v\|_p\|\sigma\|_q\,, \tag{20}$$

where $p > n$ is a number fixed at the beginning of this section, and $q = \left(\frac{1}{2} - \frac{1}{p}\right)^{-1}$. Except in the case of two-dimensions, we have the multiplicative Sobolev inequality

$$\|\sigma\|_q \le c\|\sigma\|_{1,2}^{\frac{n}{p}}\|\sigma\|^{\frac{p-n}{p}} \le c\|\nabla\sigma\|^{\frac{n}{p}}\|\sigma\|^{\frac{p-n}{p}} + c\|\sigma\|\,, \tag{21}$$

which we use together with Young's inequality, to obtain

$$\gamma\|\nabla\nabla\cdot v\|_p\|\sigma\|_q \le \epsilon\|\nabla\sigma\| + C\left(\epsilon, \gamma\|\nabla\nabla\cdot v\|_p\right)\|\sigma\|\,, \tag{22}$$

for any $\epsilon > 0$, where $C\left(\epsilon, \gamma\|\nabla\nabla\cdot v\|_p\right)$ is constant depending on ϵ, $\gamma\|\nabla\nabla\cdot v\|_p$, and the Sobolev constant for the domain. The inequality (22) is proved similarly in two-dimensions, using a two-dimensional multiplicative Sobolev inequality. Thus, from (20) and (5), we get

$$\left(k-\tfrac{\gamma}{2}\|\nabla\cdot v\|_\infty-\gamma\|Dv\|_\infty-\epsilon\right)\|\nabla\sigma\|\leq\|\nabla H\|+\frac{C\left(\epsilon,\gamma\|\nabla\nabla\cdot v|_p\right)}{\left(k-\frac{\gamma}{2}\|\nabla\cdot v\|_\infty\right)}\|H\|\ , \tag{23}$$

for any $\epsilon>0$. This estimate holds also for the Galerkin approximations used in Theorem 1, with the result that we obtain a solution of (1) that satisfies (23) and belongs to $W^{1,2}(\Omega)$. As before, we obtain $\sigma\in W_0^{1,2}(\Omega)$ if $H\in W_0^{1,2}(\Omega)$, by using eigenfunctions of the Dirichlet problem. This completes the proof.

Theorem 3. *If $k>\frac{\gamma}{2}\|\nabla\cdot v\|_\infty+\gamma\|Dv\|_\infty$ and $H\in L^2(\Omega)$, then there is at most one L^2-solution of either the transport equation (1), or of its adjoint (2).*

Proof. Clearly, if there are two L^2-solutions of (1), then their difference σ satisfies

$$(\sigma,k\psi-\gamma v\cdot\nabla\psi)=0\ ,\qquad\forall\psi\in W_0^{1,2}(\Omega)\ .$$

This implies that $\sigma=0$, if functions of the form $k\psi-\gamma v\cdot\nabla\psi$, for $\psi\in W_0^{1,2}(\Omega)$, are dense in $L^2(\Omega)$. This was shown in Theorem 2, in showing the solvability of the adjoint equation, for a solution in $W_0^{1,2}(\Omega)$, when $H\in W_0^{1,2}(\Omega)$. Uniqueness for (2) is proved similarly.

Theorem 4. *Suppose that $k\geq\frac{\gamma}{2}\|\nabla\cdot v\|_\infty+\gamma\|Dv\|_\infty$, and also that $k>\gamma\|\nabla\cdot v\|_\infty$. Let σ be an L^2-solution of either the transport equation (1) or its adjoint (2), with right side $H\in L^q(\Omega)$, for some q satisfying $2\leq q\leq\infty$.*

Let $c_q=\left(k-\frac{\gamma}{q'}\|\nabla\cdot v\|_\infty\right)^{-1}$ and $c_{q'}=\left(k-\frac{\gamma}{q}\|\nabla\cdot v\|_\infty\right)^{-1}$, where $q'=q/(q-1)$. Also set $c_\infty=(k-\gamma\|\nabla\cdot v\|_\infty)^{-1}$ and $c_{\infty'}=1/k$. Then the solution of the transport equation satisfies

$$\|\sigma\|_q\leq c_q\|H\|_q\ , \tag{24}$$

while the solution of its adjoint equation satisfies

$$\|\sigma\|_q\leq c_{q'}\|H\|_q\ . \tag{25}$$

Proof. We remark that the only purpose of the hypothesis $k>\gamma\|\nabla\cdot v\|_\infty$ is to ensure that $c_q,c_{q'}$ and c_∞ are well defined (in fact it implies $1/k\leq c_{q'}\leq c_q\leq c_\infty<\infty$). This hypothesis is superfluous if the number of spatial dimensions $n\leq 4$, since $\sqrt{n}\|Dv\|_\infty\geq\|\nabla\cdot v\|_\infty$.

Now, given any $f\in C_0^\infty(\Omega)$, Theorem 2 provides a solution $\psi^f\in W_0^{1,2}(\Omega)$ of the adjoint equation

$$k\psi^f-\gamma v\cdot\nabla\psi^f=f\ . \tag{26}$$

This solution satisfies the estimate

$$\|\psi^f\|_{q'}\leq c_q\|f\|_{q'}\ , \tag{27}$$

for any q' satisfying $1 < q' < 2$. Indeed, for such q' and for any functions $f, \psi \in W^{1,2}(\Omega)$ which satisfy $k\psi - \gamma v \cdot \nabla\psi = f$, we can multiply the equation through by $\psi |\psi|^{q'-2}$ and integrate over Ω to obtain

$$k \|\psi\|_{q'}^{q'} - \tfrac{\gamma}{q'} \int_\Omega v \cdot \nabla |\psi|^{q'} \, dx = \left(f, \psi |\psi|^{q'-2}\right), \tag{28}$$

and hence

$$\left(k - \tfrac{\gamma}{q'} \|\nabla \cdot v\|_\infty\right) \|\psi\|_{q'}^{q'} \le \|f\|_{q'} \|\psi\|_{q'}^{q'-1}, \tag{29}$$

and therefore $\|\psi\|_{q'} \le c_q \|f\|_{q'}$.

Now, taking q', f, and ψ^f as above, we can use (6) and (27) to obtain

$$\begin{aligned} \|\sigma\|_q &= \sup_{f \in C_0^\infty(\Omega)} \frac{|(\sigma, f)|}{\|f\|_{q'}} = \sup_{f \in C_0^\infty(\Omega)} \frac{\left|\left(\sigma, k\psi^f - \gamma v \cdot \nabla\psi^f\right)\right|}{\|f\|_{q'}} \\ &= \sup_{f \in C_0^\infty(\Omega)} \frac{\left|\left(H, \psi^f\right)\right|}{\|f\|_{q'}} \le \sup_{f \in C_0^\infty(\Omega)} \frac{\|H\|_q \left\|\psi^f\right\|_{q'}}{\|f\|_{q'}} \le c_q \|H\|_q \, . \end{aligned} \tag{30}$$

This completes the proof of (24) for $2 < q < \infty$. For $q = \infty$ and $q' = 1$, we obtain (24) and (25) by passing to the limit. The case $q = q' = 2$ was proven in Theorem 1.

The adjoint equation is treated similarly. For any $f \in C_0^\infty(\Omega)$, Theorem 2 provides a solution $\psi^f \in W_0^{1,2}(\Omega)$ of the transport equation

$$k\psi^f + \gamma\nabla \cdot \left(\psi^f v\right) = f \, . \tag{31}$$

For q' satisfying $1 < q' < 2$, and $f, \psi \in W^{1,2}(\Omega)$ satisfying the transport equation $k\psi + \gamma\nabla \cdot (\psi v) = f$, we again multiply through by $\psi |\psi|^{q'-2}$ and integrate over Ω. Writing $\nabla \cdot (\psi v) = v \cdot \nabla\psi + \psi\nabla \cdot v$, it is clear that in place of the second term in (28) we now have

$$\tfrac{\gamma}{q'} \int_\Omega v \cdot \nabla |\psi|^{q'} \, dx + \gamma \int_\Omega |\psi|^{q'} \nabla \cdot v \, dx = \gamma \left(1 - \frac{1}{q'}\right) \int_\Omega |\psi|^{q'} \nabla \cdot v \, dx \, , \tag{32}$$

which results in the estimate

$$\left\|\psi^f\right\|_{q'} \le c_{q'} \|f\|_{q'} \, . \tag{33}$$

The constant $c_{q'}$ therefore replaces c_q in the analogue of (30) for the adjoint equation. This completes the proof.

The duality argument used in Theorem 4 requires a stronger condition on the size of k than is needed to ensure that the constant c_q in the estimate (24) is well defined and finite. This is a drawback of our present L^2-based methods which can be addressed through an L^p-approach.

3. Existence in $W^{1,2} \cap L^q$ for the vector transport equation

Taking the gradient of the transport equation (1), the components of $\Phi = \nabla\sigma$ are seen to formally satisfy the system $(l = 1, \cdots, n)$ of equations

$$k\Phi_l + \gamma \frac{\partial}{\partial x_i} (\Phi_l v_i) + \gamma \Phi_i \frac{\partial v_i}{\partial x_l} = E_l \, , \tag{34}$$

with right side

$$E_l = \frac{\partial H}{\partial x_l} - \gamma\sigma \frac{\partial^2 v_i}{\partial x_l \partial x_i} \, . \tag{35}$$

We call Φ an L^2*-solution* of (34) if $\Phi \in L^2(\Omega)$ and

$$\left(\Phi_l \, , k\Psi_l - \gamma v_i \frac{\partial \Psi_l}{\partial x_i} + \gamma \frac{\partial v_l}{\partial x_i} \Psi_i \right) = (E_l \, , \Psi_l) \; , \qquad \forall \Psi \in W_0^{1,2}(\Omega) \, . \tag{36}$$

Here, we are using the usual summation convention.

If σ is a solution of the transport equation (1) belonging to $\in W^{1,2}$, then $\nabla\sigma$ is an L^2-solution of (34). To show this, let Ψ be an arbitrary vector valued function in $C_0^\infty(\Omega)$, with scalar components Ψ_l. By taking $\partial\Psi_l/\partial x_l$ as the test function ψ in (6), integrating by parts, and summing over l, we obtain (36) for all test functions in $C_0^\infty(\Omega)$, and then taking a limit, for all test functions in $W_0^{1,2}(\Omega)$.

In this section we treat the equation (34) independently of its origin. In fact, the preceding section has been included in this paper mainly for pedagogical reasons, as Theorems 5–8 are only slight generalizations of Theorems 1–4. The main difference is the inclusion of an additional term on the left side of (34), with coefficient matrix $\nabla v^\top$. In what follows, we will study a slightly more general system of equations, (38) below, in which the coefficient matrix $\nabla v^\top$ is replaced by a general coefficient matrix M. Of course, we are particularly interested in the case

$$M_l^i = \frac{\partial v_i}{\partial x_l} \, , \tag{37}$$

and therefore the estimates we will assume for M are just those that are valid for ∇v. Our assumptions regarding v are the same here as in the last section, namely that v belongs to $W^{2,p}(\Omega)$, for some $p > n$, and satisfies $v \cdot n|_{\partial\Omega} = 0$. Thus we assume that

$$\|\mathcal{M}\|_\infty < \infty \, , \quad \|\nabla M\|_p < \infty \, ,$$

for some $p > n$, where

$$\mathcal{M} = \tfrac{1}{2}\left(M + M^\top\right), \quad |\mathcal{M}| = \left(\sum_{i,j} \mathcal{M}_{i,j}^2 \right)^{1/2} , \qquad \|\mathcal{M}\|_\infty = \sup_\Omega |\mathcal{M}| \, ,$$

and

$$|\nabla M| = \left(\sum_{i,j,l} \left(\frac{\partial}{\partial x_j} M_i^l \right)^2 \right)^{1/2} , \quad \|\nabla M\|_p = \left(\int_\Omega |\nabla M|^p \right)^{1/p} .$$

Rewriting (34), we now consider the system $(l = 1, \cdots, n)$ of equations

$$k\Phi_l + \gamma \frac{\partial}{\partial x_i} (\Phi_l v_i) + \gamma M_l^i \Phi_i = E_l \, , \tag{38}$$

for arbitrary $E \in L^2(\Omega)$, and refer to Φ as an L^2-*solution* of (38) if and only if $\Phi \in L^2(\Omega)$ and

$$\left(\Phi_l \, , k\Psi_l - \gamma v_i \frac{\partial \Psi_l}{\partial x_i} + \gamma M_i^l \Psi_i \right) = (E_l \, , \Psi_l) \; , \qquad \forall \Psi \in W_0^{1,2}(\Omega) \, . \tag{39}$$

It is evident from (39) that the adjoint of (38) consists of the system $(l = 1, \cdot\cdot, n)$ of equations

$$k\Phi_l - \gamma v_i \frac{\partial \Phi_l}{\partial x_i} + \gamma M_i^l \Phi_i = E_l \, , \tag{40}$$

and that we should refer to Φ as an L^2-*solution* of (40) if and only if $\Phi \in L^2(\Omega)$ and

$$\left(\Phi_l, k\Psi_l + \gamma \frac{\partial}{\partial x_i} (\Psi_l v_i) + \gamma M_l^i \Psi_i \right) = (E_i, \Psi_l) \; , \qquad \forall \Psi \in W_0^{1,2}(\Omega) \, . \tag{41}$$

Similarly to (3), for $\Phi \in W^{1,2}(\Omega)$ we have the identity

$$\left(\frac{\partial}{\partial x_i} (\Phi_l v_i) \, , \Phi_l \right) = - \left(v_i \frac{\partial \Phi_l}{\partial x_i}, \Phi_l \right) = \tfrac{1}{2} \int_\Omega \Phi^2 \nabla \cdot v \, dx \, . \tag{42}$$

Therefore, analogously to (5), solutions $\Phi \in W^{1,2}(\Omega)$ of either (38) or (40) satisfy

$$\left(k - \tfrac{\gamma}{2} \|\nabla \cdot v\|_\infty - \gamma \|\mathcal{M}\|_\infty\right) \|\Phi\| \le \|E\| \; , \tag{43}$$

and are consequently unique if $k \ge \frac{\gamma}{2} \|\nabla \cdot v\|_\infty + \gamma \|\mathcal{M}\|_\infty$.

Theorem 5. *If* $k > \frac{\gamma}{2} \|\nabla \cdot v\|_\infty + \gamma \|\mathcal{M}\|_\infty$, *then, for every vector valued* $E \in L^2(\Omega)$, *the equations (38) and (40) possess* L^2*-solutions. The solutions which are constructed here satisfy the estimate (43). They also satisfy the identities (39) and (41), in the weak formulations of the equations, for all test functions* $\Psi \in W^{1,2}(\Omega)$.

Proof. We argue similarly to the proof of Theorem 1, using the R^n-valued eigenfunctions $\{A^1, A^2, \cdots\}$ and corresponding eigenvalues $\{\lambda_1, \lambda_2, \cdots\}$ of the vector-valued Neumann problem

$$-\triangle A = \lambda A \quad , \quad \left. \frac{\partial A}{\partial n} \right|_{\partial\Omega} = 0 \, . \tag{44}$$

In dealing with the equation (38), we take $\Phi^n = \sum_{k=1}^{n} c_k A^k$ to be the unique solution of the system of linear equations

$$\left(k\Phi_l^n + \gamma \frac{\partial}{\partial x_i} (\Phi_l^n v_i) + \gamma M_l^i \Phi_i^n, A_l^k \right) = \left(E_l \, , A_l^k\right), \qquad k = 1, 2, \cdots, n \, , \tag{45}$$

which, integrating by parts, can be rewritten as

$$\left(\Phi_l^n\,, kA_l^k - \gamma v_i \frac{\partial A_l^k}{\partial x_i} + \gamma M_i^l A_i^k\right) = \left(E_l\,, A_l^k\right), \qquad k = 1, 2, \cdots, n. \tag{46}$$

Taking an appropriate linear combination of these equations, we obtain the estimate (43) for Φ^n, and complete the proof just as in Theorem 1. However, we note that the mean of Φ need not vanish when that of E does.

The existence of an L^2-solution of the adjoint equation (40) is proved similarly, taking $\Phi^n = \sum_{k=1}^{n} c_k A^k$ to be the unique solution of the system of linear equations

$$\left(k\Phi_l^n - \gamma v_i \frac{\partial \Phi_l^n}{\partial x_i} + \gamma M_i^l \Phi_i^n, A_l^k\right) = \left(E_l, A_l^k\right)\,, \qquad k = 1, 2, \cdots, n, \tag{47}$$

and proceeding as before. This completes the proof.

Theorem 6. *If $k > \frac{\gamma}{2}\|\nabla\cdot v\|_\infty + \gamma\|Dv\|_\infty + \gamma\|\mathcal{M}\|_\infty$ and $\|\nabla M\|_p < \infty$, then, for every $E \in W^{1,2}(\Omega)$, the equations (38) and (40) each possess a unique solution $\Phi \in W^{1,2}(\Omega)$. For either equation, $\Phi \in W_0^{1,2}(\Omega)$ if $E \in W_0^{1,2}(\Omega)$. For either equation, the weaker assumption that $k > \frac{\gamma}{2}\|\nabla\cdot v\|_\infty + \gamma\|\mathcal{M}\|_\infty$ suffices to imply uniqueness within the class of solutions that belong to $W^{1,2}(\Omega)$. The solutions of the equations (38) and (40) satisfy, respectively, the estimates (55) and (52) below, which are both of the form*

$$\|\nabla\Phi\| \le c\,\|E\|_{1,2}\,. \tag{48}$$

Proof. We consider first the adjoint equation (40), and begin by seeking a formal estimate for $\nabla\Phi$, by taking the $\partial/\partial x_j$ derivative of (40), multiplying through by $\partial\Phi_l/\partial x_j$, and integrating over Ω. Using the identity

$$\left(\frac{\partial}{\partial x_j}\left(v_i\frac{\partial\Phi_l}{\partial x_i}\right), \frac{\partial\Phi_l}{\partial x_j}\right) = -\tfrac{1}{2}\int_\Omega |\nabla\Phi|^2\,\nabla\cdot v\,dx + \left(\frac{\partial\Phi_l}{\partial x_i}\frac{\partial v_i}{\partial x_j}, \frac{\partial\Phi_l}{\partial x_j}\right), \tag{49}$$

we obtain

$$\begin{aligned} k\|\nabla\Phi\|^2 = {} & (\nabla E, \nabla\Phi) - \tfrac{\gamma}{2}\textstyle\int_\Omega |\nabla\Phi|^2\,\nabla\cdot v\,dx + \gamma\left(\dfrac{\partial\Phi_l}{\partial x_i}\dfrac{\partial v_i}{\partial x_j}, \dfrac{\partial\Phi_l}{\partial x_j}\right) \\ & + \gamma\left(\frac{\partial\Phi_i}{\partial x_j}M_j^l, \frac{\partial\Phi_l}{\partial x_j}\right) + \gamma\left(\Phi_i\frac{\partial}{\partial x_j}M_i^l, \frac{\partial\Phi_l}{\partial x_j}\right), \end{aligned} \tag{50}$$

and hence

$$\begin{aligned} k\|\nabla\Phi\|^2 \le {} & \|\nabla E\|\,\|\nabla\Phi\| + \tfrac{\gamma}{2}\|\nabla\cdot v\|_\infty\|\nabla\Phi\|^2 + \gamma\|Dv\|_\infty\|\nabla\Phi\|^2 \\ & + \gamma\|\mathcal{M}\|_\infty\|\nabla\Phi\|^2 + \gamma\|\nabla M\|_p\|\Phi\|_q\|\nabla\Phi\|\,, \end{aligned}$$

and

$$\left(k - \tfrac{\gamma}{2}\|\nabla\cdot v\|_\infty - \gamma\|Dv\|_\infty - \gamma\|\mathcal{M}\|_\infty\right)\|\nabla\Phi\| \le \|\nabla E\| + \gamma\|\nabla M\|_p\|\Phi\|_q\,, \tag{51}$$

where $p > n$ is a number fixed at the beginning of the last section, and $q = \left(\frac{1}{2} - \frac{1}{p}\right)^{-1}$. Exactly as in deriving (23) from (20) - (22), we now obtain

$$\left(k - \tfrac{\gamma}{2}\|\nabla \cdot v\|_\infty - \gamma\|Dv\|_\infty - \gamma\|\mathcal{M}\|_\infty - \epsilon\right)\|\nabla\Phi\| \leq \|\nabla E\| + \frac{C\left(\epsilon, \gamma\|\nabla M\|_p\right)}{\left(k - \frac{\gamma}{2}\|\nabla \cdot v\|_\infty - \gamma\|\mathcal{M}\|_\infty\right)}\|E\| \,, \tag{52}$$

for any $\epsilon > 0$. This estimate can be proven for the Galerkin approximations used in Theorems 5, since an appropriate linear combination of the equations (47) gives

$$\left(k\Phi_l^n - \gamma v_i \frac{\partial \Phi_l^n}{\partial x_i} + \gamma M_i^l \Phi_i^n, -\triangle\Phi_l^n\right) = \left(E_l, -\triangle\Phi_l^n\right)\,, \qquad k = 1, 2, \cdots, n, \tag{53}$$

which can be integrated by parts to yield analogues of (50) - (52). Again, this integration by parts is possible because Φ^n satisfies the boundary condition for the Neuman problem. Thus we obtain $\Phi \in W^{1,2}(\Omega)$ if $E \in W^{1,2}(\Omega)$. Again, if $E \in W_0^{1,2}(\Omega)$, we obtain $\Phi \in W_0^{1,2}(\Omega)$ by using the eigenfunctions of the Dirichlet problem, as in Theorems 2. This completes the proof for the adjoint equation.

We turn now to the transport equation (38), seeking first a formal estimate for $\nabla\Phi$ by taking the $\partial/\partial x_j$ derivative of the equation, multiplying through by $\partial\Phi_l/\partial x_j$, and integrating over Ω. Using the identity

$$\left(\frac{\partial}{\partial x_j}\frac{\partial}{\partial x_i}(\Phi_l v_i), \frac{\partial\Phi_l}{\partial x_j}\right) = \tfrac{1}{2}\int_\Omega |\nabla\Phi|^2\,\nabla\cdot v\,dx + \left(\frac{\partial\Phi_l}{\partial x_i}\frac{\partial v_i}{\partial x_j}, \frac{\partial\Phi_l}{\partial x_j}\right) + \left(\Phi_l\frac{\partial^2 v_i}{\partial x_j\partial x_i}, \frac{\partial\Phi_l}{\partial x_j}\right),$$

we obtain

$$\begin{aligned} k\|\nabla\Phi\|^2 = \; & -\tfrac{\gamma}{2}\int_\Omega |\nabla\Phi|^2\nabla\cdot v\,dx - \gamma\left(\frac{\partial\Phi_l}{\partial x_i}\frac{\partial v_i}{\partial x_j}, \frac{\partial\Phi_l}{\partial x_j}\right) - \gamma\left(\Phi_l\frac{\partial^2 v_i}{\partial x_j\partial x_i}, \frac{\partial\Phi_l}{\partial x_j}\right) \\ & -\gamma\left(\frac{\partial\Phi_i}{\partial x_j}M_l^i, \frac{\partial\Phi_l}{\partial x_j}\right) - \gamma\left(\Phi_i\frac{\partial}{\partial x_j}M_l^i, \frac{\partial\Phi_l}{\partial x_j}\right) + \left(\frac{\partial E_l}{\partial x_j}, \frac{\partial\Phi_l}{\partial x_j}\right). \end{aligned} \tag{54}$$

Thus,

$$\begin{aligned} k\|\nabla\Phi\|^2 \leq \; & \tfrac{\gamma}{2}\|\nabla\cdot v\|_\infty\|\nabla\Phi\|^2 + \gamma\|Dv\|_\infty\|\nabla\Phi\|^2 + \gamma\|\nabla\nabla\cdot v\|_p\|\Phi\|_q\|\nabla\Phi\| \\ & \gamma\|\mathcal{M}\|_\infty\|\nabla\Phi\|^2 + \gamma\|\nabla M\|_p\|\Phi\|_q\|\nabla\Phi\| + \|\nabla E\|\,\|\nabla\Phi\|\,, \end{aligned}$$

where as before $p > n$ and $q = \left(\frac{1}{2} - \frac{1}{p}\right)^{-1}$. Hence

$$\left(k - \tfrac{\gamma}{2}\|\nabla\cdot v\|_\infty - \gamma\|Dv\|_\infty - \gamma\|\mathcal{M}\|_\infty\right)\|\nabla\Phi\| \leq \gamma\|\nabla\nabla\cdot v\|_p\|\Phi\|_q + \gamma\|\nabla M\|_p\|\Phi\|_q + \|\nabla E\|\,,$$

and therefore, using again (21) and (22), we obtain

$$\left(k - \tfrac{\gamma}{2}\left\|\nabla\cdot v\right\|_\infty - \gamma\left\|Dv\right\|_\infty - \gamma\left\|\mathcal{M}\right\|_\infty - \epsilon\right)\left\|\nabla\Phi\right\| \\ \leq \left\|\nabla E\right\| + \frac{C\left(\epsilon,\, \gamma\left\|\nabla\nabla\cdot v\right\|_p,\, \gamma\left\|\nabla M\right\|_p\right)}{\left(k - \frac{\gamma}{2}\left\|\nabla\cdot v\right\|_\infty - \gamma\left\|\mathcal{M}\right\|_\infty\right)}\left\|E\right\| . \tag{55}$$

The proof is now completed as in the case of the adjoint equation, by showing that this estimate holds for the Galerkin approximations of Theorem 5, and proceeding as in Theorem 2.

Theorem 7. *If $k \geq \frac{\gamma}{2}\left\|\nabla\cdot v\right\|_\infty + \gamma\left\|Dv\right\|_\infty + \gamma\left\|\mathcal{M}\right\|_\infty$ and $\left\|\nabla M\right\|_p < \infty$, then, for any $E \in L^2(\Omega)$, there is at most one L^2-solution of either the transport equation (38), or its adjoint (40).*

Proof. In view of (39), the difference Φ of two L^2-solutions of (38) satisfies

$$\left(\Phi_l\,, k\Psi_l - \gamma v_i\frac{\partial\Psi_l}{\partial x_i} + \gamma M_i^l\Psi_i\right) = 0\,, \qquad \forall\Psi \in W_0^{1,2}(\Omega)$$

Thus, $\Phi = 0$ if functions with l^{th} components of the form $k\Psi_l - \gamma v_i\left(\partial\Psi_l/\partial x_i\right) + \gamma M_i^l\Psi_i$, for $\Psi \in W_0^{1,2}(\Omega)$, are dense in the vector-valued space $L^2(\Omega)$. This is guaranteed by Theorem 6. Uniqueness for the adjoint equation (40) is proven similarly.

Theorem 8. *Suppose that $k \geq \frac{\gamma}{2}\left\|\nabla\cdot v\right\|_\infty + \gamma\left\|Dv\right\|_\infty + \gamma\left\|\mathcal{M}\right\|_\infty$, that $\left\|\nabla M\right\|_p < \infty$, and that $k > \gamma\left\|\nabla\cdot v\right\|_\infty + \gamma\left\|\mathcal{M}\right\|_\infty$. Let Φ be an L^2-solution of either the transport equation (38) or its adjoint (40), with right side $E \in L^q(\Omega)$, for some q satisfying $2 \leq q \leq \infty$. Let $c_q = \left(k - \frac{\gamma}{q'}\left\|\nabla\cdot v\right\|_\infty - \gamma\left\|\mathcal{M}\right\|_\infty\right)^{-1}$, where $q' = q/(q-1)$. Also let $c_{q'} = \left(k - \frac{\gamma}{q}\left\|\nabla\cdot v\right\|_\infty - \gamma\left\|\mathcal{M}\right\|_\infty\right)^{-1}$, $c_\infty = \left(k - \gamma\left\|\nabla\cdot v\right\|_\infty - \gamma\left\|\mathcal{M}\right\|_\infty\right)^{-1}$, and $c_{\infty'} = \left(k - \gamma\left\|\mathcal{M}\right\|_\infty\right)^{-1}$. Then the solution of the transport equation satisfies*

$$\left\|\Phi\right\|_q \leq c_q\left\|E\right\|_q\,, \tag{56}$$

while the solution of its adjoint equation satisfies

$$\left\|\Phi\right\|_q \leq c_{q'}\left\|E\right\|_q\,. \tag{57}$$

Proof. As remarked in Theorem 4, the second hypothesis on k, which ensures that the coefficients $c_q, c_{q'}, c_\infty$ are well defined, is superfluous if $n \leq 4$, since $\sqrt{n}\left\|Dv\right\|_\infty \geq \left\|\nabla\cdot v\right\|_\infty$.

Now, given any vector valued $F \in C_0^\infty(\Omega)$, Theorem 6 provides a solution $\Psi^F \in W_0^{1,2}(\Omega)$ of the adjoint equations

$$k\Psi_l^F - \gamma v_i\frac{\partial\Psi_l^F}{\partial x_i} + \gamma M_i^l\Psi_i^F = F_l\,. \tag{58}$$

For q' satisfying $1 < q' < 2$, this solution satisfies the estimate

$$\left\|\Psi^F\right\|_{q'} \leq c_q \left\|F\right\|_{q'} . \tag{59}$$

Indeed, given any functions $F, \Psi \in W^{1,2}(\Omega)$ which satisfy the adjoint equations, we can multiply the equations through by $\Psi_l \left|\Psi\right|^{q'-2}$, and integrate over Ω, to obtain

$$k \left\|\Psi\right\|_{q'}^{q'} - \frac{\gamma}{q'} \int_\Omega v \cdot \nabla \left|\Psi\right|^{q'} dx + \gamma \int_\Omega M_i^l \Psi_i \Psi_l \left|\Psi\right|^{q'-2} dx = \left(F, \Psi \left|\Psi\right|^{q'-2}\right),$$

and hence

$$\left(k - \frac{\gamma}{q'} \left\|\nabla \cdot v\right\|_\infty - \gamma \left\|\mathcal{M}\right\|_\infty\right) \left\|\Psi\right\|_{q'}^{q'} \leq \left\|F\right\|_{q'} \left\|\Psi\right\|_{q'}^{q'-1},$$

and therefore $\left\|\Psi\right\|_{q'} \leq c_q \left\|F\right\|_{q'}$.

Now, taking q', F and Ψ^F as above, and using (39) and (59), we obtain

$$\begin{aligned} \left\|\Phi\right\|_q &= \sup_{F \in C_0^\infty(\Omega)} \frac{\left|(\Phi, F)\right|}{\left\|F\right\|_{q'}} = \sup_{F \in C_0^\infty(\Omega)} \frac{\left|\left(\Phi_l, k\Psi_l^F - \gamma v \cdot \nabla \Psi_l^F + \gamma M_i^l \Psi_i^F\right)\right|}{\left\|F\right\|_{q'}} \\ &= \sup_{F \in C_0^\infty(\Omega)} \frac{\left|\left(E, \Psi^F\right)\right|}{\left\|F\right\|_{q'}} \leq \sup_{F \in C_0^\infty(\Omega)} \frac{\left\|E\right\|_q \left\|\Psi^F\right\|_{q'}}{\left\|F\right\|_{q'}} \leq c_q \left\|E\right\|_q . \end{aligned} \tag{60}$$

This completes the proof for $2 < q < \infty$. For $q = \infty$ and $q' = 1$, we obtain (56) and (57) by passing to the limit. The case $q = q' = 2$ was proven in Theorem 5.

The adjoint equation is treated similarly. The only difference is the matter we dealt with in proving (33), in the scalar case.

4. Estimates for $\left\|\sigma\right\|_{2,2}$, $\left\|\triangle\sigma\right\|$ and $\left\|\triangle\sigma\right\|_{-1,2}$

The following theorem completes (together with Theorem 2) the preliminary results that are needed in [11] for the L^2 version of our work on steady compressible viscous flow. Until now, we have been working under the assumption that $v \in W^{2,p}(\Omega)$ for some $p > n$, without any restriction on number of spatial dimensions. But in this theorem we need to strengthen our assumptions, to show, for instance, that $\sigma\nabla\nabla\nabla v \in L^2(\Omega)$. The assumptions made in Lemma 3 of [11], that $n \leq 3$ and $v \in W^{3,2}(\Omega)$, suffice for this purpose, and will be made here. In the next section we will reprove the first part of this theorem, for arbitrary $n \geq 1$, under the assumption that $v \in W^{3,p}(\Omega)$ for some $p > n$. This assumption on v, however, would be too strong for our application to [11].

Theorem 9. *Suppose that $n \leq 3$, $v \in W^{3,2}(\Omega)$, and $k > \frac{\gamma}{2} \left\|\nabla \cdot v\right\|_\infty + 2\gamma \left\|Dv\right\|_\infty$. Then, for any prescribed right side $H \in W^{2,2}(\Omega)$, there exists a unique solution $\sigma \in W^{2,2}(\Omega)$ of either the scalar transport equation (1) or its adjoint (2), and it satisfies*

$$\left\|\sigma\right\|_{2,2} \leq c \left\|H\right\|_{2,2}, \tag{61}$$

$$\|\triangle\sigma\| \le c\|\triangle H\| + c\|v\|_{3,2}\|H\|_{2,2}\,, \tag{62}$$

$$\|\triangle\sigma\|_{-1,2} \le c\|\triangle H\|_{-1,2} + c\|v\|_{3,2}\|H\|_{1,2}\,, \tag{63}$$

with constants c that can be bounded above (monotonically) in terms of various Sobolev constants for Ω, γ, $\|v\|_{3,2}$, *and* $\left(k - \frac{\gamma}{2}\|\nabla\cdot v\|_\infty - 2\gamma\|Dv\|_\infty\right)^{-1}$.

Proof. Let us consider first the transport equation (1). Theorem 2 ensures the existence of a unique solution $\sigma \in W^{1,2}(\Omega)$. Since $n \le 3$ and $H \in W^{2,2}(\Omega)$, we have $H \in L^\infty(\Omega)$. Therefore $\sigma \in L^\infty(\Omega)$, by Theorem 4. Hence E defined by (35) satisfies $\nabla E = \nabla\nabla H + \gamma\nabla\sigma\nabla\nabla v + \gamma\sigma\nabla\nabla\nabla v \in L^2(\Omega)$, and hence $E \in W^{1,2}(\Omega)$. Consequently Theorem 6, with M defined by (37), provides a solution $\Phi \in W^{1,2}(\Omega)$ of the vector transport equation (34). Since $\nabla\sigma$ is an L^2-solution of (34), as was proven following the definition (36) of L^2-solutions, we may conclude that $\nabla\sigma = \Phi$ by the uniqueness of L^2-solutions proven in Theorem 7. Thus $\nabla\sigma \in W^{1,2}(\Omega)$ and $\sigma \in W^{2,2}(\Omega)$. Explicit estimates accompany the forgoing arguments, which together imply (61).

Consider now the adjoint equation (2). Again, Theorem 2 provides a solution $\sigma \in W^{1,2}(\Omega)$. Its gradient is an L^2-solution of (40), if we take

$$M_i^l = -\frac{\partial v_i}{\partial x_l} \quad \text{and} \quad E_l = \frac{\partial H}{\partial x_l}\,. \tag{64}$$

This right side E clearly belongs to $W^{1,2}(\Omega)$, and this M, which is the negative transpose of that previously considered, satisfies the same estimates as before. The proof of (61) is now completed for the adjoint equation as it was for the transport equation.

To prove (62) for the transport equation, having proved that $\sigma \in W^{2,2}(\Omega)$, we observe that $\triangle\sigma$ is an L^2-solution of the transport equation (1), with right side $F = \triangle H - \gamma\nabla\cdot(\sigma\triangle v + 2\nabla\sigma\cdot\nabla v) \in L^2(\Omega)$. This is verified by taking $\triangle\psi$ as the test function in (6), with $\psi \in C_0^\infty(\Omega)$, obtaining

$$(\triangle\sigma, k\psi - \gamma v\cdot\nabla\psi) = (\triangle H - \gamma\nabla\cdot(\sigma\triangle v + 2\nabla\sigma\cdot\nabla v), \psi)\,,$$

which is then seen to hold for all test functions $\psi \in W_0^{1,2}(\Omega)$ by taking a limit. By Theorem 3, $\triangle\sigma$ can be identified with the solution constructed in Theorem 1, which satisfies the estimate (5) with right side F. Thus

$$\begin{aligned}\left(k - \tfrac{\gamma}{2}\|\nabla\cdot v\|_\infty\right)\|\triangle\sigma\| &\le \|\triangle H - \gamma\nabla\cdot(\sigma\triangle v + 2\nabla\sigma\cdot\nabla v)\| \\ &\le \|\triangle H\| + \gamma c\|v\|_{3,2}\|\sigma\|_{2,2} \\ &\le \|\triangle H\| + \gamma c\|v\|_{3,2}\|H\|_{2,2}\,,\end{aligned}$$

using Sobolev's inequalities and (61). This proves (62) for the transport equation. If σ is a solution of the adjoint equation, then $\triangle\sigma$ is an L^2-solution of the adjoint equation with right side $F = \triangle H + \gamma\triangle v\cdot\nabla\sigma + 2\gamma\nabla v : \nabla\nabla\sigma$, and we proceed as before.

To prove (63) for the transport equation, notice first that $\nabla\sigma$ actually satisfies (36) for all test functions $\Psi \in W^{1,2}(\Omega)$, since it can be identified by Theorem 7 with the solution Φ constructed in Theorem 5.

For $\Phi = \nabla\sigma$ and test functions of the form $\Psi = \nabla\psi$, with $\psi \in W_0^{1,2}(\Omega) \cap W^{2,2}(\Omega)$, we can rewrite (36) as

$$\begin{aligned}\left(\frac{\partial\sigma}{\partial x_l}, \frac{\partial}{\partial x_l}\left[k\psi - \gamma v_i \frac{\partial\psi}{\partial x_i}\right]\right) = &\left(\frac{\partial H}{\partial x_l}, \frac{\partial\psi}{\partial x_l}\right) - \gamma\left(\sigma\frac{\partial^2 v_i}{\partial x_l \partial x_i}, \frac{\partial\psi}{\partial x_l}\right) \\ &- \gamma\left(\frac{\partial\sigma}{\partial x_l}, \left(\frac{\partial v_i}{\partial x_l} + \frac{\partial v_l}{\partial x_i}\right)\frac{\partial\psi}{\partial x_i}\right).\end{aligned} \tag{65}$$

Now, for arbitrary $h \in C_0^\infty(\Omega)$, the solution of the scalar adjoint equation

$$k\psi^h - \gamma v_i \frac{\partial\psi^h}{\partial x_i} = h$$

satisfies $\psi^h \in W_0^{1,2}(\Omega) \cap W^{2,2}(\Omega)$ and

$$\|\nabla\psi^h\| \le c\,\|\nabla h\|\ , \tag{66}$$

in virtue of Theorem 2 and the first part of this present theorem. Hence we can insert it into (65), and use (5), (14) and (66) to obtain

$$\begin{aligned}|(\nabla\sigma, \nabla h)| &\le \left|(\nabla H, \nabla\psi^h)\right| + c\gamma\,\|\sigma\|_{1,2}\,\|v\|_{3,2}\,\|\nabla\psi^h\| \\ &\le \left|(\nabla H, \nabla\psi^h)\right| + c\gamma\,\|H\|_{1,2}\,\|v\|_{3,2}\,\|\nabla h\|\ ,\end{aligned}$$

and therefore

$$\begin{aligned}\|\triangle\sigma\|_{-1,2} &= \sup_{h\in C_0^\infty(\Omega)} \frac{|(\nabla\sigma, \nabla h)|}{\|\nabla h\|} \\ &\le c \sup_{h\in C_0^\infty(\Omega)} \frac{\left|(\nabla H, \nabla\psi^h)\right|}{\|\nabla\psi^h\|} + c\gamma\,\|H\|_{1,2}\,\|v\|_{3,2} \\ &\le c\,\|\triangle H\|_{-1,2} + c\gamma\,\|v\|_{3,2}\,\|H\|_{1,2}\ .\end{aligned}$$

This proves (63) for the transport equation. The proof for the adjoint equation is similar.

5. Existence in $W^{m,2}(\Omega)$, for any fixed m

We wish to show here that the proof of the first part of Theorem 9 can be iterated, to give an existence theorem for solutions belonging to $W^{m,2}(\Omega)$, for any fixed $m \ge 0$. This result is not needed in [11], and therefore we prefer to return to the consideration of an arbitrary number of spatial dimensions, although this requires a further strengthening of our assumption about v, namely to $v \in W^{m+1,p}(\Omega)$, for some $p > n$. The first conclusion of Theorem 9, that σ belongs to $W^{2,2}(\Omega)$ and satisfies (61), is easily obtained under these modified assumptions. In particular, for $\sigma \in W^{1,2}(\Omega)$ and $v \in W^{3,p}(\Omega)$, we obtain $\sigma\nabla\nabla\nabla v \in L^2(\Omega)$ using standard Sobolev imbeddings.

In Theorem 9, we began by showing that if $\sigma \in W^{1,2}(\Omega)$ is a solution of the transport equation, then $\nabla\sigma$ is an L^2-solution of the vector transport equation (38). Now, we need to see that if $\Phi \in W^{1,2}(\Omega)$ is a solution of the vector transport equation (38), then $\nabla\Phi$ is an L^2-solution of another vector transport equation of the same form as (38). This requires a slight generalization of our previous definition of vector transport equation. In first writing the vector transport equation (38), with a special interest in the case $\Phi = \nabla\sigma$, we considered Φ to be a vector with n components Φ_l. In fact, nothing of the theory of the vector transport equation given in Theorems 5–8 is changed by allowing Φ to be a vector with N components, for any $N \geq 1$. From now on, in referring to the vector transport equation, it will be understood that Φ can have an arbitrary number of components.

If Φ is a solution of a vector transport equation (38) with N components, and sufficiently regular, then $\Gamma = \nabla\Phi$ is a solution of a vector transport equation with Nn components. Setting $\Gamma_{lj} = \partial\Phi_l/\partial x_j$, $l = 1,\cdots,N$, $j = 1,\cdots,n$, this second equation is

$$\begin{aligned} k\Gamma_{lj} + \gamma\frac{\partial}{\partial x_i}(\Gamma_{lj}v_i) + \gamma M_l^i\Gamma_{ij} + \gamma\frac{\partial v_i}{\partial x_j}\Gamma_{li} \\ = \frac{\partial E_l}{\partial x_j} - \gamma\Phi_l\frac{\partial^2 v_i}{\partial x_i \partial x_k} - \gamma\frac{\partial M_l^i}{\partial x_j}\Phi_i\,. \end{aligned} \tag{67}$$

This is again of the form (38), since Γ_{li} can be regarded as a vector with Nn components, in which case the third and fourth terms on the left can be combined into a single matrix-vector multiplication, with a $Nn \times Mn$ matrix $\overline{M}$. As before, we make rigorous that $\Gamma = \nabla\Phi$ is an L^2-solution if $\Phi \in W^{1,2}(\Omega)$, and that by uniqueness this solution coincides with a solution $\Gamma \in W^{1,2}(\Omega)$ constructed in Theorem 6. It is desirable in these applications of Theorems 6 and 7 to give the hypotheses directly in terms of v and M, rather than $\overline{M}$. By reconsidering the basic estimates for (67), considering (67) as an $N \times n$ matrix equation, it is easily seen that the required smallness condition is

$$k > \frac{\gamma}{2}\|\nabla \cdot v\|_\infty + 2\gamma\|Dv\|_\infty + \gamma\|\mathcal{M}\|_\infty\,, \tag{68}$$

which is used along with the assumption that $\|\nabla M\|_p < \infty$. A seemingly unfortunate implication of this is that we will need to assume successively stronger smallness conditions to obtain solutions in $W^{m,2}(\Omega)$ for successively greater values of m.

If $E \in W^{2,2}(\Omega)$, then the right side of (67) will belong to $W^{1,2}(\Omega)$ when $\Phi \in W^{1,2}(\Omega)$, provided we assume that $v \in W^{3,p}(\Omega)$ and $M \in W^{2,p}(\Omega)$, for some $p > n$. We therefore have the following theorem.

Theorem 10. *Consider the vector transport equation (38), or its adjoint (40), in an n-dimensional domain Ω, for an N-component solution Φ, for any $n, N \geq 1$.*

Suppose that $v \in W^{3,p}(\Omega)$ and $M \in W^{2,p}(\Omega)$, for some $p > n$. Suppose also that the condition (68) is satisfied. Then, for any prescribed right side $E \in W^{2,2}(\Omega)$, there exists a unique solution $\Phi \in W^{2,2}(\Omega)$, and it satisfies

$$\|\Phi\|_{2,2} \leq c\|E\|_{2,2} ,$$

where c depends only $\left(k - \frac{\gamma}{2}\|\nabla \cdot v\|_\infty - 2\gamma\|Dv\|_\infty - \gamma\|\mathcal{M}\|_\infty\right)^{-1}$, $\|v\|_{3,p}$, $\|M\|_{2,p}$, γ and several Sobolev constants for Ω.

It is a straightforward matter to apply this theorem recursively to the original scalar transport equation (1), to prove that $\sigma \in W^{m,2}(\Omega)$ when $H \in W^{m,2}(\Omega)$, provided certain other hypotheses are satisfied. As pointed out above, the smallness condition becomes stronger with each iteration. Therefore, we can only consider fixed finite values of m. In proving $\sigma \in W^{m,2}(\Omega)$, the most troublesome term on the right side of the equation to be dealt with is of the form $\sigma\nabla^{m+1}v$, which will belong to $L^2(\Omega)$ if $v \in W^{m+1,p}(\Omega)$, for some $p > n$. Alternative hypotheses are also possible, as in Theorem 9. Through these reasonings we have, for example, the following theorem.

Theorem 11. *Consider the scalar transport equation (1) in an n-dimensional domain Ω. Suppose that $k > \frac{\gamma}{2}\|\nabla \cdot v\|_\infty + m\gamma\|Dv\|_\infty$, and that $v \in W^{m+1,p}(\Omega)$, for some $p > n$. Then, for any prescribed right side $H \in W^{m,2}(\Omega)$, there exists a unique solution $\sigma \in W^{m,2}(\Omega)$, and it satisfies*

$$\|\sigma\|_{m,2} \leq c\|H\|_{m,2} ,$$

where c depends only on $\left(k - \frac{\gamma}{2}\|\nabla \cdot v\|_\infty - m\gamma\|Dv\|_\infty\right)^{-1}$, $\|v\|_{m+1,p}$, γ and certain Sobolev constants for Ω.

6. Integration along characteristics

In this section we make some observations of a heuristic nature, stemming from the theory of characteristics. We will consider the transport equation here in the normalized form

$$v \cdot \nabla\sigma + (\nabla \cdot v + k)\,\sigma = H . \tag{69}$$

Wherever v is smooth and $|v| > 0$, this amounts to an ordinary differential equation for σ along the integral curves of v, referred to as characteristics. If we divide by $|v|$, and denote arc length along such curves by s, then the equation (69) can be written as

$$\frac{\partial\sigma}{\partial s} = -\frac{\nabla \cdot v + k}{|v|}\sigma + \frac{H}{|v|} , \tag{70}$$

the solution of which, along a given characteristic, is

$$\sigma(s) = \exp(-G(s))\left\{\sigma(s_0) + \int_{s_0}^{s} \exp(G(\tau))\frac{H(\tau)}{|v(\tau)|}\,d\tau\right\}, \tag{71}$$

where

$$G(s) = \int_{s_0}^{s} \frac{\nabla \cdot v + k}{|v|}\, d\tau \tag{72}$$

is an integral along the characteristic from an arbitrarily chosen point on it, where $s = s_0$. The problems caused by possible zeros of v will not be discussed here, but they seem not insurmountable.

Let us consider several situations. First, it is clear that if Ω is a "flow through domain", completely covered by characteristics entering the domain along a portion Γ of the boundary where $v \cdot n|_\Gamma < 0$, then σ is determined everywhere by prescribing σ along Γ. More interesting is the case of what might be described as a "flow through step domain", with some characteristics entering and leaving as before, but also with an eddy behind the step where the characteristics form closed curves. It may seem surprising at first, but the differential equation (70) can also be used to determine σ along these closed characteristics. In place of an initial condition at the point where a characteristic enters the domain, one uses the requirement that σ must be single valued. To determine σ at a point s_0 on a closed characteristic curve of length L, we observe that the requirement that $\sigma(s_0) = \sigma(s_0 + L)$ is expressed by

$$\sigma(s_0) = \exp(-G(s_0 + L)) \left\{ \sigma(s_0) + \int_{s_0}^{s_0+L} \exp(G(\tau)) \frac{H(\tau)}{|v(\tau)|} d\tau \right\} ,$$

which is uniquely solvable for $\sigma(s_0)$ if $G(s_0 + L) \neq 0$. In view of (72), this can be guaranteed by assuming that $\nabla \cdot v + k > 0$ everywhere along the curve, a condition similar to the smallness hypotheses in previous sections. It corresponds to $c_\infty < \infty$, in the estimate $\|\sigma\|_\infty \leq c_\infty \|H\|_\infty$ of Theorem 4. The condition required in Theorem 1 is weaker, being essentially that $\nabla \cdot v + 2k > 0$ should hold everywhere. Apparently L^2-solutions exist even in cases where the velocity field includes closed curves along which $G(s_0 + L) < 0$, and $\sigma = \infty$.

If the velocity field v is that of a compressible fluid flow, with density $\rho = 1+\sigma$, then, from the equation of continuity $v \cdot \nabla\rho = -\rho \nabla \cdot v$, we deduce that

$$\rho(s) = \exp\left(-\int_{s_0}^{s} \frac{\nabla \cdot v}{|v|} d\tau \right) \rho(s_0) ,$$

and therefore, since $\rho(s_0 + L) = \rho(s_0)$, that

$$\int_{s_0}^{s_0+L} \frac{\nabla \cdot v}{|v|} d\tau = 0 .$$

Hence, there must always hold $G(s_0 + L) > 0$ around closed integral curves, for the velocity field of a smooth steady compressible fluid.

The determination of σ by integration along characteristics may appear difficult to generalize to the case of characteristics which neither cross the boundary nor close upon themselves. However, if $\inf (\nabla \cdot v + k) / |v| > 0$, then, by (70), the influence of the value of σ at any given point decays exponentially as one goes forward along a characteristic. Hence, the value of σ at a specified point along a characteristic can be approximated by first tracing the characteristic backwards some distance, where one may choose a point s_0 at which to assign a value of $\sigma(s_0)$ more or less arbitrarily. Then integrating (70) forward from that point, the effect of this arbitrary value will be exponentially lost, resulting in a good value of σ at the originally specified point.

In previous sections, our theorems required successively stronger hypotheses concerning the velocity field, in order to prove successively higher regularity of the solution. To understand the necessity for this, consider the following 2-dimensional example. Let $v = (-\sin x, 1)$ in the square $\Omega = (-\pi, \pi) \times (0, L)$. Then the velocity field satisfies $v \cdot n|_\Gamma = 0$ along the lateral boundaries at $x = \pm\pi$, and is periodic at the upper and lower boundaries, nearly simulating flow around an annulus. We are interested in whether solutions σ of the transport equation

$$(k + \nabla \cdot v)\,\sigma + v \cdot \nabla\sigma = H \tag{73}$$

will be finite along the central characteristic at $x = 0$, and whether their derivatives σ_x, σ_{xx}, etc., will be finite there. For the given velocity field v, (73) becomes

$$(k - \cos x)\,\sigma + (-\sin x, 1) \cdot \nabla\sigma = H\,.$$

Along the y-axis, this reduces to

$$\frac{\partial \sigma}{\partial y} = -\,(k - \cos x)\,\sigma + H\,,$$

which is unstable (indicating that generally $\sigma = \infty$, depending on H), unless

$$k > \cos x = -\nabla \cdot v. \tag{74}$$

Now, the derivative σ_x satisfies the transport equation

$$(k - 2\cos x)\,\sigma_x + (-\sin x, 1) \cdot \nabla\sigma_x = H_x - (\sin x)\,\sigma\ ,$$

which reduces along the y-axis to

$$\frac{\partial \sigma_x}{\partial y} = -\,(k - 2\cos x)\,\sigma_x + (H_x - k\sigma)\ ,$$

which is unstable unless

$$k > 2\cos x = -2\nabla \cdot v. \tag{75}$$

Similarly for σ_{xx}, etc.. The progressively stronger hypotheses (74), (75), etc., are clearly related to those of our theorems in previous sections, especially Theorem 11.

Finally, we mention a heuristic argument for the density in $L^2(\Omega)$ of test functions of the form $k\psi - v \cdot \nabla\psi$, for $\psi \in W_0^{1,2}(\Omega)$. This was proven in Theorem 2, and subsequently used in proving the uniqueness of L^2-solutions in Theorem 3. To establish this density, it is enough to consider nonnegative test functions χ with very small supports. Consider such a test function χ with support so small that v

is almost constant in an extended neighborhood of it. Then we would like to find $\psi \in W_0^{1,2}(\Omega)$ such that $k\psi - v \cdot \nabla\psi = \chi$. By reversing the direction of v, we can consider instead the equation $v \cdot \nabla\psi = -k\psi + \chi$. Upstream of the support of χ, we can set $\psi = 0$. After passing through the support of χ, ψ decays exponentially along characteristics, so that ψ has sign and effective support similar to χ. If v goes through zero along a characteristic, then a heuristic argument shows that ψ tends to zero at the zero of v. Thus, while further arguments would have to be made, it seems possible that Theorem 2 could be approached through the theory of characteristics.

References

[1] Beirão da Veiga, H., An L^p-theory for the n-dimensional, stationary, compressible Navier-Stokes equations, and the incompressible limit for compressible fluids. The equilibrium solutions, Commun. Math. Phys., 109(1987), 229–248.

[2] Beirao da Veiga, H., On a stationary transport equation, Ann. Univ. Ferrara, 32(1986), 79–91.

[3] Beirao da Veiga, H., Existence results in Sobolev spaces for a stationary transport equation, Ricerche di Matematica suppl. 36(1987), 173–184.

[4] Beirao da Veiga, H., Boundary value problems for a class of first order partial differential equations in Sobolev spaces and applications to the Euler flow, Rend. Sem. Mat. Univ. Padova 79(1988), 247–273.

[5] Coscia, V. & G.P. Galdi, Existence, uniqueness and stability of regular motions of a second grade fluid, Int. J. Nonlinear Mechanics, 29(1994), 493–506.

[6] Farwig, R., Stationary solutions of compressible Navier-Stokes equations with slip boundary condition, Commun. In Partial Differential Equations, 14(11), (1989), 1579–1606.

[7] Fichera. G., Sulle equazioni differenziali lineari elliptico-paraboliche del secondo ordine, Atti Acad. Naz. Lincei, Mem. Cl. Sc. Fis. Mat. Nat., Sez. I, 5(1956), 1–30.

[8] Fichera. G., On a unified theory of boundary value problems for elliptic-parabolic equations of second order, in Boundary Problems. Differential Equations, Univ. of Wisconsin Press, Madison, Wisconsin, (1960), 97–120.

[9] Friedrichs, K.O., Symmetric positive linear differential equations, Comm. Pure Appl. Math., 11(1958), 333–418.

[10] Galdi, G.P., Mathematical theory of second-grade fluids, CISM Courses and Lectures No. 344, Stability And Wave Propagation In Fluids And Solids, G.P. Galdi editor, Springer-Verlag 1995, 67–104.

[11] Heywood, J.G. & M. Padula, On the existence and uniqueness theory for steady compressible viscous flow, this book, 1999.

[12] Kohn, J.J. & L. Nirenberg, Degenerate elliptic-parabolic equations of second order, Comm. Pure Appl. Math., 20(1967), 797–872.

[13] Lax, P.D. & R.S. Phillips, Local boundary conditions for dissipative symmetric linear differential operators, Comm. Pure Appl. Math., 13(1960), 427–455.

[14] Mizohata, S., The Theory of Partial Differential Equations, Cambridge Univ. Press, 1973.

[15] Novotný, A., Steady flows of viscous compressible flows L^2 approach, SAACM, 3(1993), 181–199.

[16] Novotný, A., Some remarks to the compactness of steady compressible isentropic Navier-Stokes equations via the decomposition method, Comment. Math. Univ. Carolina, 37(1996), 1–38.

[17] Novotný, A., About the steady transport equation, L^p approach in domains with smooth boundaries, Comm. Mat. Univ. Carolinae, 37(1996), 43–89.

[18] Novotný, A. & M. Padula, L^p-approach to steady flows of viscous compressible fluids in exterior domains, Arch. Rational Mech. Anal. 126(1994), 234–297.

[19] Oleinik, O.A., Linear equations of second order with nonnegative characteristic form, Amer. Math. Soc. Transl. (2) 65(1967), 167–199.

[20] Padula, M., An existence theorem for steady state compressible Navier-Stokes equations, Proc. Meeting "Waves and Stability in Continuous Media", Univ. Catania, 1981, 276–280.

[21] Padula, M., Existence and uniqueness for steady compressible flow, Proc. Meeting "Dinamica dei Continui Fluidi e dei gas ionizzati, Univ. Trieste, 1982, 237–258.

[22] Padula, M., Existence and uniqueness for viscous steady compressible motions, Arch. Rat. Mech. Anal., 97(1987), 89–102.

[23] Padula, M., A representation formula for steady solutions of a compressible fluid moving at low speed, Transport Theory Statistical Physics, 21(1992), 593–614.

[24] Padula, M., On the exterior steady problem for the equations of a viscous isothermal gas, Comm. Math. Carolinae, 34(1993), 275–293.

[25] Temam, R., On the Euler equations of incompressible perfect fluids, J. Fun. Anal., 20(1975), 32–43.

[26] Valli, A., On the existence of stationary solutions to compressible Navier-Stokes equations, Ann. Inst. H. Poincaré, Anal. Non Linéaire 4(1987), 99–113.

John G. Heywood
Department of Mathematics
University of British Columbia

Mariarosaria Padula
Dipartimento di Matematica
Università di Ferrara

On the Existence and Uniqueness Theory for Steady Compressible Viscous Flow*

John G. Heywood and Mariarosaria Padula

Abstract. We give a simplified approach to the existence theory for compressible viscous flow, based on a new iterative scheme. This paper introduces the scheme in the context of steady isothermal flow in a bounded domain with homogeneous boundary conditions. The solution that is obtained is shown to be locally unique, using estimates similar to those used in proving its existence. A second direct approach to uniqueness is also investigated. The analysis presented here is based on estimates for the velocity and density in $W^{3,2} \times W^{2,2}$. It appears to be possible to give a similar analysis based on estimates in $W^{2,p} \times W^{1,p}$, for $p > n$.

1. Introduction

In this paper we give a simplified approach to the existence theory for steady compressible viscous flow, based on a new iterative scheme. We intend to use it in future extensions of the theory, and also as the foundation for new computational procedures. To keep the presentation as simple as possible in the present paper, we have restricted our considerations to isothermal flow, and to homogeneous boundary value problems in bounded domains. It is also for simplicity that we have based the analysis on L^2-estimates, although we could have worked with derivatives one order lower by using L^p-estimates with $p > n$.

The scheme introduced here is a modification and refinement of that which was used in proving the first existence theorem for steady compressible viscous flow, in the papers [15], [16], [17] of Padula. The novelty of approach in these papers lay in linking a steady transport equation together with a steady Stokes problem. There have been a number of subsequent improvements in the theory, but every approach has involved some similar use of the steady transport equation. Particularly important among these works are the papers of Beirao da Veiga [2], Valli [22], Farwig [7], Padula [18], [19], Novotny [12], [13], and Novotny and Padula [14].

* This work has been supported by the Natural Sciences and Engineering Research Council of Canada, contracts MURST and the GNFM of Italian CNR, and the Deutsche Forschungsgemeinschaft.

One aspect of our approach, that we would draw special attention to, is that it is based on a single iteration between a Stokes problem and a transport equation. The schemes used by other authors have been based on nested double iterations between these and other related problems. We expect that the use of a single iteration will prove to be more practical for computational procedures. Recent numerical experiments of Golay and Helluy [10] are already indicative of this.

We have also had a pedagogical goal in writing this paper. It has been our intention to make it self contained and readable at a fairly elementary level, to provide an introduction to this subject which is suitable for students and for scientists from other fields. We particularly hope that it will be useful for specialists in computational fluid dynamics.

For this pedagogical purpose, it is clearly important to have a simple self contained existence theory for the steady transport equation that provides the specialized results which are needed here and in other works on compressible viscous flow. The papers [3], [4], [5] of Beirao da Veiga were written with this intention, and provide most of the results that we need. So does Farwig's approach to the transport equation given in [7]. Nevertheless, the approaches taken in these papers still involve some difficult points. Subsequently, a much simpler approach to the steady transport equation was given by Galdi [8] and by Coscia and Galdi [6], based on the use of a special orthonormal system that had been introduced for the study of the Euler equation by Temam [21]. This approach, however, is not well suited for our purposes as the estimates it is based on are altogether different from those that we need. Therefore, while writing this paper we have sought and found yet another approach to the steady transport equation. It provides the results we need by elementary arguments, with certain improvements with respect to our applications. Because of its independent interest, with applications to several fields, we are presenting this material on the steady transport equation in an accompanying self contained paper [11].

The main result of this paper is the proof of the convergence of the new iterative scheme. As a corollary, we obtain a new proof of existence for steady compressible viscous flow. This and an associated uniqueness theorem are described further in Section 3. Another direct approach to uniqueness is investigated in Section 9.

Our notation is standard. We mention only that L^2-norms are denoted simply by $\|\cdot\|$, while all other norms are distinguished by subscripts. For example, $\|\cdot\|_p$ denotes the norm for $L^p(\Omega)$, and $\|\cdot\|_{m,p}$ the norm for the Sobolev space $W^{m,p}(\Omega)$, which consists, of course, of all functions that possess derivatives in $L^p(\Omega)$ up to order m. Also, we do not distinguish through our notation between scalar and vector valued functions, or function spaces, or norms.

2. Poisson-Stokes equations for isothermal flow

The pressure in an ideal gas depends linearly on its mass density and temperature, $p = R\rho\theta$, where R is a constant. What follows is written under the "isothermal" assumption that $p = k\rho$, where k is a constant. Stokes' hypotheses imply that the linearization of the stress tensor for a compressible viscous fluid must be of the form $T = (-p + \lambda\nabla \cdot v) I + 2\mu Dv$, where $Dv = \frac{1}{2}\left(\nabla v + (\nabla v)^{\perp}\right)$. From the laws of thermodynamics it follows that the net dissipation of mechanical energy into heat must be positive in any cyclic deformation. It can be shown that this implies that the viscous coefficients must satisfy $\mu > 0$ and $3\lambda + 2\mu > 0$. What will be required below is somewhat less, namely that $\mu > 0$ and $\lambda + 2\mu > 0$. We will assume further that the fluid under consideration adheres to the boundary of the region it occupies, and that this region is a bounded subdomain of R^n, $n \le 3$, denoted by Ω. The initial boundary value problem for flow under these assumptions is

$$\begin{aligned} &\rho_t + \nabla \cdot (\rho v) = 0 \\ &\rho (v_t + v \cdot \nabla v) = -\nabla p + \mu \Delta v + (\lambda + \mu) \nabla\nabla \cdot v + \rho f \\ &p = k\rho \\ &v|_{\partial\Omega} = 0, \quad v(x,0) = v_0(x), \quad \rho(x,0) = \rho_0(x). \end{aligned} \tag{1}$$

It is appropriate to assume here that the initial density ρ_0 is everywhere positive. The corresponding stationary problem is

$$\begin{aligned} &\nabla \cdot (\rho v) = 0 \\ &\rho v \cdot \nabla v = -\nabla p + \mu \Delta v + (\lambda + \mu) \nabla\nabla \cdot v + \rho f \\ &p = k\rho \\ &v|_{\partial\Omega} = 0, \quad \textstyle\int_\Omega \rho\, dx = \overline{\rho}\, |\Omega| \;, \end{aligned} \tag{2}$$

where the mean density $\overline{\rho}$ must be prescribed, and must be positive.

3. Main result

Our main purpose in this paper is to present a new iterative scheme that we believe will be useful in the future as the basis of a general theory. Our main result is the proof of its convergence, which is given in Lemmas 6 and 7. By related methods, we also prove the local uniqueness of the resulting solution in Lemma 8. Together, these lemmas provide a simplified proof of existence and local uniqueness, for small data, which is summarized in the following theorem.

Theorem 1. *Let Ω be a bounded subdomain of R^n, $n \le 3$, with a boundary of class $C^{2,1}$. Let $f \in W^{1,2}(\Omega)$. Then, if $\|f\|_{1,2}$ is sufficiently small, depending on Ω, k, μ, λ and $\overline{\rho}$, there exists a locally unique solution v, ρ of problem (2) with $v \in W^{3,2}(\Omega) \cap W_0^{1,2}(\Omega)$, $\rho \in W^{2,2}(\Omega)$, and $\inf\limits_{\Omega} \rho > 0$.*

In Section 9 we offer another approach to the question of uniqueness, with results that are summarized there in Theorem 2.

The norms in Theorem 1 are rather heavy, requiring that $f \in W^{1,2}(\Omega)$ and yielding $v \in W^{3,2}(\Omega)$. However, we have structured most of the proof so that it can be easily modified by substituting other appropriate choices of the basic function spaces. We plan to present the details of these modifications in the future. The weakest spaces we can substitute result in the requirement that $f \in L^p(\Omega)$, for some $p > n$, where n is the dimension of Ω, and yield a velocity $v \in W^{2,p}(\Omega)$. Similar substitutions of stronger function spaces yield smoother solutions, provided the force is correspondingly smooth. We mention, however, that it remains an open problem to show that a solution which is obtained in one of these function spaces, with the regularity properties of the function space, is actually as smooth as would seem to be allowed by the force, if the force has some further higher order regularity; see [20]. The matter of proving the positivity of the density is another vexing problem in the theory of compressible viscous flow. Its assertion in our theorem is simply due to our assumption that the data is small, combined with an estimate for the perturbation of the density in a space of continuous functions.

4. Iterative scheme

Let $\sigma = \rho - \overline{\rho}$ denote the perturbation in the density ρ from the mean density $\overline{\rho}$. Normalizing the density by dividing ρ, p, μ and λ by $\overline{\rho}$, and relabeling, we may assume that $\overline{\rho} = 1$, so that $\sigma = \rho - 1$. Then the steady problem (2) can be rewritten as

$$\begin{aligned} &\nabla \cdot v = -\nabla \cdot (\sigma v) \\ &-\mu \Delta v = -\nabla p + (\lambda + \mu)\nabla \nabla \cdot v - (1+\sigma)(v \cdot \nabla v - f) \\ &p = k(1+\sigma) \\ &v|_{\partial\Omega} = 0\,, \quad \textstyle\int_\Omega \sigma\, dx = 0\,. \end{aligned} \tag{3}$$

This suggests a Stokes problem for iteration, with a prescribed nonzero divergence $g = -\nabla \cdot (\sigma v)$, a prescribed force $F = -(1+\sigma)(v \cdot \nabla v - f)$, and an unknown pressure $\pi = p - k - (\lambda + \mu)\nabla \cdot v$. We have included the constant $-k$ in this expression for π so that the normalization $\int_\Omega \pi\, dx = 0$ made in (5) below will be equivalent to the normalization of σ in (3).

We propose to solve (3) by an iterative scheme taking $v_n, \sigma_n \to v_{n+1}, \sigma_{n+1}$, beginning with $v_0 = 0$, $\sigma_0 = 0$. Given v_n, σ_n, we set

$$g_n = -\nabla \cdot (\sigma_n v_n)\,, \qquad F_n = -(1+\sigma_n)(v_n \cdot \nabla v_n - f)\,, \tag{4}$$

and find v_{n+1} and an auxiliary pressure like function π_{n+1} by solving the Stokes problem

$$\begin{aligned} &\nabla \cdot v_{n+1} = g_n \\ &-\mu \Delta v_{n+1} = -\nabla \pi_{n+1} + F_n \\ &v_{n+1}|_{\partial\Omega} = 0\,, \quad \textstyle\int_\Omega \pi_{n+1}\, dx = 0\,. \end{aligned} \tag{5}$$

We then set

$$H_{n+1} = \pi_{n+1} - \mu \nabla \cdot v_{n+1} \,, \tag{6}$$

and obtain σ_{n+1} by solving the steady transport equation

$$k\sigma_{n+1} + (\lambda + 2\mu) \nabla \cdot (\sigma_{n+1} v_{n+1}) = H_{n+1}. \tag{7}$$

Even simpler schemes may seem to suggest themselves upon rewriting the problem (2) as (3). In particular, after defining g_n and F_n by (4), and finding v_{n+1} and π_{n+1} by solving (5), one could try to determine σ_{n+1} by setting

$$k\sigma_{n+1} = \pi_{n+1} + (\lambda + \mu) \nabla \cdot v_{n+1} \,, \tag{8}$$

as suggested by combining the relations $\pi \equiv p - k - (\lambda + \mu) \nabla \cdot v$ and $p = k(1 + \sigma)$. However, the iteration of this scheme appears to lose regularity through the term $g_n = -\nabla \cdot (\sigma_n v_n)$. For instance, if $f \in W^{1,2}(\Omega)$, and if $v_0 = 0$, $\sigma_0 = 0$, then the regularity estimates for the Stokes problem (5) give $v_1 \in W^{3,2}(\Omega)$, $\pi_1 \in W^{2,2}(\Omega)$, and hence $\sigma_1 \in W^{2,2}(\Omega)$. But this is not enough to imply that $\nabla \cdot (\sigma_1 v_1) \in W^{2,2}(\Omega)$, as would be needed in order to apply the same estimates again to get $v_2 \in W^{3,2}(\Omega)$, $\pi_2 \in W^{2,2}(\Omega)$, etc..

In order to maintain the same regularity at each step, one could modify (8) as suggested by the relation $\nabla \cdot v = -\nabla \cdot (\sigma v)$, to convert it into the steady transport equation

$$k\sigma_{n+1} + (\lambda + \mu) \nabla \cdot (\sigma_{n+1} v_{n+1}) = \pi_{n+1} \,. \tag{9}$$

Since the right-hand side belongs to $W^{2,2}(\Omega)$, one obtains the regularity $\nabla \cdot (\sigma_{n+1} v_{n+1}) \in W^{2,2}(\Omega)$ needed for iteration, using known results for the transport equation. This scheme was used in [15], [16], [17]. However, coming from the Stokes problem, the right-hand side π_{n+1} is estimated in terms of F_n and g_n. The dependence of F_n on the size of the solution is quadratic, but that of g_n is essentially linear, since $g_n = \nabla \cdot v_{n+1}$. This linear dependence on g_n carries through to the estimate for $\|\nabla \cdot (\sigma_{n+1} v_{n+1})\|_{2,2}$. As a consequence, it seems to be impossible to prove the convergence of this scheme without assuming that the value of $(\lambda + \mu)^{-1}$ is small, regardless of how small the data might be assumed to be. This will be explained further in connection with the inequality (29) below. In any case, we consider this condition on the viscosity coefficients to be unacceptable, as it is not suggested by physical considerations, and is not satisfied by most fluids.

In our scheme, the expression $H_{n+1} = \pi_{n+1} - \mu \nabla \cdot v_{n+1}$ that we put on the right-hand side of the transport equation (7) admits an estimate for its Laplacian that is independent of g_n. This leads to an estimate for $\|\nabla \cdot (\sigma_{n+1} v_{n+1})\|_{2,2}$ in which the dependence on g_n comes only in multiplicative combinations that become quadratically small with the size of the solution and the data, making possible a proof of convergence without any unnatural restrictions on the parameters k, μ, λ and $\overline{\rho}$. The estimate for $\triangle H_{n+1}$ just referred to is obtained from the identity

$$\mu \nabla \times \nabla \times v_{n+1} + \nabla H_{n+1} = F_n \,, \tag{10}$$

which is derived by rewriting the second of the equations (5) using the vector identity $-\triangle v = \nabla\times\nabla\times v - \nabla\nabla\cdot v$. This explains why it is precisely the quantity H_{n+1} that we set on the right-hand side of (7). Of course, our scheme (4), (5), (6), (7) is consistent with the problem to be solved, in that if v_n, σ_n converges to a limit v, σ, and if we set $p \equiv k\,(1+\sigma)$, then v, σ, p will satisfy (3).

From (10), it is also apparent that H_{n+1} is the unique solution of the Neumann problem

$$\begin{aligned} \triangle H_{n+1} &= \nabla\cdot[F_n - \mu\,(\nabla\times\nabla\times v_{n+1})] \\ \left.\frac{\partial H_{n+1}}{\partial n}\right|_{\partial\Omega} &= [F_n - \mu\,(\nabla\times\nabla\times v_{n+1})]\cdot n\,|_{\partial\Omega} \\ \textstyle\int_\Omega H_{n+1}\,dx &= 0\,. \end{aligned} \tag{11}$$

In numerical schemes, it may be preferable to determine H_{n+1} through this Neumann problem, eliminating any use of the function π_{n+1}.

5. Regularity lemmas

To prove Theorem 1, we will show that the iteration $v_n, \sigma_n \to v_{n+1}, \sigma_{n+1}$ remains bounded in $W^{3,2}(\Omega)\times W^{2,2}(\Omega)$, and is a contraction in $W^{2,2}(\Omega)\times W^{1,2}(\Omega)$. This will be based on some estimates for subsidiary problems that are provided below in Lemmas 1–5. It is tacitly assumed in these lemmas that the boundary has the regularity assumed in Theorem 1. The numbered constants $c_1, c_2, \ldots, c_{11}$ and the generic constant c which are introduced in these lemmas depend at most on Ω, k, λ, and μ in the normalized problem (3). The reason for numbering some of the constants is for clarity in the proof of Lemma 6, to dispel any question of circular reasoning. Only the estimates with numbered constants are needed in Sections 5 and 6. In Sections 7 and 8, all the constants are treated as generic.

Lemmas 1–5 will be stated in the course of tracing through the regularity obtained at each intermediate step of our iterative procedure, to be sure that the next step is well defined. To that end, we will take the conditions

$$v_n \in W^{3,2}(\Omega)\,,\ \sigma_n \in W^{2,2}(\Omega)\,,\ \nabla\cdot(\sigma_n v_n) \in W^{2,2}(\Omega)\,,\ \nabla\cdot(\sigma_n v_n) \in W_0^{1,2}(\Omega) \tag{12}$$

as induction hypotheses, and show that they are preserved by our iterative procedure, provided that the norms $\|v_n\|_{3,2}$ remain bounded by a constant c_7 introduced in Lemma 3.

In the next section we will add several estimates to the forgoing induction hypotheses and prove the boundedness of the iterates, including the bound $\|v_n\|_{3,2} \le c_7$ that is assumed in this section.

Turning now to the main point of this section, we note that the conditions (12) are certainly satisfied by $v_0 = 0$ and $\sigma_0 = 0$. Taking the conditions (12) as induction hypotheses, the first three of them are easily seen to guarantee that the functions g_n and F_n defined by (4) satisfy

$$g_n \in W^{2,2}(\Omega) , \quad F_n \in W^{1,2}(\Omega) .$$

Here, in verifying that $F_n \in W^{1,2}(\Omega)$, we use the inequalities

$$\begin{aligned} \|\sigma f\|_{1,2} &\leq c_1 \|\sigma\|_{2,2} \|f\|_{1,2} , \quad \text{for } \sigma \in W^{2,2}(\Omega) , \ f \in W^{1,2}(\Omega) \\ \|v \cdot \nabla v\|_{1,2} &\leq c_2 \|v\|_{2,2}^2 , \quad \text{for } v \in W^{2,2}(\Omega) , \end{aligned} \tag{13}$$

which hold in virtue of Sobolev's inequalities, with constants c_1, c_2 that depend only on Ω.

Lemma 1. *In the steady Stokes problem (5), let $g_n \in W^{2,2}(\Omega)$ and $F_n \in W^{1,2}(\Omega)$. Then there exists a unique solution v_{n+1}, π_{n+1} satisfying $v_{n+1} \in W^{3,2}(\Omega) \cap W_0^{1,2}(\Omega)$ and $\pi_{n+1} \in W^{2,2}(\Omega)$. Moreover,*

$$\begin{aligned} \|v_{n+1}\|_{2,2} &\leq c \|F_n\| + c \|g_n\|_{1,2} \\ \|v_{n+1}\|_{3,2} &\leq c_3 \|F_n\|_{1,2} + c_4 \|g_n\|_{2,2} . \end{aligned} \tag{14}$$

Lemma 2. *If $F_n \in W^{1,2}(\Omega)$, $v_{n+1} \in W^{3,2}(\Omega)$ and $\pi_{n+1} \in W^{2,2}(\Omega)$, then H_{n+1} defined by (6) belongs to $W^{2,2}(\Omega)$ and satisfies*

$$\begin{aligned} \|H_{n+1}\|_{1,2} &\leq c \left(\|F_n\| + \|v_{n+1}\|_{2,2} \right) , \\ \|H_{n+1}\|_{2,2} &\leq c_5 \left(\|F_n\|_{1,2} + \|v_{n+1}\|_{3,2} \right) , \\ \|\triangle H_{n+1}\| &\leq c_6 \|F_n\|_{1,2} , \\ \|\triangle H_{n+1}\|_{-1,2} &\leq \|F_n\| . \end{aligned} \tag{15}$$

Lemma 3. *There exists a positive constant c_7, such that if $v_{n+1} \in W^{3,2}(\Omega)$ with $n \cdot v_{n+1}|_{\partial\Omega} = 0$, and if*

$$\|v_{n+1}\|_{3,2} \leq c_7 , \tag{16}$$

then, for any prescribed right-hand side $H_{n+1} \in W^{2,2}(\Omega)$, there exists a unique solution $\sigma_{n+1} \in W^{2,2}(\Omega)$ of the transport equation (7), and it satisfies

$$\begin{aligned} \|\sigma_{n+1}\|_{1,2} &\leq c \|H_{n+1}\|_{1,2} , \\ \|\sigma_{n+1}\|_{2,2} &\leq c_8 \|H_{n+1}\|_{2,2} , \\ \|\triangle \sigma_{n+1}\| &\leq c \|\triangle H_{n+1}\| + c \|v_{n+1}\|_{3,2} \|H_{n+1}\|_{2,2} , \\ \|\triangle \sigma_{n+1}\|_{-1,2} &\leq c \|\triangle H_{n+1}\|_{-1,2} + c \|v_{n+1}\|_{3,2} \|H_{n+1}\|_{1,2} , \end{aligned} \tag{17}$$

and also, as a consequence of the third of these inequalities,

$$\|\triangle \nabla \cdot (\sigma_{n+1} v_{n+1})\| \leq c_9 \|\triangle H_{n+1}\| + c_{10} \|v_{n+1}\|_{3,2} \|H_{n+1}\|_{2,2} . \tag{18}$$

Lemma 1 is well known if the boundary is of class C^3. The extension to boundaries of class $C^{2,1}$ is due to Amrouche and Girault [1]. Lemma 2 is a trivial consequence of (10), with c_6 being just the square root of the spatial dimension. Lemma 3 is proven in our accompanying paper [11].

From the second of the estimates (17) and the transport equation (7), it follows that $\nabla\cdot(\sigma_{n+1}v_{n+1})\in W^{2,2}(\Omega)$. Thus, as long as the values of $\|v_{n+1}\|_{3,2}$ remain less than c_7, the sequence of functions v_{n+1},σ_{n+1} will continue to be well defined, and will satisfy the first three of the induction hypotheses (12).

The fourth induction hypothesis, $\nabla\cdot(\sigma_n v_n)\in W_0^{1,2}(\Omega)$, together with the condition $\nabla\cdot v_{n+1} = -\nabla\cdot(\sigma_n v_n)$ in the Stokes problem (5), implies that $\nabla\cdot v_{n+1}\in W_0^{1,2}(\Omega)$. Thus all of the hypotheses of the following lemma are satisfied. It guarantees that the fourth induction hypothesis is preserved under iteration.

Lemma 4. *The conditions $v_{n+1}\in W^{3,2}(\Omega)\cap W_0^{1,2}(\Omega)$, $\sigma_{n+1}\in W^{2,2}(\Omega)$, and $\nabla\cdot v_{n+1}\in W_0^{1,2}(\Omega)$ imply that $\nabla\cdot(\sigma_{n+1}v_{n+1})\in W_0^{1,2}(\Omega)$.*

This is easily proved writing $\nabla\cdot(\sigma_{n+1}v_{n+1}) = \sigma_{n+1}\nabla\cdot v_{n+1}+v_{n+1}\cdot\nabla\sigma_{n+1}$ and noting that the right-hand side is a sum of terms, each of which is the product of a function belonging to $W^{2,2}(\Omega)\cap W_0^{1,2}(\Omega)$ with a function belonging to $W^{1,2}(\Omega)$. The product of such functions must belong to $W_0^{1,2}(\Omega)$. Indeed, given any $\phi\in W^{2,2}(\Omega)\cap W_0^{1,2}(\Omega)$ and any $\psi\in W^{1,2}(\Omega)$, we can approximate ψ in $W^{1,2}(\Omega)$ by functions $\psi_n\in C^1(\overline{\Omega})$. The products $\phi\psi_n$ belong to $W_0^{1,2}(\Omega)$, since they belong to $W^{1,2}(\Omega)\cap C(\overline{\Omega})$ and vanish on $\partial\Omega$. Since $\phi\psi_n\to\phi\psi$ in $W^{1,2}(\Omega)$, it follows that $\phi\psi\in W_0^{1,2}(\Omega)$.

This completes the induction argument for the preservation of the conditions (12), modulo the hypothesis $\|v_{n+1}\|_{3,2}\le c_7$ of Lemma 3, which will be established in the next section along with other estimates. There we will also need the following Lemma.

Lemma 5. *The solutions σ_{n+1}, v_{n+1} of Lemmas 1 and 3 satisfy*

$$\|\nabla\cdot(\sigma_{n+1}v_{n+1})\|_{2,2}\le c_{11}\|\Delta\nabla\cdot(\sigma_{n+1}v_{n+1})\| \,. \tag{19}$$

This is a standard elliptic regularity estimate for $\nabla\cdot(\sigma_{n+1}v_{n+1})$ considered as the solution φ of a Dirichlet problem $\Delta\varphi=\psi$, $\varphi|_{\partial\Omega}=0$. We know that $\psi\in L^2(\Omega)$ by Lemma 3, and that $\varphi|_{\partial\Omega}=0$ by Lemma 4.

6. Bounds for the iterates

The following lemma completes the proof that the iterates are well defined and bounded. In the course of proving it we introduce further constants $b_1,\dots,b_{11}$, which, like the constants $c_1,\dots,,c_{11}$ of the last section, depend only on Ω, k, λ, and μ in the normalized problem (3).

Lemma 6. *If R, L, B are chosen to satisfy*

$$\begin{gathered} R\le R^* = \min\{1, c_7/(c_3b_1+c_4), 1/(2b_4), b_{10}, b_{11}\} \\ L = 2b_6R^2\,, \quad B = c_2R^2\,, \end{gathered} \tag{20}$$

and if $\|f\|_{1,2} < B$, *then the iterates* v_n, σ_n *defined in Section 3 satisfy the hypothesis (16) of Lemma 3, and the estimates*

$$\begin{aligned} \|v_n\|_{3,2} + \|\sigma_n\|_{2,2} \le R\,, \\ \|\nabla \cdot (\sigma_n v_n)\|_{2,2} \le L\,, \end{aligned} \tag{21}$$

as well as the previously considered induction hypotheses (12).

Proof: We take (21) along with (12) as induction hypotheses. During the course of the proof, we will assume a number of restrictions on the size of L, namely that

$$L \le \min\{1, R, R/(2b_5), 1/(2b_7)\}\,. \tag{22}$$

At the end of the proof, it will be shown that these restrictions are satisfied by L as defined in (20).

By (4), (13), (20), (21) we have

$$\begin{aligned} \|F_n\|_{1,2} &\le \left(1 + c_1 \|\sigma_n\|_{2,2}\right)\left(\|f\|_{1,2} + c_2 \|v_n\|_{2,2}^2\right) \\ &\le (1 + c_1 R)\left(\|f\|_{1,2} + c_2 R^2\right) \le b_1 R^2\,, \\ \|g_n\|_{2,2} &= \|\nabla \cdot (\sigma_n v_n)\|_{2,2} \le L\,, \end{aligned} \tag{23}$$

where $b_1 = 2(1 + c_1)c_2$. Therefore, by the second of the Stokes estimates (14),

$$\|v_{n+1}\|_{3,2} \le c_3 b_1 R^2 + c_4 L\,. \tag{24}$$

Using the second of the estimates (15), together with (23) and (24), gives

$$\begin{aligned} \|H_{n+1}\|_{2,2} &\le c_5\left(\|F_n\|_{1,2} + \|v_{n+1}\|_{3,2}\right) \\ &\le b_2 R^2 + b_3 L\,, \end{aligned} \tag{25}$$

where $b_2 = c_5(1 + c_3)b_1$ and $b_3 = c_5 c_4$.

The hypothesis of Lemma 3, that $\|v_{n+1}\|_{3,2} \le c_7$, is satisfied in virtue of (24) and the restrictions $L \le R$, $R \le 1$, and $R \le c_7/(c_3 b_1 + c_4)$. Therefore, from the second of the estimates (17), we get

$$\|\sigma_{n+1}\|_{2,2} \le c_8 b_2 R^2 + c_8 b_3 L. \tag{26}$$

Combining (24) and (26), we have

$$\|v_{n+1}\|_{3,2} + \|\sigma_{n+1}\|_{2,2} \le b_4 R^2 + b_5 L\,, \tag{27}$$

where $b_4 = c_3 b_1 + c_8 b_2$ and $b_5 = c_4 + c_8 b_3$. Therefore, v_{n+1}, σ_{n+1} satisfy the first of the induction estimates (21) in virtue of the restrictions $L \le R/(2b_5)$ and $R \le 1/(2b_4)$.

The second of the estimates (21) is proved by using first Lemma 5, then (18), then the third of the estimates (15) combined with (23), along with (27) and (25), and finally the restrictions $R \leq 1$, $L \leq 1$, to obtain

$$\begin{aligned}\|\nabla \cdot (\sigma_{n+1} v_{n+1})\|_{2,2} &\leq c_{11} \|\triangle \nabla \cdot (\sigma_{n+1} v_{n+1})\| \\ &\leq c_{11} c_9 \|\triangle H_{n+1}\| - c_{11} c_{10} \|v_{n+1}\|_{3,2} \|H_{n+1}\|_{2,2} \\ &\leq c_{11} c_9 c_6 b_1 R^2 + c_{11} c_{10} \left(b_4 R^2 + b_5 L\right) \left(b_2 R^2 + b_3 L\right) \\ &\leq b_6 R^2 + b_7 L^2,\end{aligned}$$

with suitable definitions of b_6 and b_7. In virtue of the restriction $L \leq 1/(2b_7)$, we have $b_6 R^2 + b_7 L^2 \leq L$, confirming that σ_{n+1}, v_{n+1} satisfy (21), provided that

$$2b_6 R^2 \leq L\,. \tag{28}$$

This last requirement is consistent with our preceding restrictions, since those relating L to R have been linear. Indeed, the restrictions used so far are $R \leq R_1 \equiv \min\{1, c_7/(c_3 b_1 + c_4), 1/(2b_4)\}$ and $L \leq b_8$ and $L \leq b_9 R$, where $b_8 = \min\{1, 1/(2b_7)\}$ and $b_9 = \min\{1, 1/(2b_5)\}$. The curve $2b_6 R^2 = L$, bounding the region (28) in the R, L-plane, intersects $L = b_8$ at $R = b_{10} \equiv \sqrt{b_8/2b_6}$, and intersects $L = b_9 R$ at $R = b_{11} \equiv b_9/2b_6$. These then present two additional restrictions, namely $R \leq b_{10}$ and $R \leq b_{11}$. If $R \leq R^* = \min\{1, c_7/(c_3 b_1 + c_4), 1/(2b_4), b_{10}, b_{11}\}$ and $L = 2b_6 R^2$, then L will satisfy all of the restrictions that have been put upon it.

Therefore, we have reached the desired conclusion that every iterate will satisfy (21), provided $\|f\|_{1,2} < B \equiv c_2 R^2$. This completes the proof.

It is instructive to conclude this section by considering the iteration (4), (5), (9), discussed in Section 3. The first of the estimates (21) can be proven as before. But in trying to prove the second estimate, without an analogue for $\triangle \pi_{n+1}$ of the estimate for $\triangle H_{n+1}$ of Lemma 2, the best we can do is as follows. From the transport equation (9), and analogues of (25) and (26), we obtain

$$\begin{aligned}\|\nabla \cdot (\sigma_{n+1} v_{n+1})\|_{2,2} &\leq (\lambda + \mu)^{-1} \left(\|\pi_{n+1}\|_{2,2} + k \|\sigma_{n+1}\|_{2,2}\right) \\ &\leq (\lambda + \mu)^{-1} \left[(1 + kc_8) b_2 R^2 + (1 + kc_8) b_3 L\right]\,.\end{aligned}$$

This, however, can not possibly be bounded by L, unless

$$(\lambda + \mu)^{-1} (1 + kc_8) b_3 < 1\,, \tag{29}$$

which is an unphysical and unacceptable condition on the viscosity coefficients.

7. Convergence of the iterates

It remains to prove the convergence of the iterative scheme. We denote by v'_{n+1}, π'_{n+1}, g'_{n+1}, F'_{n+1}, H'_{n+1}, σ'_{n+1} the differences $v'_{n+1} \equiv v_{n+1} - v_n$, etc.. For $n \geq 1$, they satisfy the Stokes problem (5) and the equation (6), both of which are linear, as well as the transport equation

$$k\sigma'_{n+1} + (\lambda + 2\mu) \nabla \cdot \left(\sigma'_{n+1} v_{n+1}\right) = G'_{n+1}\,, \tag{30}$$

with right-hand side G'_{n+1} defined by

$$G'_{n+1} = H'_{n+1} - (\lambda + 2\mu)\,\nabla \cdot (\sigma_n v'_{n+1})\,. \tag{31}$$

Therefore, v'_{n+1}, H'_{n+1}, and σ'_{n+1}, satisfy the estimates of Lemmas 1, 2, and 3, with G'_{n+1} replacing H'_{n+1} in Lemma 3.

For convenience, let us also recall from Lemma 6, that if $R \le R^*$, and if f satisfies

$$\|f\|_{1,2} \le B = cR^2, \tag{32}$$

then, for all $n \ge 0$,

$$\|v_n\|_{3,2} + \|\sigma_n\|_{2,2} \le R\,. \tag{33}$$

Lemma 7. *If R and f satisfy the hypotheses of Lemma 6, then*

$$\left\|v'_{n+1}\right\|_{2,2} + \left\|\sigma'_{n+1}\right\|_{1,2} \le cR\left(\|v'_n\|_{2,2} + \|\sigma'_n\|_{1,2} + \left\|v'_{n-1}\right\|_{2,2} + \left\|\sigma'_{n-1}\right\|_{1,2}\right), \tag{34}$$

for all $n \ge 1$, where c is a constant that depends only on Ω, k, λ and μ, and where we have set $v'_0 \equiv 0$ and $\sigma'_0 \equiv 0$. Under the additional assumption that $cR < 1/2$, the inequality (34) implies the geometric convergence of the iterates v_n, σ_n in $W^{2,2}(\Omega) \times W^{1,2}(\Omega)$. The limit v, σ belongs to $W^{3,2}(\Omega) \times W^{2,2}(\Omega)$ and satisfies all of the equations and conditions of problem (3).

Proof: From Lemmas 1 and 3, we have

$$\begin{aligned} \left\|v'_{n+1}\right\|_{2,2} &\le c\,\|F'_n\| + c\,\|g'_n\|_{1,2}\,, \\ \left\|\sigma'_{n+1}\right\|_{1,2} &\le c\left\|G'_{n+1}\right\|_{1,2}\,, \end{aligned} \tag{35}$$

for $n \ge 1$. To prove (34) we need to bound the terms on the right-hand sides of (35) by expressions of the form $cR\left(\|v'_n\|_{2,2} + \|\sigma'_n\|_{1,2} + \left\|v'_{n-1}\right\|_{2,2} + \left\|\sigma'_{n-1}\right\|_{1,2}\right)$.

To estimate F'_n we note that

$$F'_n = -(1+\sigma_n)\,(v_n \cdot \nabla v'_n + v'_n \cdot \nabla v_{n-1}) - \sigma'_n v_{n-1} \cdot \nabla v_{n-1} + \sigma'_n f \tag{36}$$

and hence, using (33) and (32),

$$\|F'_n\| \le cR\,\|v'_n\|_{1,2} + cR^2\,\|\sigma'_n\|_{1,2}\,. \tag{37}$$

In estimating $g'_n = -\nabla \cdot (\sigma_n v_n) + \nabla \cdot (\sigma_{n-1} v_{n-1})$, we need to treat separately the case $n = 1$. Since $v_0 = 0$, and $v'_1 = v_1$, it follows from (33) that

$$\|g'_1\|_{1,2} \le \|\nabla \cdot (\sigma_1 v_1)\|_{1,2} \le c\,\|\sigma_1\|_{2,2}\,\|v_1\|_{2,2} \le cR\,\|v'_1\|_{2,2}\,,$$

which provides a suitable estimate for g'_1 on the right-hand side of (35).

To estimate g'_n for $n \ge 2$, we take the difference between the transport equation (7) for σ_n and for σ_{n-1} to get

$$g'_n = \frac{k}{\lambda + 2\mu}\sigma'_n - \frac{1}{\lambda + 2\mu}H'_n\,. \tag{38}$$

Arguing as in the proof of Lemma 5, and then using (38), and finally Lemma 2, we have

$$\begin{aligned} \|g'_n\|_{1,2} &\le c\,\|\triangle g'_n\|_{-1,2} \\ &\le c\,\|\triangle\sigma'_n\|_{-1,2} + c\,\|\triangle H'_n\|_{-1,2} \\ &\le c\,\|\triangle\sigma'_n\|_{-1,2} + c\,\|F'_{n-1}\| \;. \end{aligned} \tag{39}$$

The term $\|F'_{n-1}\|$ is estimated appropriately by (37), but the term $\|\triangle\sigma'_n\|_{-1,2}$ needs further consideration. From Lemma 3 we have

$$\|\triangle\sigma'_n\|_{-1,2} \le c\,\|\triangle G'_n\|_{-1,2} + c\,\|v_n\|_{3,2}\,\|G'_n\|_{1,2} \;. \tag{40}$$

By first substituting (31) into (40), and then using Lemma 2 and the inequality $\|\nabla\cdot\varphi\|_{-1,2} \le \|\varphi\|$ along with (33), and finally (33) and the restriction $R \le 1$, we obtain

$$\begin{aligned} \|\triangle\sigma'_n\|_{-1,2} &\le c\,\|\triangle H'_n\|_{-1,2} + c\,\|\triangle\nabla\cdot(\sigma_{n-1}v'_n)\|_{-1,2} \\ &\quad + c\,\|v_n\|_{3,2}\,\|H'_n\|_{1,2} + c\,\|v_n\|_{3,2}\,\|\nabla\cdot(\sigma_{n-1}v'_n)\|_{1,2} \\ &\le c\,\|F'_{n-1}\| + c\,\|\sigma_{n-1}\|_{2,2}\,\|v'_n\|_{2,2} \\ &\quad + cR\left(\|F'_{n-1}\| + \|v'_n\|_{2,2}\right) + cR\,\|\sigma_{n-1}\|_{2,2}\,\|v'_n\|_{2,2} \\ &\le c\,\|F'_{n-1}\| + cR\,\|v'_n\|_{2,2} \;. \end{aligned} \tag{41}$$

Substituting (41) into (39) and using (37) gives the desired estimate

$$\begin{aligned} \|g'_n\|_{1,2} &\le c\,\|F'_{n-1}\| + cR\,\|v'_n\|_{2,2} \\ &\le cR\,\|v'_{n-1}\|_{1,2} + cR^2\,\|\sigma'_{n-1}\|_{1,2} + cR\,\|v'_n\|_{2,2} \;. \end{aligned} \tag{42}$$

We now obtain the estimate

$$\|v'_{n+1}\|_{2,2} \le cR\left(\|v'_n\|_{2,2} + R\,\|\sigma'_n\|_{1,2} + \|v'_{n-1}\|_{1,2} + R\,\|\sigma'_{n-1}\|_{1,2}\right) , \tag{43}$$

by substituting (37) and (42) into the first of the inequalities (35).

It remains to estimate G'_{n+1} on the right-hand side of (35). Recalling its definition (31), and using Lemma 2, and then finally (37) and (33), we obtain

$$\begin{aligned} \|G'_{n+1}\|_{1,2} &\le \|H'_{n+1}\|_{1,2} + c\,\|\nabla\cdot(\sigma_n v'_{n+1})\|_{1,2} \\ &\le c\left(\|F'_n\| + \|v'_{n+1}\|_{2,2}\right) + c\,\|\sigma_n\|_{2,2}\,\|v'_{n+1}\|_{2,2} \\ &\le c\,\|v'_{n+1}\|_{2,2} + cR\left(\|v'_n\|_{1,2} + R\,\|\sigma'_n\|_{1,2}\right) . \end{aligned} \tag{44}$$

This provides the estimate we need, in view of (43). Together, (35), (44) and(43) imply (34).

To show that (34) implies the geometric convergence of the iterates v_n, σ_n in $W^{2,2}(\Omega) \times W^{1,2}(\Omega)$, let $\alpha_n \equiv \|v'_n\|_{2,2} + \|\sigma'_n\|_{1,2}$ for $n \ge 0$, and let $\gamma \equiv cR$. Then

$$\begin{aligned} &\alpha_0 = 0 \;, \\ &\alpha_1 = \|v_1\|_{2,2} + \|\sigma_1\|_{1,2} \;, \text{ and} \\ &\alpha_{n+1} \le \gamma\,(\alpha_n + \alpha_{n-1}) \;, \text{ for } n \ge 1. \end{aligned}$$

The terms of this sequence are majorized, after a normalization of α_1, by those of the sequence

$$\begin{aligned} a_0 &= 0 \\ a_1 &= 1\,, \text{ and} \\ a_{n+1} &= \gamma\,(a_n + a_{n-1})\ , \text{ for } n \geq 1. \end{aligned}$$

We will now show this sequence to be geometrically decreasing if $\gamma < 1/2$. Letting

$$r_{n+1} = \frac{a_{n+1}}{a_n} = \gamma\left(1 + \frac{1}{r_n}\right), \tag{45}$$

we note that

$$|r_{n+2} - r_{n+1}| = \frac{\gamma}{r_{n+1} r_n}\,|r_{n+1} - r_n|\ ,$$

and that

$$\frac{\gamma}{r_{n+1} r_n} = \frac{1}{1 + r_n} < \frac{1}{1+\gamma}\,.$$

Hence, the sequence of ratios r_{n+1} are geometrically convergent to some limit r. By passing to the limit in (45), we find that

$$r = \frac{\gamma + \sqrt{\gamma^2 + 4\gamma}}{2}\,,$$

which is less than 1 for $\gamma < 1/2$. Thus, for $\gamma < 1/2$, the sequence a_n is also geometrically convergent, and, by comparison, so is the sequence α_n. This completes the proof of the convergence of v_n, σ_n in $W^{2,2}(\Omega) \times W^{1,2}(\Omega)$, to a limit v, σ. Since the iterates are bounded in $W^{3,2}(\Omega) \times W^{2,2}(\Omega)$, their limit must also belong to this space, by a standard argument.

The geometric convergence of F_n in $L^2(\Omega)$ and of g_n in $W^{1,2}(\Omega)$, to limits F and g, now follows from (37), (42) and the convergence of v_n, σ_n. The geometric convergence of H_n and π_n in $W^{1,2}(\Omega)$, to limits H and π, then follows from Lemma 2 and (6). Thus passing to the limit in the equations (4), (5), (6) and (7), and setting $p = k\,(1+\sigma)$, it is easily seen that all of the conditions of problem (3) are satisfied.

8. Uniqueness in the ball of existence

In order to be physically reasonable, the density $\rho = 1 + \sigma$ must be everywhere positive, as claimed in Theorem 1 and needed in the following proof of uniqueness. Up to this point, we have not addressed this matter. However, Lemmas 6 and 7 allow for the setting of further smallness conditions on the constant R, i.e., on the size of the force. Therefore, we can make the additional assumption that R is sufficiently small that

$$\|\sigma\|_{2,2} < R \quad \text{implies} \quad \sup_\Omega |\sigma| < 1. \tag{46}$$

Then the positivity of the density is ensured by the conclusion (21) of Lemma 6, which implies (47) below. This completes the proof of existence claimed in Theorem 1.

The proof of convergence given in Lemma 7 suggests an analogous argument for the local uniqueness of the solution that was obtained. The resulting uniqueness theorem is independent of the existence theorem, in that it applies to any solutions satisfying the stated hypotheses. The assumption (46) can be dropped by simply assuming that the solutions under consideration have positive densities.

Lemma 8. *If R and f satisfy the hypotheses of Lemmas 6 and 7, then there can be at most one solution of the problem (3) satisfying* $\inf_\Omega (1+\sigma) > 0$ *and*

$$\|v\|_{3,2} + \|\sigma\|_{2,2} \le R\,. \tag{47}$$

Proof: Given one solution v, σ, such as that obtained in Lemma 7 when (46) is assumed, we set

$$\begin{aligned} g &= -\nabla\cdot(\sigma v)\ , \quad F = -(1+\sigma)(v\cdot\nabla v - f)\ ,\\ p &= k(1+\sigma)\ , \quad \pi = p - k - (\lambda+\mu)\nabla\cdot v\ . \end{aligned} \tag{48}$$

Then the Stokes problem

$$\begin{aligned} &\nabla\cdot v = g\\ &-\mu\triangle v = -\nabla\pi + F\\ &v\,|_{\partial\Omega} = 0\ , \quad \textstyle\int_\Omega \pi\,dx = 0 \end{aligned} \tag{49}$$

is satisfied, and also, setting

$$H = \pi - \mu\nabla\cdot v\ , \tag{50}$$

the transport equation

$$k\sigma + (\lambda+2\mu)\nabla\cdot(\sigma v) = H \tag{51}$$

is satisfied.

If $\overline{v}, \overline{\sigma}$ is a second solution satisfying the hypotheses of the lemma, we define $\overline{g}, \overline{F}, \overline{p}, \overline{\pi}, \overline{H}$ similarly as in (48) and (50), and then define $v', \sigma', g', F', p', \pi', H'$ by setting $v' = \overline{v} - v$, etc.

Analogously to (34), we will now show that

$$\|v'\|_{2,2} + \|\sigma'\|_{1,2} < 2cR\left(\|v'\|_{2,2} + \|\sigma'\|_{1,2}\right), \tag{52}$$

if $R \le R^*$, and hence that $v' = 0, \sigma' = 0$ if R also satisfies $cR < 1/2$. The constant c here is the same as in (34), because the proofs of (52) and (34) are based, step by step, on the same estimates.

The differences v', g', F', π', H' satisfy obvious analogues of the linear equations (49) and (50). Also, σ' satisfies the transport equation

$$k\sigma' + (\lambda+2\mu)\nabla\cdot(\sigma'\overline{v}) = G'\ , \tag{53}$$

with right-hand side

$$G' = H' - (\lambda+2\mu)\nabla\cdot(\sigma v')\ . \tag{54}$$

From Lemmas 1 and 3 we have

$$\begin{aligned} \|v'\|_{2,2} &\le c\|F'\| + c\|g'\|_{1,2} , \\ \|\sigma'\|_{1,2} &\le c\|G'\|_{1,2} . \end{aligned} \tag{55}$$

Similarly to (36), we have

$$F' = -(1+\overline{\sigma})(\overline{v}\cdot\nabla v' + v'\cdot\nabla v) - \sigma' v\cdot\nabla v + \sigma' f \tag{56}$$

and hence, using the hypotheses (47) and (32),

$$\|F'\| \le cR\|v'\|_{1,2} + cR^2\|\sigma'\|_{1,2} . \tag{57}$$

Similarly to (38), we have

$$g' = \frac{k}{\lambda+2\mu}\sigma' - \frac{1}{\lambda+2\mu}H'. \tag{58}$$

From our assumption that $\inf\rho > 0$, combined with the boundary condition $v|_{\partial\Omega} = 0$, the equation of continuity $0 = \nabla\cdot(\rho v) = v\cdot\nabla\rho + \rho\nabla\cdot v$, and our regularity assumptions, it follows that $\nabla\cdot v|_{\partial\Omega} = 0$. Hence $g \in W_0^{1,2}$, and similarly $\overline{g} \in W_0^{1,2}$. Therefore $g' \in W_0^{1,2}$, and we may argue as in (39) that

$$\begin{aligned} \|g'\|_{1,2} &\le c\|\triangle g'\|_{-1,2} \\ &\le c\|\triangle\sigma'\|_{-1,2} + c\|\triangle H'\|_{-1,2} \\ &\le c\|\triangle\sigma'\|_{-1,2} + c\|F'\| . \end{aligned} \tag{59}$$

To estimate $\|\triangle\sigma'\|_{-1,2}$ we note that Lemma 3 implies

$$\|\triangle\sigma'\|_{-1,2} \le c\|\triangle G'\|_{-1,2} + c\|v\|_{3,2}\|G'\|_{1,2} , \tag{60}$$

and therefore, using (54), Lemma 2, and the hypothesis (47), we obtain

$$\begin{aligned} \|\triangle\sigma'\|_{-1,2} &\le c\|\triangle H'\|_{-1,2} + c\|\triangle\nabla\cdot(\sigma v')\|_{-1,2} \\ &\quad + c\|v\|_{3,2}\|H'\|_{1,2} + c\|v\|_{3,2}\|\nabla\cdot(\sigma v')\|_{1,2} \\ &\le c\|F'\| + c\|\sigma\|_{2,2}\|v'\|_{2,2} \\ &\quad + cR\left(\|F'\| + \|v'\|_{2,2}\right) + cR\|\sigma\|_{2,2}\|v'\|_{2,2} \\ &\le c\|F'\| + cR\|v'\|_{2,2} . \end{aligned} \tag{61}$$

Thus, by substituting (61) into (59) and using (57) we obtain

$$\|g'\|_{1,2} \le cR\|v'\|_{1,2} + cR^2\|\sigma'\|_{1,2} + cR\|v'\|_{2,2} . \tag{62}$$

Substituting (57) and (62) into (55) gives

$$\|v'\|_{2,2} \le cR\left(\|v'\|_{2,2} + R\|\sigma'\|_{1,2} + \|v'\|_{1,2} + R\|\sigma'\|_{1,2}\right) . \tag{63}$$

Finally, similarly to (44), but using the hypothesis (47), we obtain

$$\begin{aligned} \|G'\|_{1,2} &\le \|H'\|_{1,2} + c\,\|\nabla\cdot(\sigma v')\|_{1,2} \\ &\le c\left(\|F'\| + \|v'\|_{2,2}\right) + c\,\|\sigma\|_{2,2}\,\|v'\|_{2,2} \\ &\le c\,\|v'\|_{2,2} + cR\left(\|v'\|_{1,2} + R\,\|\sigma'\|_{1,2}\right). \end{aligned} \tag{64}$$

Together, (55), (63), and (64) imply that

$$\|v'\|_{2,2} + \|\sigma'\|_{1,2} \le 2cR\left(\|v'\|_{2,2} + \|\sigma'\|_{1,2}\right).$$

Therefore $v' = 0$ and $\sigma' = 0$, if $cR < 1/2$. Throughout this section, all of the constants c are the same as in the corresponding estimates of the last section. Thus the condition $cR < 1/2$ which ensures uniqueness here is the same as that which guarantees existence there.

9. Uniqueness reconsidered directly

The current uniqueness theory for steady compressible viscous flow is far from satisfactory. Generally speaking, it is desirable in a uniqueness theorem that the assumptions on the second solution should be as weak as possible. It is natural, therefore, to try to improve the result of Lemma 8 by seeking another approach to the proof of uniqueness. Let us try to find a direct proof, again in the case of a bounded domain, that there can be no more than one solution u, ρ of the problem

$$\begin{aligned} &\nabla\cdot(\rho u) = 0 \\ &\rho u\cdot\nabla u = -\nabla p + \mu\Delta u + (\lambda+\mu)\,\nabla\nabla\cdot u + \rho f \\ &p = k\rho \\ &v|_{\partial\Omega} = 0, \quad \textstyle\int_\Omega \rho\,dx = \bar{\rho}\,|\Omega|\,, \quad \inf_\Omega \rho > 0\,. \end{aligned} \tag{65}$$

To set ideas, let us consider first the problem of showing that $u \equiv 0$ if $f \equiv 0$. Multiplying (65) by u and integrating, etc., one easily obtains

$$\mu\,\|\nabla u\|^2 + (\lambda+\mu)\,\|\nabla\cdot u\|^2 = -k\,(\nabla\rho, u) = -k\,(\nabla\log\rho, \rho u) = 0\,, \tag{66}$$

and thus the result. The main point to be noticed here is the use of the equation of continuity in showing that $(\nabla\rho, u) = 0$.

Now consider two solutions, u_1, ρ_1 and u_2, ρ_2 of (65), and their difference $w = u_2 - u_1$, $\sigma = \rho_2 - \rho_1$. The perturbation equation for w, σ may be written as

$$-\mu\Delta w - (\lambda+\mu)\,\nabla\nabla\cdot w = -k\nabla\sigma - \rho_2 u_2\cdot\nabla w - \rho_2 w\cdot\nabla u_1 - \sigma u_1\cdot\nabla u_1 + \sigma f\,, \tag{67}$$

since

$$\begin{aligned} \rho_2 u_2\cdot\nabla u_2 - \rho_1 u_1\cdot\nabla u_1 &= \rho_2 u_2\cdot\nabla(u_2-u_1) + \rho_2 u_2\cdot\nabla u_1 - \rho_1 u_1\cdot\nabla u_1 \\ &= \rho_2 u_2\cdot\nabla w + \rho_2(u_2-u_1)\cdot\nabla u_1 + (\rho_2-\rho_1)u_1\cdot\nabla u_1\,. \end{aligned}$$

Multiplying (67) by w, integrating, and noting that $(\rho_2 u_2 \cdot \nabla w, w) = 0$, etc., we obtain

$$\begin{aligned}\mu \|\nabla w\|^2 + (\lambda+\mu)\|\nabla\cdot w\|^2 &= -k(\nabla\sigma, w) + (\sigma f, w)\\ &\quad - (\rho_2 w\cdot\nabla u_1, w) - (\sigma u_1\cdot\nabla u_1, w)\\ &\le k(\sigma,\nabla\cdot w) + c\|f\|_3\|\sigma\|\|\nabla w\|\\ &\quad + c\|\rho_2\|_\infty\|\nabla u_1\|_{3/2}\|\nabla w\|^2 + c\|u_1\cdot\nabla u_1\|_3\|\sigma\|\|\nabla w\|\,. \end{aligned} \tag{68}$$

Here, we could not write $(\nabla\sigma, w) = (\nabla\log\sigma, \sigma w) = 0$, as in (66), because σw is not solenoidal. But the treatment of the term $k(\sigma,\nabla\cdot w)$ must still be based on the equation of continuity. To get an expression for $\nabla\cdot w$, we can write

$$\begin{aligned}0 = \nabla\cdot(\rho_2 u_2 - \rho_1 u_1) &= \nabla\cdot(\sigma u_2 + \rho_1 w)\\ &= \sigma\nabla\cdot u_2 + u_2\cdot\nabla\sigma + \rho_1\nabla\cdot w + w\cdot\nabla\rho_1\end{aligned}$$

and isolate $\nabla\cdot w$, to obtain

$$\nabla\cdot w = -\sigma\frac{\nabla\cdot u_2}{\rho_1} - \frac{u_2}{\rho_1}\cdot\nabla\sigma - \frac{w\cdot\nabla\rho_1}{\rho_1}\,. \tag{69}$$

Hence

$$\begin{aligned}k(\sigma,\nabla\cdot w) &= -k\left(\sigma^2, \frac{\nabla\cdot u_2}{\rho_1}\right) - \frac{k}{2}\left(\sigma^2, \nabla\cdot\left(\frac{u_2}{\rho_1}\right)\right) - k\left(\sigma, \frac{w\cdot\nabla\rho_1}{\rho_1}\right)\\ &\le \frac{k}{2}\sup_\Omega\left|\frac{3\nabla\cdot u_2}{\rho_1} - \frac{u_2\cdot\nabla\rho_1}{\rho_1^2}\right|\|\sigma\|^2 + ck\left\|\frac{\nabla\rho_1}{\rho_1}\right\|_3\|\sigma\,|\,\|\nabla w\|\,.\end{aligned} \tag{70}$$

We now return to the perturbation equation (67) to get an estimate for the perturbation σ in the density. By Bogovskii's formula (see [9],p.121), there exists ϕ such that

$$\nabla\cdot\phi = \sigma\,,\quad \phi|_{\partial\Omega} = 0,\quad \|\nabla\phi\| \le \gamma\|\sigma\|\,.$$

Multiplying (67) by ϕ and isolating the term $-k(\nabla\sigma,\phi) = k(\sigma,\nabla\cdot\phi) = k\|\sigma\|^2$ gives

$$\begin{aligned}k\|\sigma\|^2 &= \mu(\nabla w,\nabla\phi) + (\lambda+\mu)(\nabla\cdot w,\sigma) - (\sigma f,\phi)\\ &\quad + (\rho_2 u_2\cdot\nabla w,\phi) + (\rho_2 w\cdot\nabla u_1,\phi) + (\sigma u_1\cdot\nabla u_1,\phi)\\ &\le \mu\gamma\|\nabla w\|\|\sigma\| + \sqrt{3}(\lambda+\mu)\|\nabla w\|\|\sigma\| + \gamma\|f\|_3\|\sigma\|^2\\ &\quad + c\gamma\|\rho_2 u_2\|_3\|\nabla w\|\|\sigma\| + c\gamma\|\rho_2\|_\infty\|\nabla u_1\|_{3/2}\|\nabla w\|\|\sigma\|\\ &\quad + c\gamma\|u_1\cdot\nabla u_1\|_3\|\sigma\|^2\end{aligned}$$

and hence

$$\begin{aligned}&(k - \gamma\|f\|_3 - c\gamma\|u_1\nabla u_1\|_3)\|\sigma\|\\ &\qquad \le \left(\mu\gamma + \sqrt{3}(\lambda+\mu) + c\gamma\|\rho_2 u_2\|_3 + c\gamma\|\rho_2\|_\infty\|\nabla u_1\|_{3/2}\right)\|\nabla w\|\,.\end{aligned} \tag{71}$$

Let us write the estimates (68), (70) and (71) in the abbreviated form

$$\begin{aligned} \mu \|\nabla w\|^2 \leq k(\sigma, \nabla \cdot w) + m_1 \|\sigma\| \|\nabla w\| + m_2 \|\nabla w\|^2 + m_3 \|\sigma\| \|\nabla w\| \\ k(\sigma, \nabla \cdot w) \leq m_4 \|\sigma\|^2 + m_5 \|\sigma\| \|\nabla w\| \\ \overline{k} \|\sigma\| \leq M \|\nabla w\| \ , \end{aligned} \tag{72}$$

where the precise forms of the m_i, and of $\overline{k}$ and M, can be read from the full original estimates. The value of $\overline{k}$ will be positive, and the values of the m_i will all be small, if the force and the two solutions are small. While M is not small, it will be bounded if the solutions are. Combining these estimates gives

$$\begin{aligned} \mu \|\nabla w\|^2 & \leq m_4 \|\sigma\|^2 + m_5 \|\sigma\| \|\nabla w\| + m_1 \|\sigma\| \|\nabla w\| \\ & \quad + m_2 \|\nabla w\|^2 + m_3 \|\sigma\| \|\nabla w\| \\ & \leq \left(\frac{m_4 M^2}{\overline{k}^2} + \frac{(m_5 + m_1 + m_3) M}{\overline{k}} + m_2 \right) \|\nabla w\|^2 . \end{aligned} \tag{73}$$

Evidently, if the force is small, then any two small solutions must be identical. More precisely, we have the following result.

Theorem 2. *The coefficient on the right of (73) will be less than μ, and consequently the two solutions under consideration will be identical, if the force and the solutions are suitably restricted in terms of the quantities*

$$\gamma \|f\|_3 \ , \quad \|\nabla u_1\|_3 \ , \quad \|\nabla \rho_1\|_\infty \ , \quad \inf_\Omega \rho_1 \ , \quad \|\rho_2\|_\infty \ , \quad \|u_2\|_\infty \ , \quad \|\nabla \cdot u_2\|_\infty \ .$$

It remains a striking open problem to prove the uniqueness of a given solution without restrictions on the size of the second solution, when the size of the force is assumed to be small. A related problem, equally striking, is that even when the force is assumed to be small, we have no satisfactory a priori estimate that applies to all possible solutions.

References

[1] Amrouche, C. & V. Girault, On the existence and regularity of the solutions of Stokes problem in arbitrary dimension, Proc. Japan Acad., 67(1991), 171–175.

[2] Beirão da Veiga, H., An L^p-theory for the n-dimensional, stationary, compressible Navier-Stokes equations, and the incompressible limit for compressible fluids. The equilibrium solutions, Commun. Math. Phys., 109(1987), 229–248.

[3] Beirao da Veiga, H., On a stationary transport equation, Ann. Univ. Ferrara, 32(1986), 79–91.

[4] Beirao da Veiga, H., Existence results in Sobolev spaces for a stationary transport equation, Ricerche di Matematica suppl. 36(1987), 173–184.

[5] Beirao da Veiga, H., Boundary value problems for a class of first order partial differential equations in Sobolev spaces and applications to the Euler flow, Rend. Sem. Mat. Univ. Padova 79(1988), 247–273.

[6] Coscia, V. & G.P. Galdi, Existence, uniqueness and stability of regular motions of a second grade fluid, Int. J. Nonlinear Mechanics, 29(1994), 493–506.

[7] Farwig, R., Stationary solutions of compressible Navier-Stokes equations with slip boundary condition, Commun. In Partial Differential Equations, 14(11), (1989), 1579–1606.

[8] Galdi, G.P., Mathematical theory of second-grade fluids, CISM Courses and Lectures No. 344, Stability And Wave Propagation In Fluids And Solids, G.P. Galdi editor, Springer-Verlag 1995, 67–104.

[9] Galdi, G.P., An Introduction to the Mathematical Theory of the Navier-Stokes Equations, Vol.1, Springer-Verlag, corrected second printing 1998.

[10] Golay, F. & P. Helluy, Numerical simulation of a viscous compressible fluid based on a decomposition method, Université de Toulon, preprint 1998.

[11] Heywood, J.G. & M. Padula, On the steady transport equation, this book, 1999.

[12] Novotný, A., Steady flows of viscous compressible flows L^2 approach, SAACM, 3(1993), 181–199.

[13] Novotný, A., Some remarks to the compactness of steady compressible isentropic Navier-Stokes equations via the decomposition method, Comment. Math. Univ. Carolina, 37(1996), 1–38.

[14] Novotný, A. & M. Padula, L^p-approach to steady flows of viscous compressible fluids in exterior domains, Arch. Rational Mech. Anal. 126(1994), 234–297.

[15] Padula, M., An existence theorem for steady state compressible Navier-Stokes equations, Proc. Meeting "Waves and Stability in Continuous Media", Univ. Catania, 1981, 276–280.

[16] Padula, M., Existence and uniqueness for steady compressible flow, Proc. Meeting "Dinamica dei Continui Fluidi e dei gas ionizzati, Univ. Trieste, 1982, 237–258.

[17] Padula, M., Existence and uniqueness for viscous steady compressible motions, Arch. Rat. Mech. Anal., 97(1987), 89–102.

[18] Padula, M., A representation formula for steady solutions of a compressible fluid moving at low speed, Transport Theory Statistical Physics, 21(1992), 593–614.

[19] Padula, M., On the exterior steady problem for the equations of a viscous isothermal gas, Comm. Math. Carolinae, 34(1993), 275–293.

[20] Padula, M., On the regularity of steady flows of viscous compressible fluids, Int. Conf. on Navier-Stokes Equations: Theory and Numerical Methods, ed. R. Salvi, Addison Wesley Longman Limited 1997.

[21] Temam, R., On the Euler equations of incompressible perfect fluids, J. Fun. Anal., 20(1975), 32–43.

[22] Valli, A., On the existence of stationary solutions to compressible Navier-Stokes equations, Ann. Inst. H. Poincaré, Anal. Non Linéaire 4(1987), 99–113.

John G. Heywood
Department of Mathematics
University of British Columbia

Mariarosaria Padula
Dipartimento di Matematica
Università di Ferrara

Finite Element Methods for the Incompressible Navier-Stokes Equations

Rolf Rannacher*

Abstract. These notes are based on lectures given in a Short Course on Theoretical and Numerical Fluid Mechanics in Vancouver, British Columbia, Canada, July 27–28, 1996, and at several other places since then. They provide an introduction to recent developments in the numerical solution of the Navier-Stokes equations by the finite element method. The material is presented in eight sections:

1. Introduction: Computational aspects of laminar flows
2. Models of viscous flow
3. Spatial discretization by finite elements
4. Time discretization and linearization
5. Solution of algebraic systems
6. A review of theoretical analysis
7. Error control and mesh adaptation
8. Extension to weakly compressible flows

Theoretical analysis is offered to support the construction of numerical methods, and often computational examples are used to illustrate theoretical results. The variational setting of the finite element Galerkin method provides the theoretical framework. The goal is to guide the development of more efficient and accurate numerical tools for computing viscous flows. A number of open theoretical problems will be formulated, and many references are made to the relevant literature.

1. Introduction

In the following sections, we will discuss a computational methodology for simulating viscous incompressible laminar flows. The description of the numerical algorithms will be accompanied by a heoretical analysis so far as it is relevant to understanding the performance of the method. In this sense, these notes are meant as a contribution of Mathematics to "CFD" (Computational Fluid Dynamics).

The author acknowledges the support by the German Research Association (DFG) through the SFB 359 "Reactive Flow, Diffusion and Transport" at the University of Heidelberg, Im Neuenheimer Feld 294, D-69120 Heidelberg, Germany.

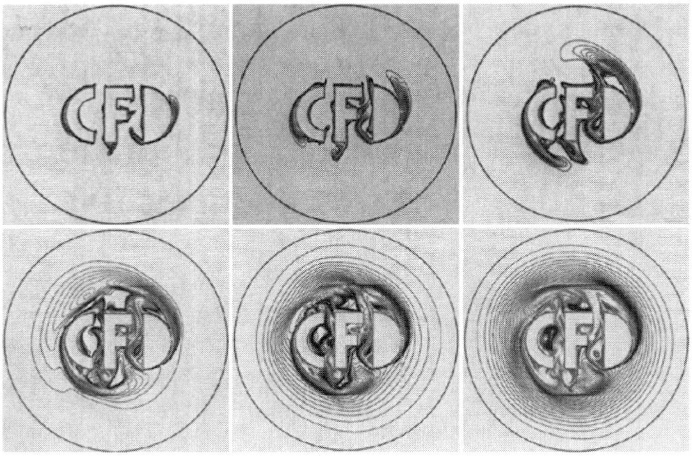

FIGURE 1. Nonstationary flow around "CFD" for $\mathrm{Re} = 500$, driven by rotation of the outer circle and visualized by temperature isolines; from Turek [98].

The established model for viscous Newtonian incompressible flow is the the system of Navier-Stokes equations,

$$\partial_t v - \nu \Delta v + v \cdot \nabla v + \nabla p = f, \quad \nabla \cdot v = 0, \tag{1.1}$$

in some region $\Omega \times (0, T)$ with appropriate initial and boundary conditions. We concentrate on "laminar" flows, i.e., on flows with Reynolds number in the range $1 \leq \mathrm{Re} \leq 10^5$, where $\mathrm{Re} \sim \bar{v}\bar{l}/\nu$. The numerical solution of this system involves several typical difficulties:

- Complicated flow structure $\Rightarrow$ fine meshes!
- $\mathrm{Re} \gg 1$ $\Rightarrow$ locally refined and anisotropic meshes in boundary layers!
- Dominant nonlinear effects $\Rightarrow$ stability problems!
- Constraint $\nabla \cdot v = 0$ $\Rightarrow$ implicit solution!
- Sensitive quantities $\Rightarrow$ solution-adapted meshes!

Accurate flow prediction requires the use of large computer power, particularly for the extension from 2D to 3D, from stationary to nonstationary flows, and from qualitative results to quantitatively accurate results. The key goals in the developing tools for computing laminar flows are:

- *fast* (nonstationary calculations in minutes or hours),
- *cheap* (simulations on workstations),
- *flexible* (general purpose solver),
- *accurate* (adaptive error control).

1.1. Solution method

The method of choice in these notes is the "Finite Element Method" (FEM) for computing the primitive variables v (velocity) and p (pressure). This special Galerkin method is based an a variational formulation of the Navier-Stokes problem in appropriate function spaces, and determines "discrete" approximations in certain finite dimensional subspaces ("trial spaces") consisting of piecewise polynomial functions. By this approach the discretization inherits most of the rich structure of the continuous problem, which, on the one hand provides a *high computational flexibility* and on the other hand facilitates a *rigorous mathematical error analysis.* These are the main aspects which make the FEM increasingly attractive in CFD. For completeness, we briefly comment on the essential features of the main competitors of the FEM:

- **Finite difference methods (FDM):** Approximation of the Navier-Stokes equations in their "strong" form by finite differences:
  \+ easy implementation,
  − problems along curved boundaries,
  − difficult stability and convergence analysis,
  − mesh adaptation difficult.
- **Finite volume methods (FVM):** Approximation of the Navier-Stokes equations as a system of (cell-wise) conservation equations:
  \+ based on "physical" conservation properties,
  − problems on unstructured meshes,
  − difficult stability and convergence analysis,
  − only heuristic mesh adaptation.
- **Spectral Galerkin methods:** Approximation of the Navier-Stokes equations in their variational form by a Galerkin method with "high-order" polynomial trial functions:
  \+ high accuracy,
  − treatment of complex domains difficult,
  − mesh and order (hp)-adaptation difficult.

This brief classification must be superficial and is based on personal taste. The details are the subject of much controversial discussion concerning the pros and cons of the various methods and their variants. However, this conflict is partially resolved in many cases, as the differences between the methods, particularly between FEM and FVM, often disappear on general meshes. In fact, some of the FVMs can be interpreted as variants of certain "mixed" FEMs.

1.2. Examples of computable viscous flows

Below, we give some examples of flows which can be computed by the methods described in these notes. More examples including movies of nonstationary flows can be seen on our homepage: `http://gaia.iwr.uni-heidelberg.de/`.
Some of the computer codes are available for research purposes:

- FEATFLOW Code (FORTRAN 77) by S. Turek [96], [97]: `http://gaia.iwr.uni-heidelberg.de/~featflow/`.
- deal.II Code (C++) by W. Bangerth and G. Kanschat [5]: `http://gaia.iwr.uni-heidelberg.de/~deal/`.

A collection of experimental photographs of such "computable" flows can be found in Van Dyke's book *"An Album of Fluid Motion"* [99]. In the following, we present some examples of viscous flows which have been computed by the methodology described in these notes. Most of these results emerged as side products in the course of developing the numerical solvers and testing them for standard benchmark problems. All computations were done on normal work stations.

Example 1. Cavity flow: The first example is stationary and nonstationary flow in a cavity driven by flow along the upper boundary ("driven cavity").

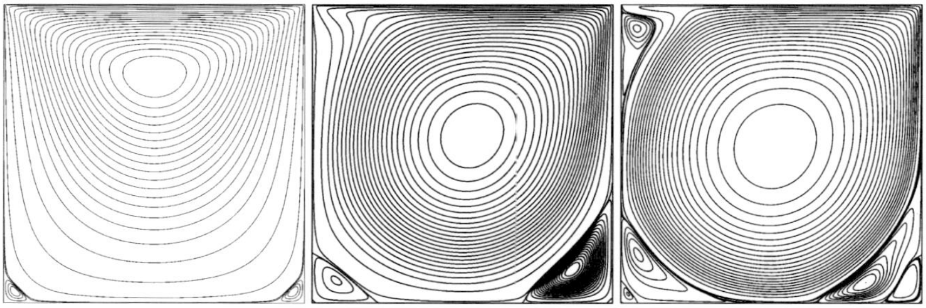

FIGURE 2. Stationary driven cavity flow in 2D for Re $=$ $1, 1000, 9000$ (from left to right); for Re > 10000 the flow becomes nonstationary; from Turek [98].

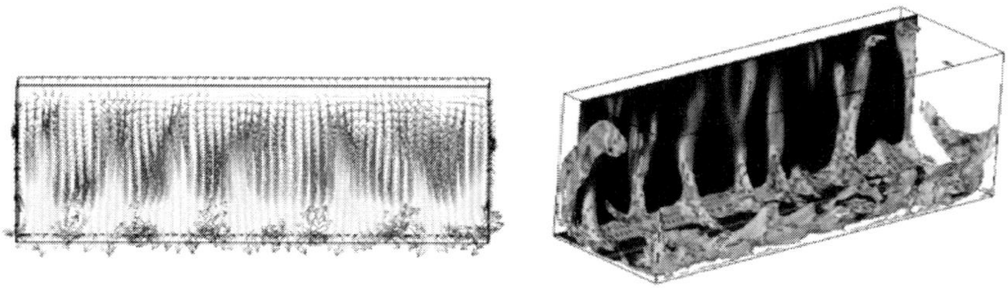

FIGURE 3. Simulation of nonstationary 3D driven cavity flow for Re $= 100$; from Oswald [66].

Example 2. Vortex shedding: The second example is nonstationary flow around a circular cylinder ("von Kármán vortex street")

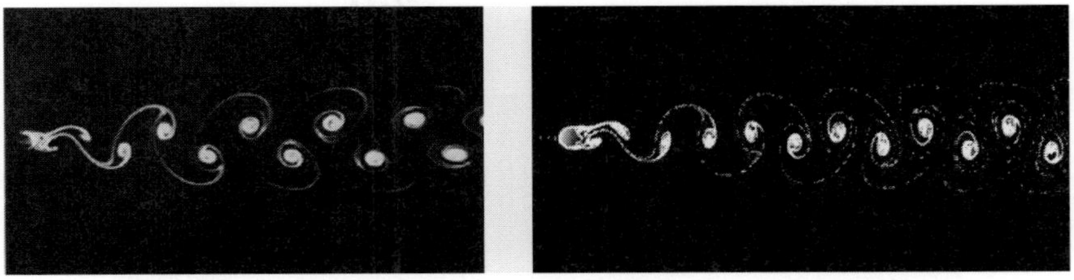

FIGURE 4. Von Kármán vortex street; experiment with Re = 105 (left; from Van Dyke [99]), and 2D computation with Re = 100 (right; from Turek [98]).

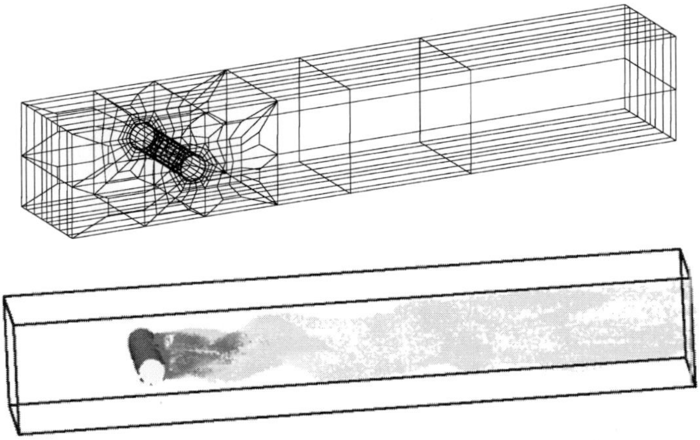

FIGURE 5. 3D simulation of vortex shedding behind a cylinder for Re = 100, coarse grid and flow visualized by particle tracing; from Oswald [66].

Example 3. Leapfrogging of vortex rings: Two successive puffs of fluid are injected through a hole and develop into vortex rings dancing around each other. The second ring travels faster in the induced wake of the first and slips through it. Then the first ring slips through the second, and so on.

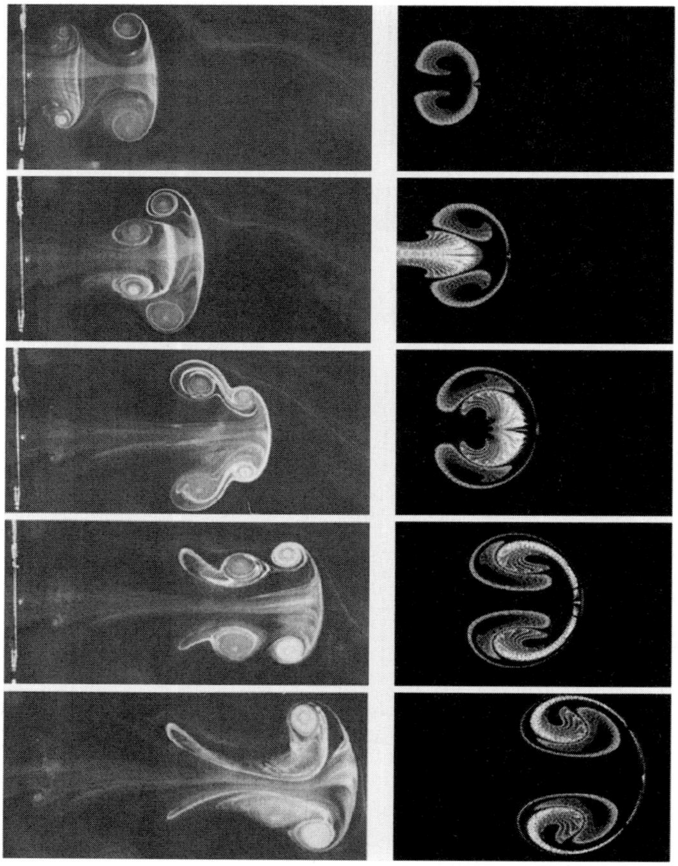

FIGURE 6. Leapfrogging of two vortex rings; experiment for Re ≈ 1600 (left; from Van Dyke [99]) and 2D computation for Re ≈ 500 (right; from [46]).

1.2.1. EXTENSIONS BEYOND STANDARD NAVIER-STOKES FLOW The numerical methodology described in these notes has primarily been developed for computing viscous incompressible Newtonian flows. However, extensions are possible in several directions. These include flows in regions with moving boundaries, as for example pipe flow driven by rotating propellers, and flow of non-Newtonian fluids modelled by a simple power-law rule. The extension to certain low-speed compressible flows will be discussed in more detail below in Section 8.

1) Flow regions with moving boundaries

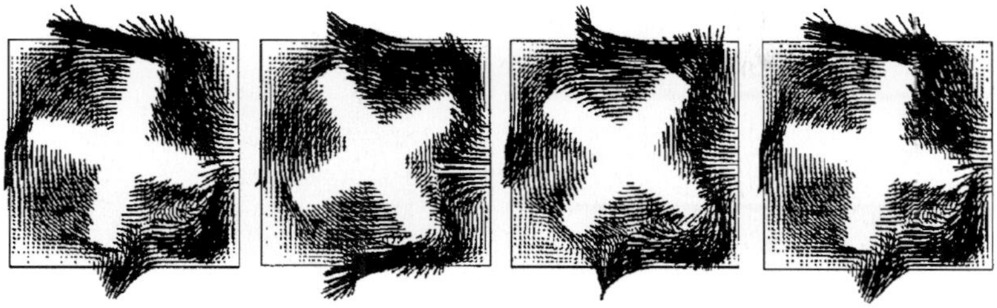

FIGURE 7. Velocity plot of 2D flow in a box driven by a rotating cross, computed by a "virtual boundary" technique; from Turek [97].

2) Flow of a non-Newtonian fluid

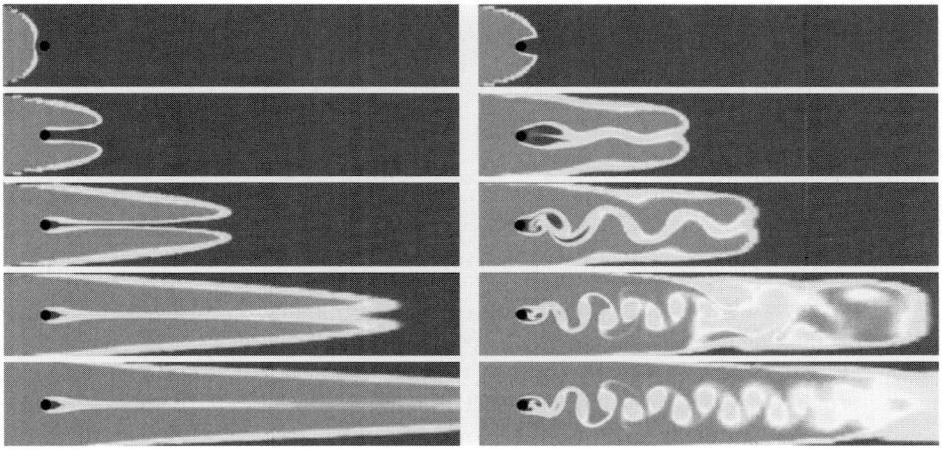

FIGURE 8. Computation of the flow of a non-Newtonian fluid around a circular obstacle in a 2D channel ("power-law" $\nu' = \nu(1+|D(v)|)^{-1}$): stationary flow in the Newtonian case (left) and nonstationary flow in the non-Newtonian case (right), both for the same Reynolds number $\mathrm{Re} = 20$; from Turek [98].

3) Low-Mach-number compressible flow

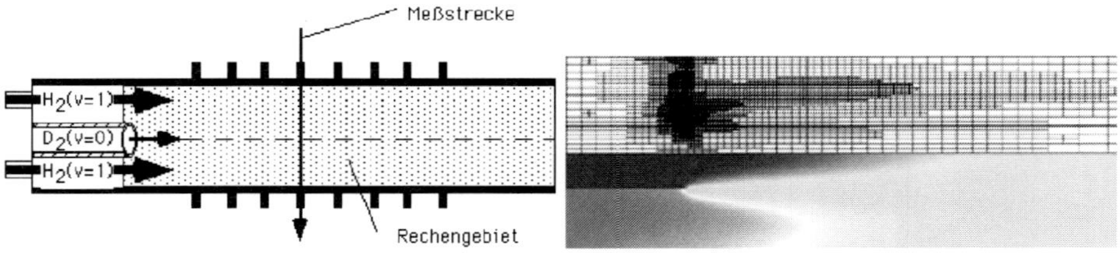

FIGURE 9. Computation of compressible flow in a chemical flow reactor: flow configuration and mass fraction of excited $H_2^{(\nu=1)}$ computed on a locally refined mesh; from Waguet [103] and [10].

2. Models of viscous flow

The mathematical model for describing viscous (Newtonian) flows is the system of Navier-Stokes equations, which are the equations of conservation of mass, momentum and energy:

$$\partial_t\rho + \nabla\cdot[\rho v] = 0\,, \tag{2.1}$$

$$\partial_t(\rho v) + \rho v\cdot\nabla v - \nabla\cdot[\mu\nabla v + \tfrac{1}{3}\mu\nabla\cdot vI] + \nabla p_{tot} = \rho f\,, \tag{2.2}$$

$$\partial_t(c_p\rho T) + c_p\rho v\cdot\nabla T - \nabla\cdot[\lambda\nabla T] = h\,. \tag{2.3}$$

Here, v is the velocity, ρ the density, p_{tot} the (total) pressure, and T the temperature of the fluid occupying a two- or three-dimensional region Ω. The parameters $\mu > 0$ (dynamic viscosity), $c_p > 0$ (heat capacity) and $\lambda > 0$ (heat conductivity) characterize the properties of the fluid. The volume force f and the heat source h are explicitly given. Since we only consider low-speed flows, the influence of stress and hydrodynamic pressure in the energy equation can be neglected. In temperature-driven flows, h may implicitly depend on the temperature and further quantities describing heat release, as for example by chemical reactions. Such "weakly compressible" flows will be briefly considered at the end of these notes in Section 8. Here, the fluid is assumed to be *incompressible* and the density to be homogeneous, $\rho \equiv \rho_0 =$ const., so that (2.1) reduces to the incompressibility constraint

$$\nabla\cdot v = 0\,. \tag{2.4}$$

In this model, we consider as the primal unknowns the velocity v, the pressure $p = p_{tot}$, and the temperature T. For most parts of the discussion, the flow is assumed to be isothermal, so that the energy equation decouples from the momentum and continuity equations, and the temperature only enters through the viscosity parameter. The system is closed by imposing appropriate initial and boundary conditions for the flow variables

$$v_{|t=0} = v^0, \quad v_{|\Gamma_{rigid}} = 0\,,\ v_{|\Gamma_{in}} = v^{in}\,,\ (\mu\partial_n v + pn)_{|\Gamma_{out}} = 0\,, \tag{2.5}$$

and corresponding ones for the temperature, where Γ_{rigid}, Γ_{in}, Γ_{out} are the rigid part, the inflow part and the outflow part of the boundary $\partial\Omega$, respectively. The role of the natural outflow boundary condition on Γ_{out} will be explained in more detail below.

In the isothermal case, assuming for simplicity that $\rho_0 = 1$, the Navier-Stokes system can be written in short as

$$\partial_t v + v\cdot\nabla v - \nu\Delta v + \nabla p = f\,, \quad \nabla\cdot v = 0\,, \tag{2.6}$$

with the kinematic viscosity parameter $\nu = \mu/\rho_0$. In this formulation the domain Ω may be taken two- or three-dimensional according to the particular requirements of the simulation. In our examples, we shall mostly refer to the two-dimensional case. Through a scaling transformation this problem is made dimensionless, with the Reynolds Number $Re = UL/\nu$ as the characteristic parameter, where U is the reference velocity (e.g., $U \approx \max|v^{\rm in}|$) and L the characteristic length (e.g., $L \approx \mathrm{diam}(\Omega)$), of the problem.

As common in the mathematical literature, we assume that $U = 1$ and $L = 1$ and consider $\nu := 1/Re$ as a *dimensionless* parameter characterizing in some sense the "complexity" of the flow problem. Then, the length of the time interval over which the solution develops its characteristic features is $T \approx 1/\nu$, and the relevant scale of its spatial structures is $\delta x \approx \sqrt{\nu}$ (width of boundary layers). This has to be kept in mind when the right spatial mesh size h and the time step k is chosen for a numerical simulation which is supposed to resolve all structures of the flow; for a more detailed discussion of the issue of reliable transient flow calculation see [54].

2.1. Variational formulation

The finite element discretization of the Navier-Stokes problem (2.6) is based on its variational formulation. To this end, we use the following sub-spaces of the usual Lebesgue function space $L^2(\Omega)$ of square-integrable functions on Ω:

$$\begin{aligned} L^2_0(\Omega) &= \{\varphi \in L^2(\Omega) :\ (v,1) = 0\}\,, \\ H^1(\Omega) &= \{v \in L^2(\Omega),\ \partial_i v \in L^2(\Omega),\ 1 \le i \le d\}\,, \\ H^1_0(\Gamma;\Omega) &= \{v \in H^1(\Omega),\ v_{|\Gamma} = 0\}\,,\quad \Gamma \subset \partial\Omega, \end{aligned}$$

and the corresponding inner products and norms

$$\begin{aligned} &(u,v) = \int_\Omega uv\,dx\,, \quad \|v\| = (v,v)^{1/2}\,, \\ &\|\nabla v\| = (\nabla v,\nabla v)^{1/2}\,, \quad \|v\|_1 = \left(\|v\|^2 + \|\nabla v\|^2\right)^{1/2}\,. \end{aligned}$$

These are all spaces of $\mathbb{R}$-valued functions. Spaces of $\mathbb{R}^d$-valued functions $v = (v_1, \dots, v_n)$ are denoted by boldface-type, but no distinction is made in the notation of norms and inner products; thus $\mathbf{H}_0^1(\Gamma;\Omega) = H_0^1(\Gamma;\Omega)^d$ has norm $\|v\|_1 = (\sum_{i=1}^d \|v_i\|_1^2)^{1/2}$, etc. All the other notation is self-evident: $\partial_t u = \partial u/\partial t$, $\partial_i u = \partial u/\partial x_i$, $\partial_n v = n\cdot\nabla v$, $\partial_\tau = \tau\cdot\nabla v$ etc., where n and τ are the normal and tangential unit vectors along the boundary $\partial\Omega$.

The pressure p in the Navier-Stokes equations is uniquely (possibly up to a constant) determined by the velocity field v. This follows from the fact that every bounded functional $F(\cdot)$ on $\mathbf{H}_0^1(\Gamma;\Omega)$ which vanishes on the subspace

$$\mathbf{J}_1(\Gamma;\Omega) = \{v \in \mathbf{H}_0^1(\Gamma;\Omega),\ \nabla\cdot v = 0\}$$

can be expressed in the form $F(\varphi) = (p, \nabla\cdot\varphi)$ for some $p \in L^2(\Omega)$. Further, there holds the stability estimate ("inf-sup" stability)

$$\inf_{q\in L^2(\Omega)} \left\{ \sup_{\phi\in\mathbf{H}_0^1(\Gamma;\Omega)} \frac{(q,\nabla\cdot\phi)}{\|\nabla\phi\|} \right\} \ge \gamma_0 > 0\,, \tag{2.7}$$

where $L^2(\Omega)$ has to be replaced by $L_0^2(\Omega)$ in the case $\Gamma = \partial\Omega$. For proofs of these facts, one may consult the first parts of the books of Ladyshenskaya [58], Temam [88] and Girault/Raviart [29]. Finally, we introduce the notation

$$a(u,v) := \nu(\nabla u, \nabla v)\,, \quad n(u,v,w) := (u\cdot\nabla v, w)\,, \quad b(p,v) := -(p, \nabla\cdot v),$$

and the abbreviations

$$\mathbf{H} := \mathbf{H}_0^1(\Gamma;\Omega), \quad L := L^2(\Omega) \quad \big(L := L_0^2(\Omega) \text{ in the case } \Gamma = \partial\Omega\big),$$

where $\Gamma = \Gamma_{in} \cup \Gamma_{rigid}$. Then, the variational formulation of the Navier-Stokes problem (2.6), reads as follows: Find functions $v(\cdot,t) \in v^{\rm in} + \mathbf{H}$ and $p(\cdot,t) \in L$, such that $v_{|t=0} = v^0$, and setting $\Gamma := \Gamma_{in} \cup \Gamma_{rigid}$,

$$(\partial_t v, \varphi) + a(v,\varphi) + n(v,v,\varphi) + b(p,v) = (f,\varphi) \quad \forall \varphi \in \mathbf{H}\,, \tag{2.8}$$

$$(\nabla\cdot v, \chi) = 0 \quad \forall \chi \in L\,. \tag{2.9}$$

It is well known that in two space dimensions the pure Dirichlet problem (2.8), (2.9), with $\Gamma_{out} = \emptyset$, possesses a unique solution on any time interval $[0,T]$, which is also a classical solution if the data of the problem are smooth enough. For small viscosity, i.e., large Reynolds numbers, this solution may be unstable. In three dimensions, the existence of a unique solution is known only for sufficiently small data, e.g., $\|v^0\|_1 \approx \nu$, or on sufficiently short intervals of time, $0 \le t \le T$, with $T \approx \nu$.

2.2. Regularity of solution

We collect some results concerning the regularity of the variational solution of the Navier-Stokes problem which are relevant for the understanding of its numerical approximation. One obtains quantitative regularity bounds from the following sequence of differential identities

$$\begin{aligned}
&\tfrac{1}{2}d_t\|v\|^2 + \nu\|\nabla v\|^2 = (f,v),\\
&\tfrac{1}{2}d_t\|\nabla v\|^2 + \nu\|\Delta v\|^2 = -(f,\Delta v) + (v\cdot\nabla v,\Delta v),\\
&\tfrac{1}{2}d_t\|\partial_t v\|^2 + \nu\|\nabla\partial_t v\|^2 = (\partial_t f,\partial_t v) - (\partial_t v\cdot\nabla v,\partial_t v),\\
&\tfrac{1}{2}d_t\|\nabla\partial_t v\|^2 + \nu\|\Delta\partial_t v\|^2 = -(\partial_t f,\Delta\partial_t v) + (\partial_t v\cdot\nabla v,\Delta\partial_t v) + \cdots,\\
&\tfrac{1}{2}d_t\|\partial_t^2 v\|^2 + \nu\|\nabla\partial_t^2 v\|^2 = (\partial_t^2 f,\partial_t^2 v) - (\partial_t^2 v\cdot\nabla v,\partial_t^2 v) - \cdots,\\
&\ldots
\end{aligned}$$

which are easily derived by standard energy arguments; see [40] and [41]. Assuming a bound on the Dirichlet norm of v,

$$\sup_{t\in(0,T]}\|\nabla v(t)\| \le M, \tag{2.10}$$

the above estimates together with the usual elliptic regularity results imply that v is smooth on $\bar\Omega$ for $0<t\le T$, if all the data and $\partial\Omega$ are smooth. However, for the purposes of numerical analysis one needs regularity estimates which hold uniformly for $t\to 0$. To get such information from the above equations requires starting values for all the quantities $\|v\|$, $\|\nabla v\|$, $\|\partial_t v\|$, $\|\nabla\partial_t v\|$, $\|\partial_t^2 v\|$, etc., at $t=0$. However, there is a problem already with $\|\nabla\partial_t v(0)\|$, as has been demonstrated in [41].

2.2.1. Compatibility conditions at $t=0$ To investigate this phenomenon, let us assume that the solution $\{v,p\}$ is uniformly smooth as $t\to 0$. Then, applying the divergence operator to the Navier-Stokes equations and letting $t\to 0$ implies:
(i) in Ω:

$$\nabla\cdot(\partial_t v + v\cdot\nabla v) = \nabla\cdot(\nu\Delta v - \nabla p) \quad\to\quad \nabla\cdot(v^0\cdot\nabla v^0) = -\Delta p^0,$$

(ii) on $\partial\Omega$:

$$\partial_t v + v\cdot\nabla v = \nu\Delta v - \nabla p \quad\to\quad \partial_t g_{|t=0} + v^0\cdot\nabla v^0 = \nu\Delta v^0 - \nabla p^0,$$

where g is the boundary data, v^0 the initial velocity and $p^0 := \lim_{t\to 0} p(t)$ the "initial pressure". Hence, in the limit $t=0$, we obtain an overdetermined Neumann problem for the initial pressure:

$$\Delta p^0 = -\nabla\cdot(v^0\cdot\nabla v^0) \text{ in } \Omega, \tag{2.11}$$

$$\nabla p^0{}_{|\partial\Omega} = \nu\Delta v^0 - \partial_t g_{|t=0} - v^0\cdot\nabla v^0, \tag{2.12}$$

including the compatibility condition

$$\partial_\tau p^0{}_{|\partial\Omega} = \tau\cdot(\nu\Delta v^0 - \partial_t g_{|t=0} - v^0\cdot\nabla v^0), \tag{2.13}$$

where τ is the tangent direction along $\partial\Omega$. If this compatibility is violated, then $\lim_{t\to 0}\{\|\nabla^3 v(t)\| + \|\nabla\partial_t v(t)\|\} = \infty$; see [41]. We emphasize that (2.13) together with (2.11) is a *global* condition which in general cannot be verified for given data. Without (2.13) being satisfied the maximum degree of regularity is right in the middle of $\mathbf{H}^2$ and $\mathbf{H}^3$; see [78]. In view of the foregoing discussion, the natural regularity assumption for the nonstationary Navier-Stokes equations (without additional compatibility condition) is

$$v^0 \in \mathbf{J}_1(\Omega) \cap \mathbf{H}^2(\Omega) \quad \Rightarrow \quad \sup_{t\in(0,T]} \big\{\|\nabla^2 v(t)\| + \|\partial_t v(t)\|\big\} < \infty. \tag{2.14}$$

Example: *Flow between two concentric spheres ("Taylor problem")*
Let the inner sphere with radius r_{in} be accelerated from rest $v^0 = 0$ with a constant acceleration ω, i.e., $v_{|\Gamma_{in}}\cdot(n,\tau_\theta,\tau_\phi)^T = r_{in}\cos(\theta)(0,0,\omega t)^T$ (in polar coordinates), while at the outer sphere, we set $v_{|\Gamma_{out}} = 0$. Accordingly, the Neumann problem for the "initial pressure" takes the form

$$\Delta p^0 = 0 \quad \text{in } \Omega, \quad \partial_n p^0{}_{|\partial\Omega} = 0,$$

which implies that $p^0 \equiv const$. However, this conflicts with the compatibility condition (2.13) which in this case reads

$$\partial_\phi p^0{}_{|\Gamma_{in}} = -\partial_t (v^0 \cdot n_\phi)_{t=0} = -r_{in}\cos(\theta)\omega \neq 0\,.$$

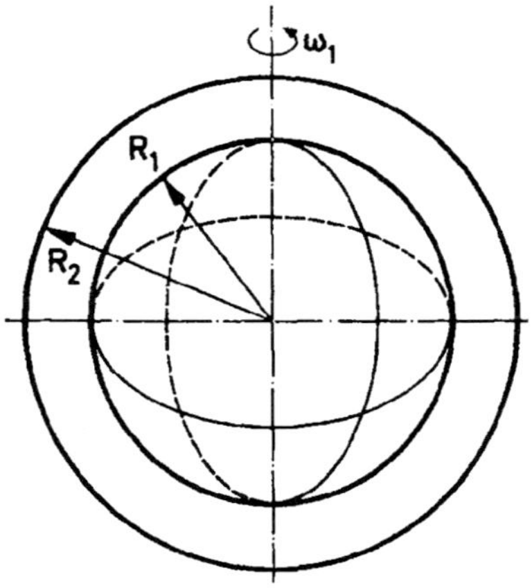

In order to describe the "natural" regularity of the solution $\{v(t), p(t)\}$, as $t \to 0$, and as $t \to \infty$, in [41] a sequence of time-weighted a priori estimates has been proven under the assumption (2.10) using the weight functions $\tau(t) = \min(t,1)$ and $e^{\alpha t}$, with fixed $\alpha > 0$:

$$\tau(t)^{2n+m-2}\Big\{\|\nabla^m \partial_t^n v(t)\| + \|\nabla^{m-1}\partial_t^n p(t)\|\Big\} \leq K, \tag{2.15}$$

and

$$e^{-\alpha t}\int_0^t e^{\alpha s}\tau(s)^{2n+m-2}\Big\{\|\nabla^m \partial_t^n v(t)\|^2 + \|\nabla^{m-1}\partial_t^n p(t)\|^2\Big\}\,dt \leq K, \tag{2.16}$$

for any $m \geq 2$, $n \geq 1$.

Open problem 2.1: *Devise a way to construct for any given initial data v^0 (e.g., fitted from experimental data) and any $\epsilon > 0$ a smooth initial data $\tilde v^0 \in \mathbf{J}_1(\Gamma;\Omega)$,* *such that* $\|v^0 - \tilde{v}^0\|_1 \le \epsilon$, *and the resulting solution of the Navier-Stokes equations satisfies the compatibility condition (2.13) at* $t = 0$.

2.3. Outflow boundary conditions

Numerical simulation of flow problems usually requires the truncation of a conceptionally unbounded flow region to a bounded computational domain, thereby introducing artificial boundaries, along which some kind of boundary conditions are needed. The variational formulation (2.8), (2.9) does not contain an explicit reference to any "outflow boundary condition". Suppose that the solution $v \in v^{in} + \mathbf{H}$, $p \in L$ is sufficiently smooth. Then, integration by parts on the terms

$$\nu(\nabla v \nabla \phi) - (p, \nabla \cdot \phi) = \int_{\Gamma_{out}} \{\nu \partial_n v - pn\}\phi \, do + (-\nu \Delta v + \nabla p, \phi)$$

yields the already mentioned "natural" condition on the outflow boundary

$$\nu \partial_n v - pn = 0 \quad \text{on } \Gamma_{out}. \tag{2.17}$$

This condition has proven to be well suited in modeling (essentially) parallel flows, see, e.g., [46], Turek [93, 97]. It naturally occurs in the variational formulation of the problem if one does not prescribe any boundary condition for the velocity at the outlet suggesting the name "do nothing" boundary condition.

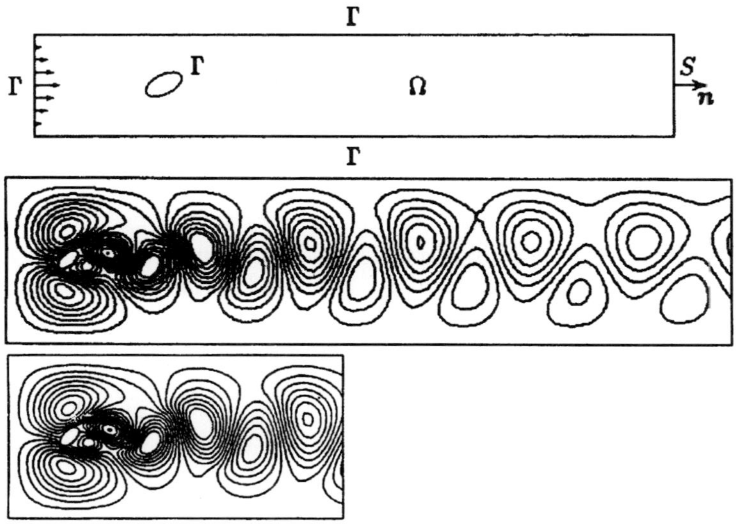

FIGURE 10. The effect of the "do nothing" outflow boundary condition shown by pressure isolines for unsteady flow around an inclined ellipse at Re$=$500; from [46].

In the following, we present some experiences in choosing the boundary conditions implicitly, through the choice of variational formulations of flow problems used in finite element computations. To fix ideas, let us begin by considering a com-

mon test problem, that of calculating nonstationary flow past an obstacle (here an inclined ellipse), situated in a rectangular channel.
We impose the usual no-slip boundary conditions on the channel walls and on the surface of the ellipse, while a parabolic "Poiseuille" inflow profile is prescribed on the upstream boundary. We denote again by Γ that portion of the boundary on which Dirichlet conditions are imposed. At the downstream boundary $S = \Gamma_{out}$, we decide to "do nothing" and leave the solution and the test space free by choosing $\mathbf{H} = \mathbf{H}_0^1(\Gamma;\Omega)$ and $L = L^2(\Omega)$. This results in the free-outflow condition (2.17). The results of computations based on (2.17) show a truly remarkable "transparency" of the downstream boundary when it is handled in this way; see Figure 10 where almost no effect of shortening the computational domain is seen.

2.3.1. PROBLEMS WITH THE "DO NOTHING" OUTFLOW CONDITION Although, the "do nothing" outflow boundary condition seems to yield very satisfactory results, one should use it with care. For example, if the flow region contains more than one outlet, like in flows through systems of pipes, undesirable effects may occur, since the "do nothing" condition contains as an additional hidden condition that the mean pressure is zero across the outflow boundary. In fact integrating (2.17) over any component S of the outflow boundary (a straight segment) and using the incompressibility constraint $\nabla\cdot v = 0$ yields

$$\int_S pn\, do = \nu \int_S \partial_n v\, do = -\nu \int_S \partial_t v\, do = v(s_2) - v(s_1) = 0\,.$$

Here s_i denote the end points of S at which $v(s_i) = 0$, due to the imposed no-slip condition along Γ. Consequently, the mean pressure over S must be zero:

$$\int_S p\, do = 0\,. \tag{2.18}$$

To illustrate this, let us consider low Reynolds number flow through a junction in a system of pipes, again prescribing a Poiseuille inflow upstream. In Figure 11, we show steady streamlines for computations based on the same variational formulations as above, each with the same inflow, but with varying lengths of pipe beyond the junction. Obviously, making one leg of the pipe longer significantly changes the flow pattern. The explanation of this effect is that by the property (2.18), in Figure 11 the pressure gradient is greater in the shorter of the two outflow sections, which explains why there is a greater flow through that section.

2.3.2. MODIFICATION OF TRANSPORT OR DIFFUSION MODEL The foregoing example suggests that one might consider formulating problems more generally, e.g., in terms of *prescribed pressure drops* or *prescribed fluxes*; we refer to [46] for a thorough development of the corresponding variational formulations. Both choices of boundary conditions lead to well posed formulations of the problem. However, the situation is less satisfactory than in the case of pure Dirichlet boundary conditions. Although the variational problem looks well set in this situation, surprisingly there is a problem with its well posedness. The related Dirichlet problem of the

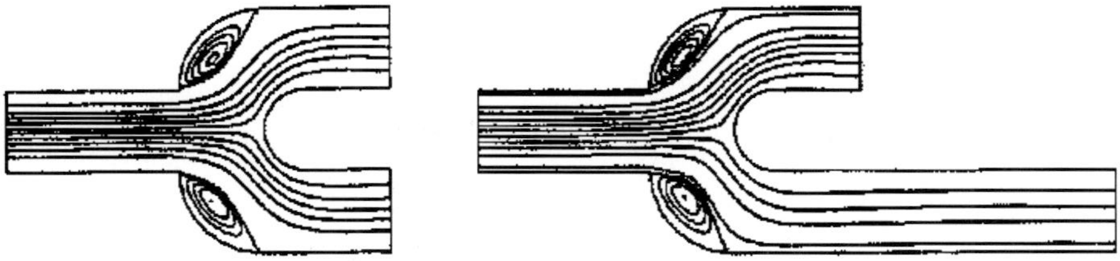

FIGURE 11. The effect of the "do nothing" outflow boundary condition shown by streamline plots for flow through a bifurcating channel for $\mathrm{Re} = 20$; from [46].

Navier-Stokes equations, stationary as well as nonstationary, is well known to possess weak solutions (not necessarily unique or stable) for any Reynolds number. The standard argument for this result is based upon the "conservation property" $(v\cdot\nabla v, v) = 0$ of the nonlinear term, which is obtained by integration by parts and using $\nabla\cdot v = 0$. In the case of a "free" boundary this relation is replaced by

$$(v\cdot\nabla v, v) = -\tfrac{1}{2}(n\cdot v, v^2)_\Gamma, \quad \Gamma = \cup_i\Gamma_i, \tag{2.19}$$

which generally does not allow to bound the energy in the system without a priori knowledge of what is an inflow and what is an outflow boundary. As a consequence, in [46] the existence of a unique solution could be shown, even in two space dimensions, only for sufficiently small data. Kracmar/Neustupa [57] have treated the case of general data by formulating the problem as a variational inequality including the energy bound as a constraint. This still leaves the question open whether one can expect existence of solutions for the original formulation with general data. A positive answer is suggested by numerical tests which do not show any unexpected instability with the discrete analogues of the formulation (2.8), (2.9) in the case of higher Reynolds numbers.

One may suspect that this theoretical difficulty can be avoided simply by changing the variational formulation of the problem, i.e., using other variational representations of the transport or diffusion terms. It has been suggested to replace in the momentum equation (2.8) the Dirichlet form $(\nabla v, \nabla\phi)$ by $(D[v], D[\phi])$, with $D[v] = \frac{1}{2}(\partial_i v_j + \partial_j v_i)_{i,j=1}^d$ being the deformation tensor. This change has no effect in the case of pure Dirichlet boundary conditions as then the two forms coincide. But in using the "do nothing" approach this modification leads to the outflow boundary condition

$$n\cdot D[v] - pn = 0 \quad \text{on } \Gamma_{out},$$

which may result in a non-physical behavior of the flow. In the case of simple Poiseuille flow the streamlines are bent outward as shown in Figure 12.

Another possible modification is to enforce the conservation property on the transport terms. Using the identity $\nabla(\frac{1}{2}|v|^2) = v\cdot(\nabla v)^T$, the transport term can be written in the form

$$v\cdot\nabla v = v\cdot\nabla v - v\cdot(\nabla v)^T + \tfrac{1}{2}\nabla|v|^2.$$

This leads to a variational formulation in which $(v\cdot\nabla v, \phi)$ is replaced by

$$\tilde{b}(v, v, \phi) := (v\cdot\nabla v, \phi) - (\phi\cdot\nabla v, v), \tag{2.20}$$

while the term $\frac{1}{2}|v|^2$ is absorbed into the pressure. An alternative form is

$$\tilde{b}(v, v, \phi) := \tfrac{1}{2}(v\cdot\nabla v, \phi) - \tfrac{1}{2}(v\cdot\nabla\phi, v), \tag{2.21}$$

which is legitimate because $\tilde{b}(v, v, \phi) = (v\cdot\nabla v, \phi)$ for $v \in \mathbf{J}_1(\Omega)$. Notice that in both cases $\tilde{b}(w, \phi, \phi) = 0$ for any w. The corresponding natural outflow boundary conditions are, for (2.20):

$$\nu\partial_n v - \overline{p}n = 0, \tag{2.22}$$

with the so-called "Bernoulli pressure" $\overline{p} = p + \frac{1}{2}|v|^2$, and for (2.21):

$$\nu\partial_n v - \tfrac{1}{2}|n\cdot v|^2 n - pn = 0. \tag{2.23}$$

Both modifications again result in a non-physical behavior across the outflow boundary; streamlines bent inward as shown in Figure 12. Hence, for physical reasons, it seems to be necessary to stay with the original formulation (2.8), (2.9). For a detailed discussion of the boundary conditions, and for an extensive list of references, we refer the reader to Gresho [34] and Gresho and Sani [35].

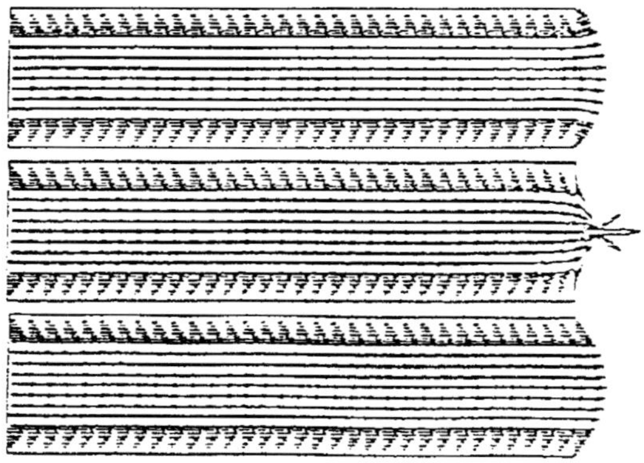

FIGURE 12. The effect of using the deformation tensor formulation (top) or the symmetrized transport formulations (middle) together with the "do nothing" outflow boundary condition compared to the correct Poiseuille flow (bottom); from [46].

Open Problem 2.2: *Prove the existence of global smooth solutions (in 2D) for the original variational Navier-Stokes equations with the "do nothing" outflow boundary condition for general data.*

3. Spatial discretization by finite elements

In this section, we recall some basics about the spatial discretization of the incompressible Navier-Stokes equations by the finite element method. The emphasis will be on those types of finite elements which are used in our codes for solving two and three dimensional flow problems, stationary as well as nonstationary. For a general discussion of finite element methods for flow problems, see to Girault/Raviart [29], Pironneau [68], and Gresho/Sani [35].

3.1. Basics of finite element discretization

We begin by a brief introduction to the basics of finite element discretization of elliptic problems, e.g., the Poisson equation in a bounded domain $\Omega \subset \mathbb{R}^d$ ($d = 2$ or 3) with a polyhedral boundary $\partial\Omega$,

$$-\nu\Delta u = f \quad \text{in } \Omega. \tag{3.1}$$

We assume homogeneous Dirichlet and Neumann boundary conditions,

$$u_{|\Gamma_D} = 0, \quad \partial_n u_{|\Gamma_N} = 0, \tag{3.2}$$

along disjoint components Γ_D and Γ_N of $\partial\Omega$, where $\partial\Omega = \overline{\Gamma_D \cup \Gamma_N}$. The starting point is the variational formulation of this problem in the natural solution space $H := H_0^1(\Gamma_D;\Omega)$: Find $u \in H$ satisfying

$$a(u,\phi) := \nu(\nabla u, \nabla\phi) = (f,\phi) \quad \forall\phi \in H. \tag{3.3}$$

To discretize this problem, we introduce decompositions, named $\mathbb{T}_h$, of $\bar{\Omega}$ into (closed) cells K (triangles or quadrilaterals in 2D, and tetrahedra or hexahedra in 3D) such that the usual regularity conditions are satisfied:

- $\bar{\Omega} = \cup\{K \in \mathbb{T}_h\}$.
- Any two cells K, K' only intersect in common faces, edges or vertices.
- The decomposition $\mathbb{T}_h$ matches the decomposition $\partial\Omega = \Gamma_D \cup \Gamma_N$.

In the following, we will also allow decompositions with "hanging nodes" in order to ease local mesh refinement. To each of the decompositions $\mathbb{T}_h$, there corresponds a mesh-size function $h = h(x)$ which is piecewise constant such that $h_{|K} =: h_K$. We set $h_K := \text{diam}(K)$ and denote by ρ_K the radius of the ball of maximal size contained in K. We will also use the notation $h := \max_{K\in\mathbb{T}_h} h_K$. The family of decompositions $\{\mathbb{T}_h\}_h$ is said to be (uniformly) "shape regular", if

$$ch_K \le \rho_K \le h_K, \tag{3.4}$$

and (uniformly) "quasi-uniform", if

$$\max_{K\in\mathbb{T}_h} h_K \le c \min_{K\in\mathbb{T}_h} h_K, \tag{3.5}$$

with some constants c independent of h; see Girault/Raviart [29] or Brenner/Scott [19] for more details of these properties. In the following, we will generally assume *shape-regularity* (unless something else is said). *Quasi-uniformity* is usually not required. Examples of admissible meshes are shown below.

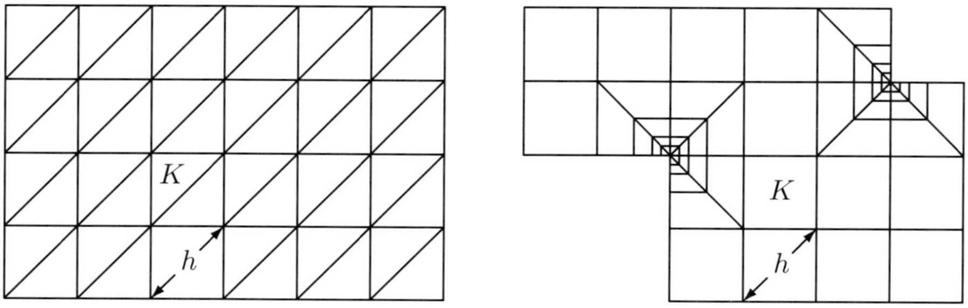

FIGURE 13. Regular finite element meshes (triangular and quadrilateral)

On the decompositions $\mathbb{T}_h$, we consider "finite element spaces" $H_h \subset H$ defined by

$$H_h := \{v_h \in H,\ v_{h|K} \in P(K),\ K \in \mathbb{T}_h\},$$

where $P(K)$ are certain spaces of elementary functions on the cells K. In the simplest case, $P(K)$ are polynomial spaces, $P(K) = P_r(K)$ for some degree $r \geq 1$. On general quadrilateral or hexahedral cells, we have to work with "parametric" elements, i.e., the local shape functions are constructed by using transformations $\psi_K : \hat{K} \to K$ between the "physical" cell K and a fixed "reference unit-cell" $\hat{K}$ by $v_{h|K}(\psi_K(\cdot)) \in P_r(\hat{K})$. This construction is necessary in general in order to preserve "conformity" (i.e., global continuity) of the cell-wise defined functions $v_h \in H_h$. For example, the use of *bilinear* shape functions $\phi \in \text{span}\{1, x_1, x_2, x_1x_2\}$ on a quadrilateral mesh in 2D employs likewise *bilinear* transformations $\psi_K : \hat{K} \to K$. We will see more examples of concrete finite element spaces below.

In a finite element discretization, "consistency" is expressed in terms of local approximation properties of the shape functions used. For example, in the case of a second-order approximation using *linear* or *d-linear* shape functions, there holds locally on each cell K:

$$\|v - I_h v\|_K + h_K\|\nabla(v - I_h v)\|_K \leq c_I h_K^2 \|\nabla^2 v\|_K, \tag{3.6}$$

and on each cell surface ∂K:

$$\|v - I_h v\|_{\partial K} + h_K\|\partial_n(v - I_h v)\|_{\partial K} \leq c_I h_K^{3/2} \|\nabla^2 v\|_K, \tag{3.7}$$

where $I_h v \in H_h$ is the natural "nodal interpolation" of a function $v \in H \cap H^2(\Omega)$, i.e., $I_h v$ coincides with v with respect to certain "nodal functionals" (e.g., point values at vertices, mean values on edges or faces, etc.). The "interpolation constant" is usually of size $c_I \sim 0.1 - 1$, depending on the shape of the cell K.

With the foregoing notation the discrete scheme reads as follows: Find $u_h \in H_h$ satisfying

$$a(u_h, \phi_h) = (f, \phi_h) \quad \forall \phi_h \in H_h. \tag{3.8}$$

Combining the two equations (3.3) and (3.8) yields the relation

$$a(u-u_h,\phi_h)=0, \quad \phi_h\in H_h, \tag{3.9}$$

which means that the error $e:=u-u_h$ is "orthogonal" to the subspace H_h with respect to the bilinear form $a(\cdot,\cdot)$. This essential feature of the finite element Galerkin scheme immediately implies the "best approximation" property

$$\|\nabla e\| = \min_{\phi_h\in H_h}\|\nabla(u-\phi_h)\|. \tag{3.10}$$

In virtue of the interpolation estimate (3.6), we obtain the (global) *a priori* convergence estimate

$$\|\nabla e\|\le c_I\, h\|\nabla^2 u\|\le c_I c_S\, h\,\|f\|, \tag{3.11}$$

provided that the solution is sufficiently regular, i.e., $u\in H^2(\Omega)$, satisfying the a priori bound

$$\|\nabla^2 v\|\le c_S\|f\|. \tag{3.12}$$

In the above model, this is the case if the polygonal domain Ω is convex. In case of reduced regularity of u due to reentrant corners, the order in the estimate is correspondingly reduced. In the case of approximation by higher-order polynomials, $r\ge 2$, and higher order of regularity of u, the estimate (3.11) shows a correspondingly increased power of h. The order of h in the "energy-error" estimate (3.11) can be improved by shifting to the L^2-norm. This is done by employing a duality argument ("Aubin-Nitsche trick"); see, e.g., Brenner/Scott [19]. Let $z\in H$ be the solution of the auxiliary problem

$$-\nu\Delta z=\|e\|^{-1}e \ \text{ in } \Omega, \quad z=0 \ \text{ on } \partial\Omega, \tag{3.13}$$

satisfying an a priori bound $\|\nabla^2 z\|\le c_S$. Then, there holds

$$\|e\|=(e,-\nu\Delta z)=a(e,z)=a(e,z-I_hz)\le c_Ic_Sh\|\nabla e\|, \tag{3.14}$$

and we conclude the improved *a priori* L^2-error estimate

$$\|e\|\le c_I^2c_S^2h^2\|f\|. \tag{3.15}$$

In order to convert the problems (3.8) into a form which is amenable to practical computation, we introduce the nodal basis $\{\phi_h^1,\dots,\phi_h^N\}$, $N=\dim H_h$, of the space H_h, defined by $\phi_h^i(a_j)=\delta_{ij}$, $i,j=1,\dots,N$, where a_j are the nodal points (e.g., the vertices) of the mesh. Then, setting

$$u_h=\sum_{i=1}^N x_i\phi_h^i,$$

problem (3.8) is equivalent to the linear algebraic system

$$Ax=b, \tag{3.16}$$

for the "nodal value" vector $x = (x_i)_{i=1}^N$. Here, the "stiffness matrix" A and the "load vector" b are defined by

$$A := \big(a(\phi_h^i, \phi_h^j)\big)_{i,j=1}^N, \quad b := \big((f, \phi_h^i)\big)_{i=1}^N.$$

In the case of variable coefficients and force the integrals have to be computed by using integration formulas; in our implementations usually Gaussian formulas are used. For the pure diffusion problem the stiffness matrix A is symmetric and positive definite. Its condition number behaves like $\kappa(A) = O(h^{-2})$, where the exponent -2 is determined by the order of the differential operator Δ (it is independent of the spatial dimension and the polynomial degree of the finite elements used).

Below, we show a sequence of hexahedral 3D meshes used for computing the "puff-puff flow" mentioned in the Introduction; observe the successively refined approximation of the curved boundary.

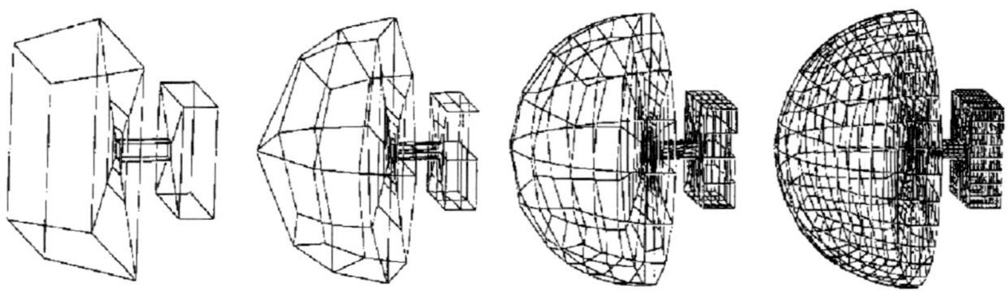

FIGURE 14. Sequence of successively refined hexahedral meshes for computing the "puff-puff" flow in 3D.

3.2. Stokes elements

We consider the *stationary* Navier-Stokes problem as specified in Section 2. In setting up a finite element model of the Navier-Stokes problem, one starts from the variational formulation of the problem: Find $v \in v^{in} + \mathbf{H}$ and $p \in L$, such that

$$a(v, \varphi) + n(v, v, \varphi) + b(p, \phi) = (f, \varphi) \quad \forall\, \phi \in \mathbf{H}, \tag{3.17}$$

$$b(\chi, v) = 0 \quad \forall\, \chi \in L. \tag{3.18}$$

The choice of the function spaces $\mathbf{H} \subset \mathbf{H}^1(\Omega))$ and $L \subset L^2(\Omega)$ depends on the specific boundary conditions imposed in the problem to be solved. On a finite element mesh $\mathbb{T}_h$ on Ω with cell width h, one defines spaces of "discrete" trial and test functions,

$$\mathbf{H}_h \text{ “}\subset\text{” } \mathbf{H}, \quad L_h \subset L.$$

The discrete analogues of (3.19), (3.20) then read as follows: Find $v_h \in v_h^{in} + \mathbf{H}_h$ and $p_h \in L_h$, such that

$$a_h(v_h, \varphi_h) + n_h(v_h, v_h, \phi_h) + b_h(p_h, \phi_h) = (f, \varphi_h) \quad \forall\, \phi_h \in \mathbf{H}_h, \tag{3.19}$$

$$b_h(\chi_h, v_h) = 0 \quad \forall\, \chi_h \in L_h, \tag{3.20}$$

where v_h^{in} is a suitable approximation of the inflow data v^{in}. The notation $\mathbf{H}_h$ "$\subset$ "$\mathbf{H}$ indicates that in this discretization the spaces $\mathbf{H}_h$ may be "nonconforming", i.e., the discrete velocities v_h are continuous across the interelement boundaries and zero along the rigid boundaries only in an approximate sense; in this case the discrete forms $a_h(\cdot,\cdot)$, $b_h(\cdot,\cdot)$, $n_h(\cdot,\cdot,\cdot)$ and the discrete "energy norm" $\|\nabla \cdot\|_h$ are defined in the piecewise sense,

$$a_h(\phi,\psi) := \sum_{K\in\mathbb{T}_h} \nu(\nabla\phi,\nabla\psi)_K\,, \quad b_h(\chi,\phi) := \sum_{K\in\mathbb{T}_h} (\chi,\nabla\cdot\phi)_K\,,$$

$$n_h(\phi,\psi,\xi) := \sum_{K\in\mathbb{T}_h} (\phi\cdot\nabla\psi,\xi)_K\,, \quad \|\nabla\phi\|_h := \Big(\sum_{K\in\mathbb{T}_h} \|\nabla\phi\|_K^2\Big)^{1/2}.$$

In order that (3.19), (3.20) is a stable approximation to (3.17), (3.18), as $h \to 0$, it is crucial that the spaces $\mathbf{H}_h \times L_h$ satisfy a compatibility condition, the so-called "inf-sup" or "Babuska-Brezzi" condition,

$$\inf_{q_h\in L_h} \left\{ \sup_{w_h\in\mathbf{H}_h} \frac{b_h(q_h,w_h)}{\|q_h\|\,\|\nabla w_h\|_h} \right\} \geq \gamma > 0\,. \tag{3.21}$$

Here, the constant γ is required to be independent of h. This ensures that the problems (3.19), (3.20) possess solutions which are uniquely determined in $\mathbf{H}_h \times L_h$ and stable. Further, for the errors $e_v := v - v_h$ and $e_p := p - p_h$, there hold *a priori* estimates of the form

$$\|\nabla e_v\|_h + \|e_p\| \leq ch\{\|\nabla^2 v\| + \|\nabla p\|\}\,. \tag{3.22}$$

A rigorous convergence analysis of spatial discretization of the Navier-Stokes problem can be found in Girault/Raviart [29] and in [41, 43].

3.2.1. EXAMPLES OF STOKES ELEMENTS Many stable pairs of finite element spaces $\{\mathbf{H}_h, L_h\}$ have been proposed in the literature (see, e.g., Girault/Raviart [29], Hughes et al. [49] and [77]). Below, two particularly simple examples of quadrilateral elements will be described which have satisfactory approximation properties and are applicable in two as well as in three space dimensions. They can be made robust against mesh degeneration (large aspect ratios) and they admit the application of efficient multigrid solvers. We note that, from the point of view of accuracy, in our context quadrilateral (hexahedral) elements are to be preferred over triangular (tetrahedral) elements because of their superior approximation properties. Both types of elements may be used in the spatial discretization underlying the discussions in the following sections.

1) The nonconforming "rotated" d-linear $\tilde{Q}_1/P_0$ Stokes element
The first example is the natural quadrilateral analogue of the well-known triangular nonconforming finite element of Crouzeix/Raviart (see [29]). It was introduced and analyzed in [77] and its two- as well as three-dimensional versions have been implemented in state-of-the-art Navier-Stokes codes (see Turek [93, 96], Schreiber/Turek [83], and Oswald [66]. In two space dimensions, this nonconforming element uses piecewise "rotated" bi-linear (reference) shape functions for the velocities, spanned by $\{1, x, y, x^2 - y^2\}$, and piecewise constant pressures. As nodal values one may take the mean values of the velocity vector over the element edges (or, alternatively, its point values at the midpoints of sides) and the mean values of the pressure over the elements. For the precise definition of this element we introduce the set $\partial\mathbb{T}_h$ of all $(d-1)$-faces S of the elements $K \in \mathbb{T}_h$. We set

$$\widetilde{Q}_1(K) = \{q \circ \psi_T^{-1} : q \in \operatorname{span}\{1, x_1, x_i^2 - x_{i+1}^2, i = 1, \dots, d\}\}.$$

The corresponding finite element spaces are

$$\mathbf{H}_h := \left\{ \begin{array}{l} v_h \in L^2(\Omega)^d : v_{h|K} \in \widetilde{Q}_1(K)^d, \ K \in \mathbb{T}_h, \\ F_S(v_{h|K}) = F_S(v_{h|K'}), \ S \subset \partial K \cap \partial K', \ F_S(v_h) = 0, \ S \subset \Gamma \end{array} \right\},$$

$$L_h := \{q_h \in L : q_{h|K} \in P_0(K), \ K \in \mathbb{T}_h\},$$

with the nodal functionals

$$F_S(v_h) = |S|^{-1} \int_S v_h \, do, \qquad F_K(p_h) = |K|^{-1} \int_K p_h \, dx.$$

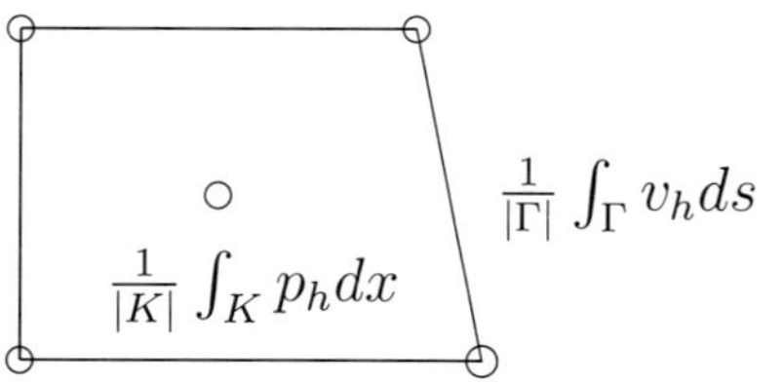

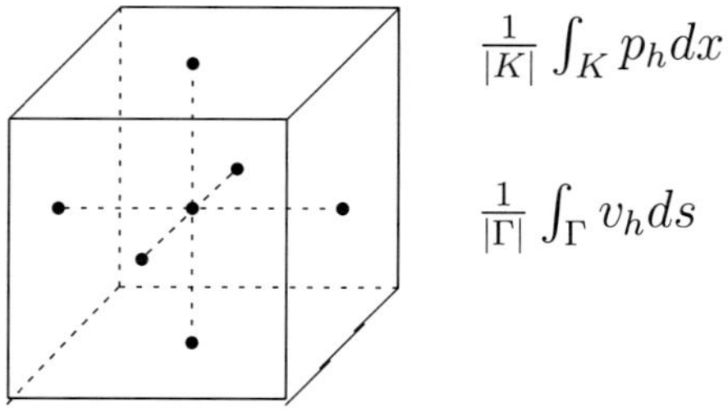

Clearly, the spaces $\mathbf{H}_h$ are non-conforming, $\mathbf{H}_h \not\subset \mathbf{H}^1(\Omega)^d$. For the pair $\{\mathbf{H}_h, L_h\}$ the discrete "inf-sup" stability condition (3.21) is known to be satisfied on fairly general meshes; see [77] and [13]. For illustration, we recall from [13] the essential steps of the argument.

Proof of the "inf-sup" stability estimate (3.21): Using the continuous "inf-sup" estimate (2.7), we conclude for an arbitrary $p_h \in L_h$ that

$$\gamma_0 \|p_h\| \le \sup_{\phi \in \mathbf{H}} \frac{|b_h(p_h, \phi)|}{\|\nabla \phi\|} \le \sup_{r_h\phi \in \mathbf{H}_h} \frac{|b_h(p_h, r_h\phi)|}{\|\nabla r_h \phi\|} \sup_{\phi \in \mathbf{H}} \frac{\|\nabla r_h \phi\|}{\|\nabla \phi\|}. \tag{3.23}$$

where $r_h\phi \in \mathbf{H}_h$ is an approximation to $\phi \in \mathbf{H}$ satisfying

$$b_h(\chi_h, \phi - r_h\phi) = 0 \quad \forall \chi_h \in L_h, \qquad \|\nabla r_h \phi\| \le c_* \|\nabla \phi\|. \tag{3.24}$$

These properties are realized for the $\tilde{Q}_1/P_0$ Stokes element by the natural nodal interpolation defined by

$$\int_S r_h\phi \, do = \int_S \phi \, do \quad \forall\, S \in \partial\mathbb{T}.$$

Then, the first relation in (3.24) is obvious, and the $\mathbf{H}^1$ stability follows from

$$\begin{aligned} \|\nabla r_h\phi\|_h^2 &= \sum_{K\in\mathbb{T}_h} \Big\{(r_h\phi, \partial_n r_h\phi)_{\partial K} - (r_h\phi, \Delta r_h\phi)_K\Big\} \\ &= (\nabla\phi, \nabla r_h\phi)_h + \sum_{K\in\mathbb{T}_h} \Big\{(r_h\phi - \phi, \partial_n r_h\phi)_{\partial K} - (r_h\phi - \phi, \Delta r_h\phi)_K\Big\}. \end{aligned}$$

The argument becomes particularly simple for parallelogram meshes. In this case $\partial_n r_h\phi_{|\Gamma} \equiv const.$ and $\Delta r_h\phi_{|K} \equiv 0$, such that the last sum vanishes. The general case requires a more involved estimation. Now, the desired "inf-sup" stability estimate follows with the constant $\gamma = \gamma_0/c_*$.

As discussed in [77], the stability and approximation properties of the $\tilde{Q}_1/P_0$ Stokes element depend very sensitively on the degree of deviation of the cells K from parallelogram shape. Stability and convergence deteriorates with increasing cell aspect ratios. This defect can be cured by using a "nonparametric" version of the element where the reference space $\tilde{Q}_1(K) := \{q \in \text{span}\{1, \xi, \eta, \xi^2 - \eta^2\}\}$ is defined for each element K independently with respect to the coordinate system (ξ, η) spanned by the directions connecting the midpoints of sides of K. This approximation turns out to be robust with respect to the shape of the elements K, and the convergence estimate (3.22) remains true. Below, we will relax this requirement even further by allowing the elements to be stretched in one or more (in 3D) directions.

Finally, we mention an important feature of the $\tilde{Q}_1/P_0$ Stokes element (see [93]): It possesses a "divergence-free" nodal-basis, which allows the elimination of the pressure from the problem resulting in a positive definite algebraic system for the velocity unknowns alone. The reduced algebraic system can be solved by specially adapted multigrid methods; see Turek [93].

2) *The conforming d-linear Q_1/Q_1 Stokes element with pressure stabilization*
The second example uses continuous isoparametric d-linear shape functions for both the velocity and the pressure approximations. The nodal values are just the function values of the velocity and the pressure at the vertices of the mesh, making this approximation particularly attractive in three dimensions. With

$$\tilde{Q}_1(K) = \{q \circ \psi_T^{-1} : \, q \in \text{span}\{1, x_i, x_i x_j, \, i,j = 1, \dots, d\}\},$$

the corresponding finite element spaces are defined by

$$\begin{aligned} \mathbf{H}_h &= \Big\{v_h \in \mathbf{H}_0^1(\Gamma;\Omega)^d : \, v_{h|K} \in \tilde{Q}_1(K)^d, \, K \in \mathbb{T}_h\Big\}, \\ L_h &= \Big\{q_h \in H^1(\Omega) : \, q_{h|K} \in \tilde{Q}_1(K), \, K \in \mathbb{T}_h\Big\}, \end{aligned}$$

with the nodal functionals (a vertex of the mesh $\mathbb{T}_h$)

$$F_a(v_h) = v_h(a)\,, \quad F_a(p_h) = p_h(a)\,.$$

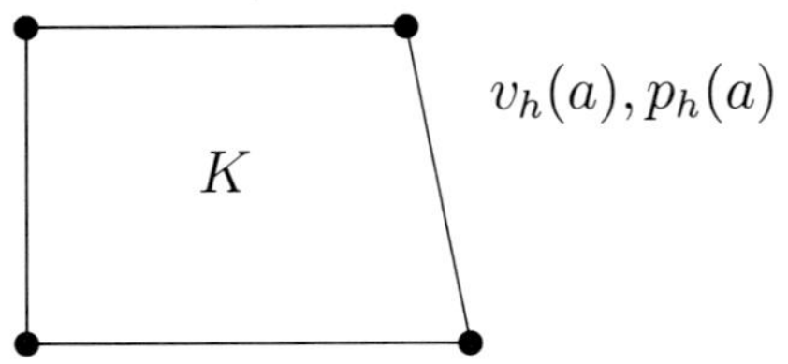

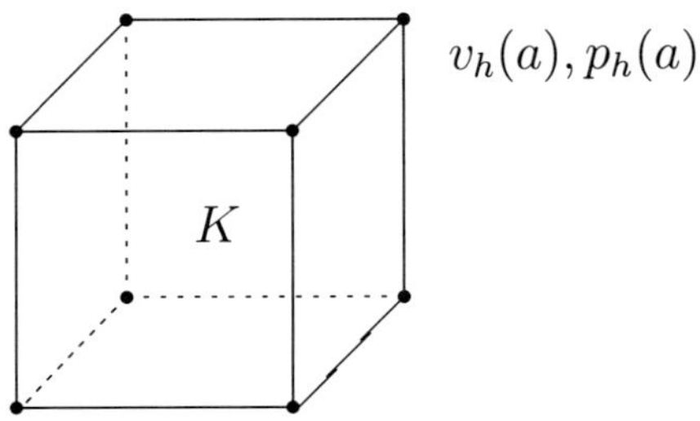

This combination of spaces, however, would be unstable, i.e., it would violate the condition (3.21), if used together with the variational formulation (3.19), (3.20). In order to get a stable discretization, it was proposed by Hughes et al. [49], to add certain least squares terms in the continuity equation (3.20) (pressure stabilization method),

$$b(\chi_h, v_h) + c_h(\chi_h, p_h) = g_h(v_h; \chi_h), \tag{3.25}$$

where

$$\begin{aligned} c_h(\chi_h, p_h) &= \frac{\alpha}{\nu} \sum_{K\in\mathbb{T}_h} h_K^2 (\nabla\chi_h, \nabla p_h)_K, \\ g_h(v_h; \chi_h) &= \frac{\alpha}{\nu} \sum_{K\in\mathbb{T}_h} h_K^2 (\nabla\chi_h, f + \nu\Delta v_h - v_h\cdot\nabla v_h)_K\,. \end{aligned}$$

The correction terms on the right-hand side have the effect that this modification is fully consistent, since the additional terms cancel out if the exact solution $\{v,p\}$ of problem (3.17), (3.18) is inserted. On regular meshes, one obtains a stable and consistent approximation of the Navier-Stokes problem (3.17), (3.18), for which a convergence estimate of form (3.22) holds true. The argument follows a slightly different track than that used above for the nonconforming $\tilde{Q}_1/P_0$ element; see [13].

Proof of the "inf-sup" stability estimate (3.21): From the continuous stability estimate (2.7) we conclude that

$$\gamma_0\|p_h\| \le \sup_{r_h\phi\in\mathbf{H}_h} \frac{|(p_h, \nabla r_h\phi)|}{\|\nabla r_h\phi\|} \sup_{\phi\in\mathbf{H}} \frac{\|\nabla r_h\phi\|}{\|\nabla\phi\|} + \sup_{\phi\in\mathbf{H}} \frac{|(\nabla p_h, \phi - r_h\phi)|}{\|\nabla\phi\|}, \tag{3.26}$$

where $r_h\phi \in \mathbf{H}_h$ is an approximation to $\phi \in \mathbf{H}$, satisfying

$$\Big(\sum_{K\in\mathbb{T}_h} h_K^{-2} \|\phi - r_h\phi\|_K^2 \Big)^{1/2} + \|\nabla r_h\phi\| \le c_* \|\nabla\phi\|\,. \tag{3.27}$$

The existence of such an approximation can be shown by employing an averaged nodal interpolation. From this, we obtain

$$\gamma\|p_h\| \leq c_* \sup_{\phi_h\in\mathbf{H}_h} \frac{|(p_h,\nabla\phi_h)|}{\|\nabla\phi_h\|} + c_*\Big(\sum_{K\in\mathbb{T}_h} h_K^2\|\nabla p_h\|_K^2\Big)^{1/2},$$

which yields the desired stability estimate with the constant $\gamma = \gamma_0/c_*$.

It was shown in Hughes et al. [49], and later on in a series of mathematical papers (see, e.g., Brezzi/Pikäranta [21], and the literature cited therein) in the context of a more general analysis of such stabilization methods, that this kind of discretization is numerically stable and of optimal order convergent for many relevant pairs of spaces $\mathbf{H}_h\times L_h$.

The stabilized Q_1/Q_1 Stokes element has several important features: With the same number of degrees of freedom it is more accurate than its triangular analogue (and also slightly more accurate than its nonconforming analogue described above). Furthermore, it has a very simple data structure due to the use of the same type of nodal values for velocities and pressure which allows for an efficient vectorization of solution processes. Thanks to the stabilization term in the continuity equation, standard multigrid techniques can be used for solving the algebraic systems with good efficiency (see the discussion in Section 5 below).

We note that the triangular analogue of this element is closely related (indeed almost algebraically equivalent) to the "inf-sup" stable MINI-element (see Brezzi/Fortin [20]) which is based on the standard Q_1/Q_1-element and stability is achieved by augmenting the velocity space by local cubic bubble functions.

The stabilized Q_1/Q_1-Stokes element has been implemented in several 2D and 3D Navier-Stokes codes (see, e.g., Harig [38], Becker [7], and Braack [17]). However, it was already reported in Harig [38] that the convergence properties of this element sensitively depend on the parameter α and may deteriorate on strongly stretched meshes. We will come back to this point below.

3.3. The algebraic problems

The discrete Navier-Stokes problem (3.19), (3.20), possibly including pressure stabilization (3.25), has to be converted into an algebraic system which can be solved on a computer. To this end, we choose appropriate local "nodal bases" $\{\phi_h^i,\, i=1,\dots,N_v\}$ of the "velocity space" $\mathbf{H}_h$, and $\{\chi_h^i,\, i=1,\dots,N_p\}$ of the "pressure space" L_h and expand the unknown solution $\{v_h,p_h\}$ in the form

$$v_h = v_h^{in} + \sum_{i=1}^{N_v} x_i\phi_h^i, \qquad p_h = \sum_{j=1}^{N_p} x_i\chi_h^j.$$

We introduce the following matrices:

$$A = \big(a_h(\phi_h^i,\phi_h^j)\big)_{i,j=1}^{N_v}, \quad B = \big(b_h(\chi_h^i,\phi_h^j)\big)_{i,j=1}^{N_p,N_v}, \quad C = \big(c_h(\chi_h^i,\chi_h^j)\big)_{i,j=1}^{N_p},$$

$$N(x) = \big(n_h(v_h^{in} + \Sigma_{k=1}^{N_v} x_k\phi_h^k, \phi_h^i, \phi_h^j) + n_h(\phi_h^i, v_h^{in}, \phi_h^j)\big)_{i,j=1}^{N_v},$$
$$b = \big((f,\phi_h^j) - a(v_h^{in},\phi_h^j) - n_h(v_h^{in}, v_h^{in}, \phi_h^j)\big)_j^{N_v}, \quad c(x) = \big(g_h(v_h;\chi_h^j)\big)_{j=1}^{N_p}.$$

Here, A is the stiffness matrix, B the "gradient matrix" with the associated "divergence matrix" $-B^T$; $N(\cdot)$ is the (nonlinear) transport matrix and b the load vector into which the nonhomogeneous inflow-boundary data have been incorporated. Further, C is the matrix arising from pressure stabilization and c the (nonlinear) correction term on the right-hand side. Occasionally, we will use the abbreviation $A(\cdot) := A + N(\cdot)$. For later use, we also introduce the velocity and pressure "mass matrices":

$$M_v = \big((\phi_h^i, \phi_h^j)\big)_{i,j=1}^{N_v}, \quad M_p = \big((\chi_h^i, \chi_h^j)\big)_{i,j=1}^{N_p}.$$

With this notation the variational problem (3.19), (3.20) can equivalently be written in form of an algebraic system for the vectors $x \in \mathbb{R}^{N_v}$ and $y \in \mathbb{R}^{N_p}$ of expansion coefficients:

$$Ax + N(x)x + By = b, \tag{3.28}$$
$$-B^T x + Cy = c(x). \tag{3.29}$$

Notice that for this system has the structure of a saddle-point problem (for $C = 0$) and is generically nonsymmetric. This poses a series of problems in solving it by iterative methods. This point will be addressed in more detail in Section 5, below.

3.4. Anisotropic meshes

In many situations it is necessary to work with (locally) *anisotropic* meshes, i.e., in some areas of the computational domain the cells are stretched in order to better resolve local solution features. Such anisotropies generically occur when tensor-product meshes are used to resolve boundary layers. In this case the mesh $\mathbb{T}_h$ is no longer "quasi-regular" and the discretization may strongly depend on the deterioration of the cells measured in terms of "cell aspect ratios". On such meshes three different phenomena occur:

- The constant c_I in the interpolation estimates (3.6), (3.7) may blow up.
- The constant γ in the "inf-sup" stability estimate (3.21) may become small.
- The conditioning of the algebraic system may deteriorate.

It is known that for most of the lower-order finite elements (including the elements considered here) the local interpolation estimates remain valid even on highly stretched elements ("maximum angle condition" versus "minimum angle condition"). Accordingly, the failure of the considered Stokes elements on stretched meshes is not so much a problem of consistency but rather one of stability. Hence, we will discuss the stability in more detail; the approximation aspects have been systematically analyzed in Apel/Dobrowolski [2], Apel [3], and the literature cited therein. The main technical difficulty arises from the deterioration of the "inverse inequality" for finite elements $\|\nabla\phi_h\|_K \le c h_K^{-1}\|\phi_h\|_K$ on stretched cells. Further,

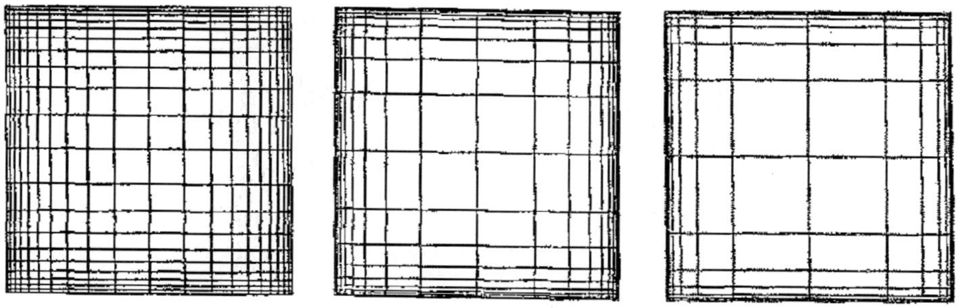

FIGURE 15. A sequence of locally anisotropic tensor-product meshes with aspect ratios $\sigma_h = 10, 100, 1000$, for computing the driven-cavity flow.

the solution of the resulting algebraic systems, e.g., by multigrid methods, becomes increasingly difficult. For simplicity, we concentrate the following discussion on the special case of cartesian tensor-product cells as shown in the figures above and below; here the "cell aspect ratio" is defined by $\sigma_K = h_x/h_y$ and the maximum "mesh aspect ratio" by $\sigma_h := \max_{K\in\mathbb{T}_h} \sigma_K$. We consider aspect ratios of size $\sigma_h \approx 1 - 10^4$.

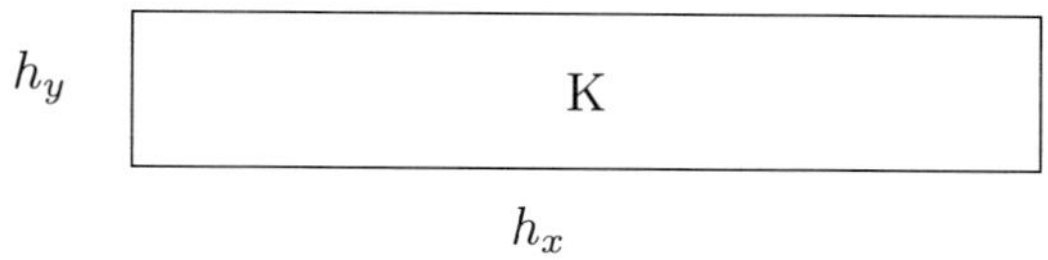

As a model problem, we consider the stationary Stokes equations

$$-\nu\Delta u + \nabla p = f, \quad \nabla\cdot u = 0, \text{ in } \Omega, \tag{3.30}$$

with homogeneous Dirichlet boundary conditions $u_{|\partial\Omega} = 0$, on a bounded polygonal domain $\Omega \in \mathbb{R}^2$. Using the notation introduced above, the finite element formulation of this problem reads as follows:

$$a_h(u_h, \phi_h) + b_h(p_h, \phi_h) = (f, v) \quad \forall \phi_h \in \mathbf{H}_h, \tag{3.31}$$

$$b_h(u_h, \chi_h) = 0 \quad \forall \chi_h \in L_h. \tag{3.32}$$

1. First, we consider the *nonconforming* $\widetilde{Q}_1/P_0$ Stokes approximation which uses "rotated" bilinear shape functions for the velocities and piecewise constants for the pressure. Above, we have introduced its "non-parametric" version where local cell coordinates $\{\xi_K, \eta_K\}$ are used for defining the local shape functions on each cell as $v_{h|K} \in \text{span}\{1, \xi_K, \eta_K, \xi_K^2 - \eta_K^2\}$. In this way one obtains a discretization which is robust with respect to deviations of the cell from parallelogram shape. It has been shown in [12] that this non-parametric element can be modified to be also

robust with respect to increasing aspect ratio. This modification employs a scaling of the local coordinate system according to the cell aspect ratio $\sigma_K := h_x/h_y$,

$$v_{h|K} \in \operatorname{span}\{1, \xi_K, \sigma_K\eta_K, \xi_K^2 - \sigma_K^2\eta_K^2\}.$$

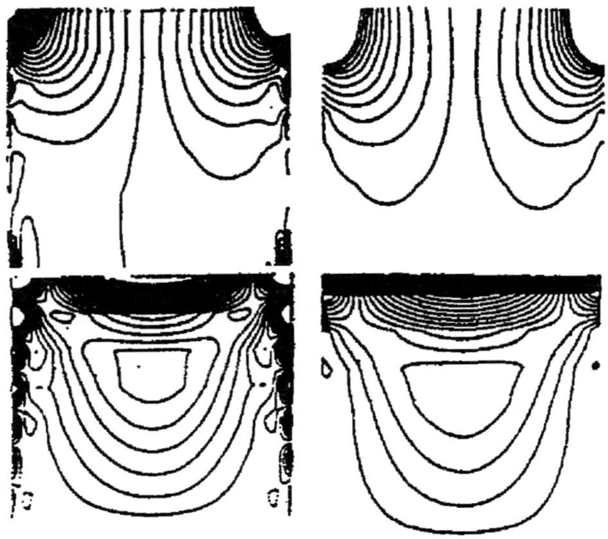

FIGURE 16. Pressure and velocity norm isolines for a driven-cavity computation with the standard isotropic $\widetilde{Q}_1/P_0$ element (left) compared to the anisotropically scaled version (right); from Becker [7].

Furthermore, the "inf-sup" stability estimate (3.21) is preserved on such meshes with a constant γ independent of the mesh aspect ratio σ_h. To demonstrate that this scaling is actually necessary for the stability of this element, we show in Figure 16 the results of a "driven cavity" calculation on meshes with $\sigma_h = 32$ using the standard isotropic approximation compared to the anisotropically scaled version. The instability caused by the large aspect ratio exhibits spurious pressure peaks and vortices along the boundary.

2. Next, we consider the stabilized Q_1/Q_1-Stokes approximation which uses continuous (isoparametric) bilinear shape functions for both the velocity and the pressure. As seen before, this discretization becomes "inf-sup" stable if the discrete model is augmented by a least-squares term of the form

$$(\nabla\cdot u_h, \chi_h) + c(p_h, \chi_h) = \text{"correction terms"}. \tag{3.33}$$

On quasi-uniform meshes, we obtain a stable and consistent approximation of the Stokes problem but this approximation sensitively depends on the choice of the form $c(\cdot,\cdot)$ and may deteriorate on strongly stretched meshes. Again, the approximation property of the Q_1/Q_1 element is not the problem. The interpolation estimates (3.6) and (3.7) remain valid also on high-aspect-ratio meshes (as defined above) with constants independent of σ_h. However, the proper design of the stabilization (3.33) is delicate. We consider the following three different choices for the stabilizing bilinear form:

$$c(p,q) = \begin{cases} c_1(p,q) &= \alpha \sum_{K\in\mathbb{T}_h} |K|(\nabla p, \nabla q)_K\,, \\ c_2(p,q) &= \alpha \sum_{K\in\mathbb{T}_h} h_K^2(\nabla p, \nabla q)_K\,, \\ c_3(p,q) &= \alpha \sum_{K\in\mathbb{T}_h} \{h_x^2(\partial_x p, \partial_x q)_K + h_y^2(\partial_y p, \partial_y q)_K\}\,. \end{cases}$$

The form $c_1(\cdot,\cdot)$ is built in analogy to the MINI-element, since condensation of the bubble functions leads directly to the cell-wise scaling factor $|K|$. We see that $c_1(\cdot,\cdot)$ gets smaller with increasing σ_h, an undesirable effect which is avoided by $c_2(\cdot,\cdot)$. Finally, $c_3(\cdot,\cdot)$ distinguishes between the different coordinate directions which requires the use of a local coordinate system in the definition of the stabilization. By a local "inverse estimate" for bilinear functions on (arbitrary) rectangles we get the stability relation $c_3(p_h,p_h) \le \|p_h\|^2$, which appears necessary for the stabilization to achieve uniformity with respect to the mesh aspect ratio. This may be seen by writing the discrete system (3.31), (3.33) in matrix notation

$$\begin{bmatrix} A & B \\ -B^T & C_i \end{bmatrix} \begin{bmatrix} x \\ y \end{bmatrix} = \begin{bmatrix} b \\ c \end{bmatrix},$$

where C_i corresponds to the stabilizing bilinear form $c_i(\cdot,\cdot)$. The Schur complement of the main diagonal block A is $\Sigma = C_i - B^T A^{-1} B$. Then, the stability constant γ in (3.33) is given by (see [13]):

$$\gamma^2 = \lambda_{min}(M^{-1}\Sigma)\,, \tag{3.34}$$

where M denotes the mass matrix of the pressure space L_h (piecewise constants in this case). This correspondence can be used in order to experimentally determine the dependence of the stability constant γ on the various parameters of the discretization, particularly the cell aspect ratio. We may detect γ by counting the number of cg-iterations (preconditioned by the mass matrix M) needed to invert the Schur complement. The convergence rate ρ of the cg-iteration applied to the preconditioned Schur complement $M^{-1}\Sigma$ is linked to the condition number $\kappa = \text{cond}(M^{-1}\Sigma)$ by the following well-known formula

$$\rho \approx 2\left(\frac{1-1/\sqrt{\kappa}}{1+1/\sqrt{\kappa}}\right).$$

These test calculations use a sequence of anisotropic grids obtained by one-directional refinements. The results are given in Table 1.

The interpretation of these observations is as follows: The increase of the stability constant for $c_1(\cdot,\cdot)$ stems from the fact, that $\gamma \approx \sigma^{-2} \to 0$ with increasing aspect ratio, whereas the bad behavior of $c_2(\cdot,\cdot)$ can be explained by the growth of $\lambda_{\max}(\Sigma) \approx \sigma^2$ due to fact that we only have $c_2(p,q) \le \sigma\|p\|\|q\|$. We also see that the anisotropic stabilization by $c_3(\cdot,\cdot)$ leads to an aspect-ratio-independent

TABLE 1. Number of cg-iterations; from Becker [7].

σ	2	4	8	16	32	64	128
c_1	8	18	39	98	559	*	*
c_2	8	18	39	88	193	*	*
c_3	8	16	29	31	29	27	24

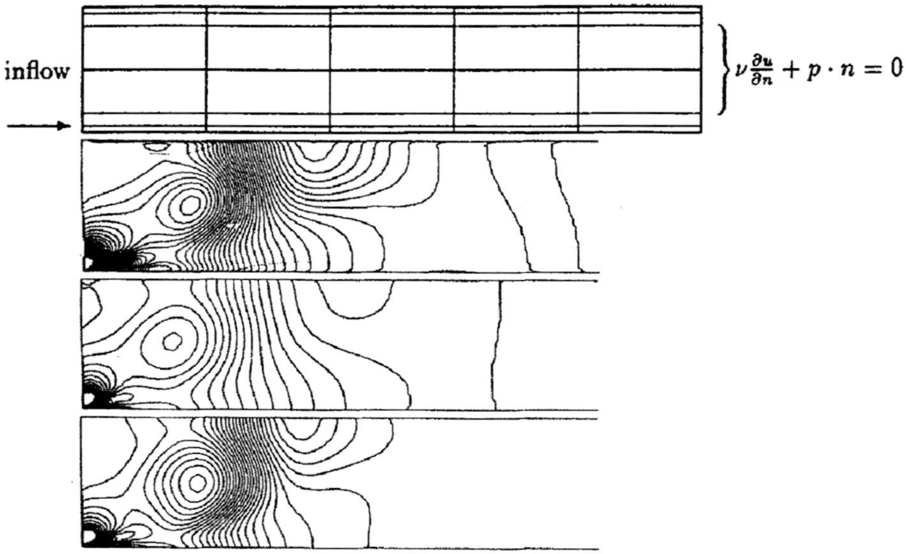

FIGURE 17. Pressure isolines for a jet flow in a channel calculation with the Q_1/Q_1 element using isotropic stabilization (top and middle) compared to the anisotropic stabilization (bottom); from Becker [7].

behavior. Similar effects are also observed for the accuracy of the different stabilizations; see Becker [7] and [13].

Open Problem 3.1: *Prove the approximation property (3.27) on general meshes with arbitrary aspect ratio σ_h. The special case of tensor-product meshes has been treated in Becker* [7] *(see also Apel* [3]*.*

3.5. Treatment of dominant transport

In the case of higher Reynolds numbers (e.g., Re > 1000 for the 2-D driven cavity, and Re > 100 for the flow around an cylinder) the finite element models (3.19),

(3.20) or (3.19), (3.25) may become unstable since they essentially use central-differences-like discretization of the advective term. This "instability" most frequently occurs in form of a drastic slow-down or even break-down of the iteration processes for solving the algebraic problems; in the extreme case the possibly existing "mathematical" solution contains strongly oscillatory components without any physical meaning. In order to avoid these effects some additional numerical damping is required. The use of simple first-order artificial viscosity is not advisable since it introduces too much numerical damping. Below, we describe two approaches used in the context of finite element discretization: i) an adaptive upwinding, and ii) the streamline diffusion method. Alternative techniques are the "characteristics Galerkin method" (for nonstationary flows) and the "discontinuous finite element method" which require major changes in the discretization and will therefore not be discussed here; for references see Pironneau [67], Morton [63] and Johnson [51].

3.5.1. UPWINDING In the finite element context "upwinding" can be defined in a quite natural way; see, e.g., Tobiska/Schieweck [90] and Turek [97], and the literature cited therein. Here, the upwinding effect is accomplished in the evaluating of the advection term through shifting integration points into the upwind direction. This modification leads to system matrices which have certain M-matrix properties and are therefore amenable to efficient and robust solution techniques. This is widely exploited in the finite element codes described in Schreiber/Turek [83], Turek [97] and Schieweck [82].

Following [97], we briefly describe the upwind strategy for the nonconforming "rotated" bilinear Stokes element. Each quadrilateral $K \in \mathbb{T}_h$ is divided into eight barycentric fragments S_{ij}, and for each edge Γ_l and mid point m_l on Γ_l the "lumping region" R_l is defined by $R_l := \cup_{k\in\Lambda_l} S_{lk}$, where $\Lambda_l = \{k, m_l$ and m_k belong to the same element $K\}$. The boundary of the lumping region R_l consists of the edges $\Gamma_{lk} := \partial S_{lk} \cap \partial S_{kl}$, i.e., $\partial R_l = \cup_{k\in\Lambda_l}\Gamma_{lk}$. In this way we obtain an edge oriented partition of the mesh domain $\Omega_h = \cup_l R_l$.

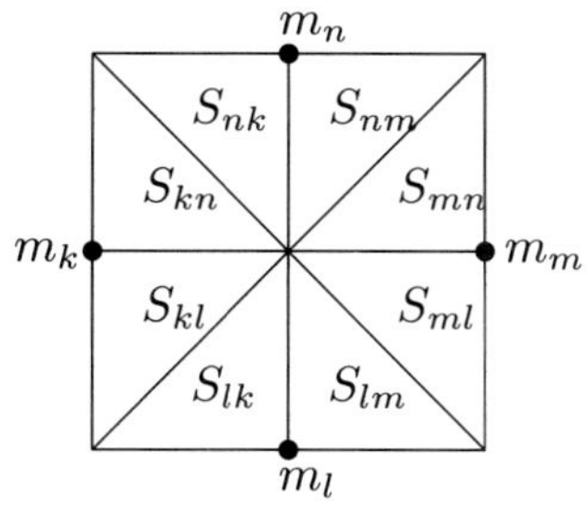

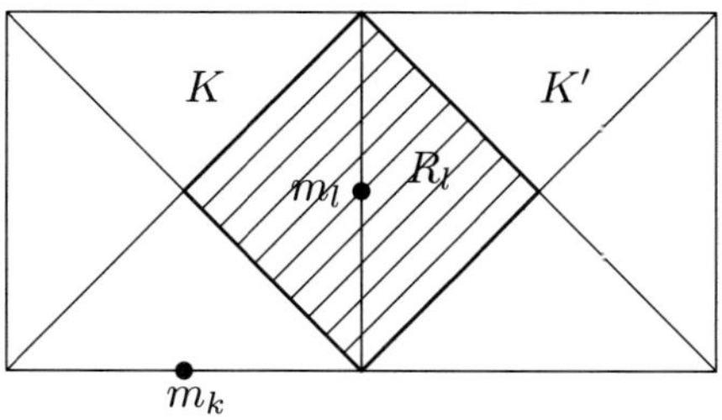

A modification of the nonlinear form $n(u_h, v_h, w_h)$ is now defined by

$$\tilde{n}_h(u_h, v_h, w_h) := \sum_{K,l} (1 - \lambda_{lk}(u_h))\,(v_h(m_k) - v_h(m_l))\,w_h(m_l) \int_{\Gamma_l} u_h{\cdot}n_{lk}\,ds\,,$$

where λ_{lk} are parameters depending on the local flux direction. Setting

$$x := \frac{1}{\nu}\int_{\Gamma_{lk}} u_h{\cdot}n_{lk}\,ds\,,$$

possible choices are

$$\lambda_{lk} \;:=\; \left\{\begin{array}{l} 1,\ \text{for } x\ge 0 \\ 0,\ \text{for } x<0 \end{array}\right\} \quad (\text{“simple upwinding”}),$$

$$\lambda_{lk} \;:=\; \left\{\begin{array}{l} (1/2+x)/(1+x),\ \text{for } x\ge 0 \\ 1/(2-2x),\ \text{for } x<0 \end{array}\right\} \quad (\text{“Samarskij upwinding”}),$$

It can be be shown (see Tobiska/Schieweck [90]) that this upwind scheme is of first order accurate and, what is most important, the main diagonal blocks of the corresponding system matrix $A+\tilde N(\cdot)$ become M-matrices. This is the key property for its inversion by fast multigrid algorithms.

The described upwind discretization can generically be extended to the three dimensional case. An analogous construction is possible for the conforming Q_1/Q_1 element with pressure stabilization also in two as well as in three dimensions; see Harig [38].

3.5.2. Streamline diffusion The idea of “streamline diffusion” is to introduce artificial diffusion acting only in the transport direction while maintaining the second-order consistency of the scheme. This can be achieved in various ways, by augmenting the test space by direction-oriented terms resulting in a “Petrov-Galerkin method”, or by adding certain least-squares terms to the discretization. For the (stationary) Navier-Stokes problem, we propose the following variant written in terms of pairs $\{\phi,\chi\}\in \mathbf{H}\times L$: Find $v_h\in v_h^{in}+\mathbf{H}_h$ and $p_h\in L_h$, such that

$$\begin{aligned} a_h(v_h,\phi_h)+n_h(v_h,v_h,\phi_h)+b_h(p_h,\nabla{\cdot}\phi_h)+s_h(\{v_h,p_h\},\{\phi_h,\chi_h\})\\ =(f,\phi_h)+r_h(\{v_h,p_h\},\{\phi_h,\chi_h\}) \end{aligned} \tag{3.35}$$

for all $\{\phi_h,\chi_h\}\in \mathbf{H}\times L_h$, where, with some “reference velocity” $\bar v_h$,

$$\begin{aligned} s_h(\{v_h,p_h\},\{\phi_h,\chi_h\}) &= \sum_{K\in\mathbb{T}_h}\delta_K\Big\{(\nabla p_h+v_h{\cdot}\nabla v_h,\nabla\chi_h+\bar v_h{\cdot}\nabla\phi_h)_K\\ &\qquad +(\nabla{\cdot}v_h,\nabla{\cdot}\phi_h)_K\Big\},\\ r_h(\{v_h,p_h\},\{\phi_h,\chi_h\}) &= \sum_{K\in\mathbb{T}_h}\delta_K(f+\nu\Delta v_h,\nabla\chi_h+\bar v_h{\cdot}\nabla\phi_h)_K. \end{aligned}$$

The stabilization parameters δ_K are chosen according to

$$\delta_K=\min\Big\{\frac{h_K^2}{\nu},\frac{h_K}{|\bar v|_K}\Big\}. \tag{3.36}$$

This discretization contains several features. The first term in the sum

$$\sum_{K\in\mathbb{T}_h}\delta_K\big\{(\nabla p_h,\nabla\chi_h)_K+(v_h{\cdot}\nabla v_h,\bar v_h{\cdot}\nabla\phi_h)_K+(\nabla{\cdot}v_h,\nabla{\cdot}\phi_h)_K\big\}$$

stabilizes the pressure-velocity coupling for the conforming Q_1/Q_1 Stokes element, the second term stabilizes the transport operator, and the third term enhances mass conservation. The other terms introduced in the stabilization are correction terms which guarantee second-order accuracy for the stabilized scheme. Theoretical analysis shows that this kind of Galerkin stabilization actually leads to an improvement over the standard upwinding scheme, namely an error behavior like $\mathcal{O}(h^{3/2})$ for the finite elements described above; see Tobiska/Verfürth [91], and also Braack [17] where the same kind of stabilization has been applied for weakly compressible flows with chemical reactions. For linear convection-diffusion problems the streamline diffusion method is known to have even $\mathcal{O}(h^2)$ convergence on fairly general meshes; see [107].

Open Problem 3.2: *Derive a strategy for choosing the stabilization parameter δ_K in the streamline diffusion method on general meshes with arbitrarily large aspect ratio σ_h.*

Open Problem 3.3: *The streamline diffusion method (like the least-squares pressure stabilization) leads to a scheme which lacks local mass conservation. Recently, for convection-diffusion problems an alternative approach has been proposed which uses a "discontinuous" Galerkin approximation on the transport term and combines (higher-order) upwinding features with local mass conservation. The extension of this method to the incompressible Navier-Stokes equations (and its practical realization) has yet to be done.*

4. Time discretization and linearization

We now consider the *nonstationary* Navier-Stokes problem: Find $v \in v^{in} + \mathbf{H}$ and $p \in L$, such that $v(0) = v^0$ and

$$(\partial_t v, \varphi) + a(v,\varphi) + n(v,v,\varphi) + b(p,\phi) = (f,\varphi), \quad \forall\, \varphi \in \mathbf{H}, \tag{4.1}$$

$$b(\chi, v) = 0, \quad \forall\, \chi \in L. \tag{4.2}$$

The choice of the function spaces $\mathbf{H} \subset \mathbf{H}^1(\Omega)^d$ and $L \subset L^2(\Omega)$ depends again on the specific boundary conditions chosen for the problem to be solved; see the discussion in Section 2.

In the past, explicit time stepping schemes have been commonly used in nonstationary flow calculations, mainly for simulating the transition to steady state limits. Because of the severe stability problems inherent to this approach (for moderately sized Reynolds numbers) the very small time steps required prohibited the accurate solution of really time dependent flows. In implicit time stepping one distinguishes traditionally between two different approaches called the "Method of Lines" and the "Rothe Method".

4.1. The Rothe Method

In the "Rothe Method", at first, the time variable is discretized by one of the common time differencing schemes; for a general account of such schemes see, e.g., Thomée [89]. For example, the backward Euler scheme leads to a sequence of Navier-Stokes-type problems of the form:

$$k_n^{-1}(v^n - v^{n-1}, \phi) + a(v^n, \phi) + n(v^n, v^n, \phi) + b(p^n, \phi) = (f^n, \phi), \tag{4.3}$$

$$b(\chi, v^n) = 0. \tag{4.4}$$

for all $\{\phi, \chi\} \in \mathbf{H} \times L$, where $k_n = t_n - t_{n-1}$ is the time step. Each of these problems is then solved by some spatial discretization method as described in the preceding section. This provides the flexibility to vary the spatial discretization, i.e. the mesh or the type of trial functions in the finite element method, during the time stepping process. In the classical Rothe method the time discretization scheme is kept fixed and only the size of the time step may change. The question of how to deal with varying spatial discretization within a time-stepping process while maintaining higher-order accuracy and conservation properties is currently subject of intensive research. It is essential to do the mesh-transfer by L^2 projection which is costly, particularly in 3D if full remeshing is used in each time step, but is easily manageable if only meshes from a family of hierarchically ordered meshes are used.

4.2. The Method of Lines

The traditional approach to solving time-dependent problems is the "Method of Lines". At first, the spatial variable is discretized, e.g. by a finite element method as described in the preceding section leading to a system of ordinary differential equations of the form:

$$M\dot{x}(t) + Ax(t) + N(x(t))x(t) + By(t) = b(t), \tag{4.5}$$

$$-B^T x(t) + Cy(t) = c(t), \quad t > 0, \tag{4.6}$$

with the initial value $x(0) = x^0$. The mass matrix M, the stiffness matrix A and the gradient matrix B are as defined above in Section 3. The matrix C and the right-hand side c stem from the pressure stabilization when using the conforming Q_1/Q_1 Stokes element. Further, the (nonlinear) matrix $N(\cdot)$ is thought to contain also all terms arising through the transport stabilization by upwinding or streamline diffusion. For abbreviation, we will sometimes use the notation $A(\cdot) := A + N(\cdot)$.

For solving this ODE system, one now applies a time differencing scheme. The most frequently used schemes are the so-called "One-Step-θ Schemes":

One-step θ-scheme: Step $t_{n-1} \to t_n$ ($k =$ time step):

$$\begin{aligned} [M + \theta k A^n]x^n + By^n &= [M - (1-\theta)kA^{n-1}]x^{n-1} + \theta k b^n + (1-\theta)kb^{n-1} \\ -B^T x^n + Cy^n &= c^n, \end{aligned}$$

where $x^n \approx x(t_n)$ and $A^n := A(x^n)$. Special cases are the "forward Euler scheme" for $\theta = 0$ (first-order explicit), the backward Euler scheme for $\theta = 1$ (first-order implicit, strongly A-stable), and the most popular Crank-Nicolson scheme for $\theta = 1/2$ (second-order implicit, A-stable). These properties can be seen by applying the method to the scalar model equation $\dot{x} = \lambda x$. In this context it is related to a rational approximation of the exponential function of the form

$$R_\theta(-\lambda) = \frac{1-(\theta-\frac{1}{2})\lambda}{1+\theta\lambda} = e^{-\lambda} + \mathcal{O}\left((\theta-\tfrac{1}{2})|\lambda|^2 + |\lambda|^3\right), \quad |\lambda| \le 1.$$

The most robust implicit Euler scheme ($\theta = 1$) is very dissipative and therefore not suitable for computing really nonstationary flow. In contrast, the Crank-Nicolson scheme has only very little dissipation but occasionally suffers from unexpected instabilities caused by the possible occurrence of rough perturbations in the data which are not damped out due to the only weak stability properties of this scheme (not *strongly* A-stable). This defect can in principle be cured by an adaptive step size selection but this may enforce the use of an unreasonably small time step, thereby increasing the computational costs. For a detailed discussion of this issue see [71]. A good time-stepping scheme of the described type should possess the following properties:

- *A-stability* ($\Rightarrow$ local convergence): $|R(-\lambda)| \le 1$
- *Global stability* ($\Rightarrow$ global convergence):
$$\overline{\lim}_{\mathrm{Re}\,\lambda\to\infty} |R(-\lambda)| \le 1 - \mathcal{O}(k).$$
- *Strong A-stability* ($\Rightarrow$ smoothing property):
$$\overline{\lim}_{\mathrm{Re}\,\lambda\to\infty} |R(-\lambda)| \le 1-\delta < 0.$$
- *Low dissipation* ($\Rightarrow$ energy preservation):
$$|R(-\lambda)| = 1 - \mathcal{O}(|Im\,\lambda|), \quad \text{for } Re\,\lambda \to 0.$$

Alternative schemes of higher order are based on the (diagonally) implicit Runge-Kutta formulas or the backward differencing multi-step formulas, both being well known from the ODE literature. These schemes, however, have not yet found wide applications in flow computations, mainly because of their higher complexity and storage requirements compared with the Crank-Nicolson scheme. Also less theoretical analysis is available for these methods when applied to large stiff systems. Some comparison of their stability and approximation properties is made in [72]; see also [65]. However, there is still another method which is an attractive alternative to the Crank-Nicolson method, the so-called "Fractional-Step-θ Scheme" originally proposed by Glowinski [30] and Bristeau et al. [22].

Fractional-Step-θ-scheme: (three substeps: $t_{n-1} \to t_{n-1+\theta} \to t_{n-\theta} \to t_n$)

$$\begin{aligned}
&(1) \quad [M+\alpha\theta k A^{n-1+\theta}]x^{n-1+\theta} + \theta k B y^{n-1+\theta} = [M-\beta\theta k A^{n-1}]x^{n-1} + \theta k b^{n-1},\\
&\qquad -B^T x^{n-1+\theta} + C y^{n-1+\theta} = c^{n-1+\theta},\\
&(2) \quad [M+\beta\theta' k A^{n-\theta}]x^{n-\theta} + \theta' k B y^{n-\theta} = [M-\alpha\theta' k A^{n-1+\theta}]x^{n-1+\theta} + \theta' k b^{n-\theta},\\
&\qquad -B^T x^{n-\theta} + C y^{n-\theta} = c^{n-\theta},\\
&(3) \quad [M+\alpha\theta k A^{n}]x^{n} + \theta k B y^{n} = [M-\beta\theta k A^{n-\theta}]x^{n-\theta} + \theta k b_h^{n-\theta},\\
&\qquad -B^T x^{n} + C y^{n} = c^{n}.
\end{aligned}$$

In the ODE context this scheme reduces to a rational approximation of the exponential function of the form

$$R_\theta(-\lambda)=\frac{(1-\alpha\theta'\lambda)(1-\beta\theta\lambda)^2}{(1+\alpha\theta\lambda)^2(1+\beta\theta'\lambda)}=e^{-\lambda}+\mathcal{O}(|\lambda|^3)\,,\ |\lambda|\le 1\,.$$

Here $\theta=1-\sqrt{2}/2=0.292893\ldots$, $\theta'=1-2\theta$, $\alpha\in(1/2,1]$, and $\beta=1-\alpha$, in order to ensure second-order accuracy, and strong A-stability,

$$\overline{\lim}_{\mathrm{Re}\,\lambda\to\infty}|R_\theta(-\lambda)|=\frac{\beta}{\alpha}<1\,.$$

For the special choice $\alpha=(1-2\theta)/(1-\theta)=0.585786\ldots$, there holds $\alpha\theta=\beta\theta'$ which is useful in building the system matrices in the three substeps. This scheme was first proposed in form of an operator splitting scheme separating the two complications "nonlinearity" and "incompressibility" within each cycle $t_n\to t_{n+\theta}\to t_{n+1}$. However, the Fractional-Step-θ scheme has also very attractive features as a pure time-stepping method. It is strongly A-stable, for any choice of $\alpha\in(1/2,1]$, and therefore possesses the full smoothing property in the case of rough initial data, in contrast to the Crank-Nicolson scheme (case $\alpha=1/2$). Furthermore, its amplification factor has modulus $|R(-\lambda)|\approx 1$, for λ approaching the imaginary axis (e.g., $|R(-0.8i)|=0.9998\ldots$), which is desirable in computing oscillatory solutions without damping out the amplitude. Finally, it also possesses very good approximation properties, i.e., one cycle of length $(2\theta+\theta')k=k$ provides the same accuracy as three steps of the Crank-Nicolson scheme with total step length $k/3$; for more details on this comparison see [72] and [65].

We mention some theoretical results on the convergence of these schemes. For the Crank-Nicolson Scheme combined with spatial discretization as described in Section 3, an optimal-order convergence estimate

$$\|v_h^n-v(\cdot,t_n)\|=\mathcal{O}(h^2+k^2) \tag{4.7}$$

has been given in [44]. This estimate requires some additional stabilization of the scheme but then holds under realistic assumptions on the data of the problem. A similar result has been shown by Müller [64] for the Fractional-Step θ-Scheme. Due to its stronger damping properties (*strong* A-stability) this scheme does not require extra stabilization.

4.2.1. COMPUTATIONAL TESTS Below, we present some results of the computational comparison between the backward Euler scheme, the Crank-Nicolson scheme and the Fractional-Step-θ scheme. The flow configuration is shown in Figure 18: flow around an inclined plate in the cross-section of a channel at Re = 500. The spatial discretization is by the nonconforming "rotated" bilinear Stokes element described in Section 3 on a uniformly refined mesh with 13,000 cells.
The first test concerns accuracy. Figure 19 shows that the backward Euler (BE) scheme is not suitable for computing time-periodic flows with acceptable time-step widths, while the Crank-Nicolson (CN) and the Fractional-Step-θ (FS) scheme show equally satisfactory results. This similar accuracy is further confirmed by

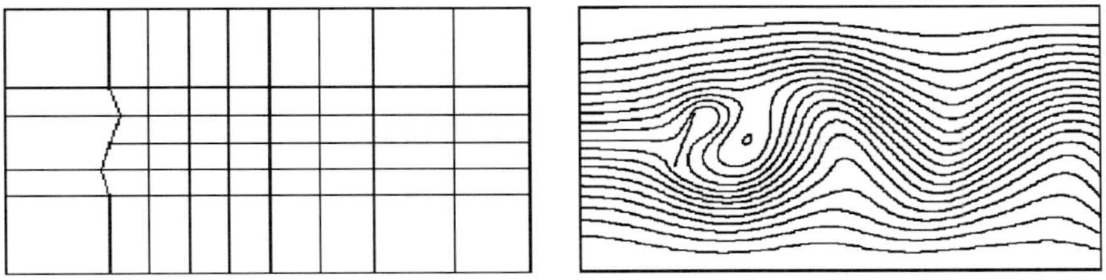

FIGURE 18. Configuration of plate-flow test, coarse mesh and streamline plot.

comparing a more sensitive error quantity (mean pressure) in Figure 20. Finally, we look at the stability of the schemes. Figure 21 demonstrates the lack of robustness of the Crank-Nicolson scheme combined with linear time-extrapolation in the nonlinearity for larger time steps.

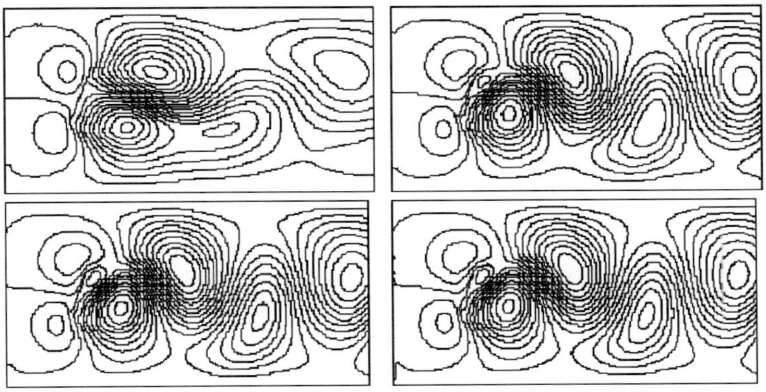

FIGURE 19. Pressure isolines of the plate-flow test: BE scheme with $3k = 1$ (top left), BE scheme with $3k = 0.1$ (top right). CN scheme with $3k = 1$ (bottom left), FS scheme with $k = 1$ (bottom right); from [65].

4.3. Splitting and projection schemes

As already mentioned, the Fractional-Step-θ scheme was originally introduced as an operator splitting scheme in order to separate the two main difficulties in solving problem (4.1) namely the nonlinearity causing nonsymmetry and the incompressibility constraint causing indefiniteness. At that time, handling both complications simultaneously was not feasible. Therefore, the use of operator splitting seemed the only way to compute nonstationary flows. Using the notation from above, the splitting scheme reads as follows (suppressing here the terms stemming from pressure stabilization).

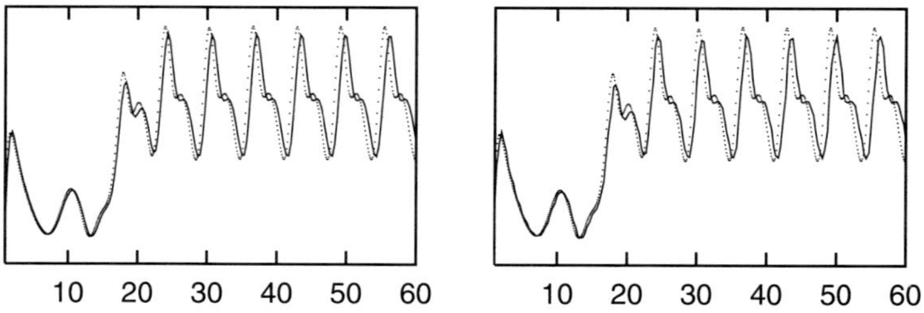

FIGURE 20. Mean pressure plots for the plate test with fully implicit treatment of the nonlinearity; left: CN scheme with $3k = 0.33$; right: FS scheme with $k = 0.33$, both compared to a reference solution (dotted line); from[65].

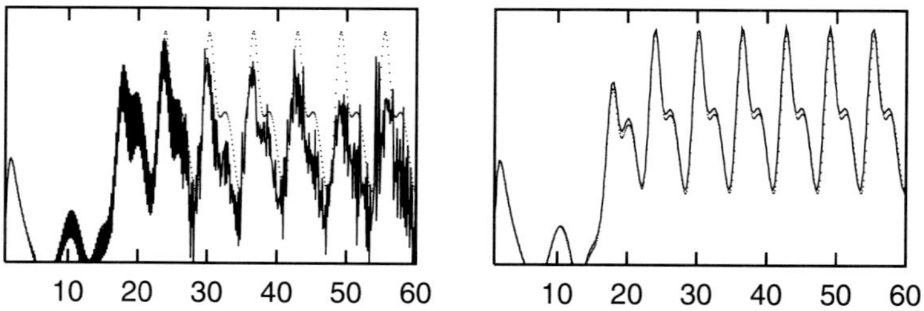

FIGURE 21. Mean pressure plots for the plate test with linear time-extrapolation; left: CN scheme with $3k = 0.11$; right: FS scheme with $k = 0.11$, both compared to a reference solution (dotted line); [65].

Splitting-Fractional-Step-θ Scheme:

$$
\begin{aligned}
(1)\quad & [M+\alpha\theta kA]x^{n-1+\theta} + \theta k B y^{n-1+\theta} = [M-\beta\theta kA]x^{n-1} + \theta k b^{n-1} - \\
& \qquad\qquad\qquad\qquad\qquad\qquad\qquad -\theta k N^{n-1} x^{n-1}, \\
& B^T x^{n-1+\theta} = 0\,, \\
(2)\quad & [M+\beta\theta' k A^{n-\theta}]x^{n-\theta} = [M-\alpha\theta' k A^{n-1+\theta}]x^{n-1+\theta} - \theta' k B y^{n-\theta} + \theta' k b^{n-\theta}, \\
& \ldots\ldots \\
(3)\quad & [M+\alpha\theta kA]x^{n} + \theta k B y^{n} = [M-\beta\theta kA]x^{n-\theta} - \theta k N^{n-\theta} x^{n-\theta} + \theta k b^{n-\theta}, \\
& B^T x^{n} = 0\,.
\end{aligned}
$$

The first and last step solve *linear* Stokes problems treating the nonlinearity explicitly, while in the middle step a nonlinear Burgers-type problem (without incompressibility constraint) is solved. The symmetric form of this scheme follows the ideas from Strang [87], in order to achieve a second-order splitting approximation. The results of Müller [64] suggest that the optimal-order convergence estimate

(4.7) remains true also for this splitting scheme. However, a complete proof under realistic assumptions is still missing.

Open Problem 4.1: *Prove that the Splitting-Fractional-Step-θ scheme is actually second order accurate for all choices of the parameter* $\alpha \in (\frac{1}{2}, 1]$.

In these days, the efficient solution of the *nonlinear* incompressible Navier-Stokes equations is standard by the use of new multigrid techniques. Hence, the splitting of nonlinearity and incompressibility is no longer an important issue. One of these new approaches uses the Fractional-Step-θ scheme in combination with the idea of a "projection method" due to Chorin [24]; for a survey see Gresho/Sani [35]. Finally, Turek [95] (see also [97]) has designed the "Discrete Projection Fractional-Step-θ scheme" as component in his solver for the nonstationary Navier-Stokes problem.

Next, we address the problem of how to deal with the incompressibility constraint $\nabla{\cdot}v = 0$. The traditional approach is to decouple the continuity equation from the momentum equation through an iterative process (again "operator splitting"). There are various schemes of this kind in the literature referred to, e.g., as "quasi-compressibility method", "projection method", "SIMPLE method", etc. All these methods are based on the same principle idea. The continuity equation $\nabla{\cdot}\mathbf{v} = 0$ is supplemented by certain stabilizing terms involving the pressure, e.g.,

$$\nabla{\cdot}v + \varepsilon p = 0, \tag{4.8}$$

$$\nabla{\cdot}v - \varepsilon\Delta p = 0, \qquad \partial_n p_{|\partial\Omega} = 0, \tag{4.9}$$

$$\nabla{\cdot}v + \varepsilon\partial_t p = 0, \qquad p_{|t=0} = 0, \tag{4.10}$$

$$\nabla{\cdot}v - \varepsilon\partial_t\Delta p = 0, \qquad \partial_n p_{|\partial\Omega} = 0,\ p_{|t=0} = 0, \tag{4.11}$$

where the small parameter ε is usually taken as $\varepsilon \approx h^\alpha$, or $\varepsilon \approx k^\beta$, depending on the purpose of the procedure. For example, (4.8) corresponds to the classical "penalty method", and (4.9) is the simplest form of the "least squares pressure stabilization" scheme (4.11) described above, with $\varepsilon \approx h^2$ in both cases. Further, (4.10) corresponds to the "quasi-compressibility method".

These approaches are closely related to the classical "projection method" of Chorin [24]. Since this method used to be particularly attractive for computing nonstationary incompressible flow, we will discuss it in a some detail. For simplicity consider the case of homogeneous Dirichlet boundary conditions, $v_{|\delta\Omega} = 0$. The projection method reads as follows. For an admissible initial value v^0, choose a time step k, and solve for $n \geq 1$:

(i) $\tilde{v}^n \in \mathbf{H}$ (implicit "Burgers step"):

$$k^{-1}(\tilde{v}^n - v^{n-1}) - \nu\Delta\tilde{v}^n + \tilde{v}^n{\cdot}\nabla\tilde{v}^n = f^n. \tag{4.12}$$

(ii) $v^n = P\tilde{v}^n \in \mathbf{J}_0(\Omega)$ ("Projection step"):

$$\nabla{\cdot}v^n = 0, \quad n{\cdot}v^n_{|\partial\Omega} = 0. \tag{4.13}$$

Here, the function space $\mathbf{J}_0(\Omega)$ is obtained through the completion of the space $\{\phi \in \mathcal{D}(\Omega), \nabla{\cdot}\phi \equiv 0\}$ of solenoidal test functions with respect to the norm of

$\mathbf{L}^2(\Omega)$. This time stepping scheme can be combined with any spatial discretization method, e.g. the finite element methods described in Section 3. The projection step (ii) can equivalently be expressed in the form

$$\text{(ii')} \qquad v^n = \tilde{v}^n + k\nabla\tilde{p}^n, \tag{4.14}$$

with some "pressure" $\tilde{p}^n \in H^1(\Omega)$, which is determined by the properties

$$\text{(ii")} \qquad \Delta\tilde{p}^n = k^{-1}\nabla\cdot\tilde{v}^n, \quad \partial_n\tilde{p}^n_{|\partial\Omega} = 0. \tag{4.15}$$

This amounts to a Poisson equation for $\tilde{p}^n$ with zero Neumann boundary conditions. It is this non-physical boundary condition, $\partial_n\tilde{p}^n{}_{|\partial\Omega} = 0$, which has caused a lot of controversial discussion about the value of the projection method. Nevertheless, the method has proven to work well for representing the velocity field in many flow problems of physical interest (see, e.g. Gresho [32] and Gresho/Chan [33]). It is very economical as it requires in each time step only the solution of a (nonlinear) advection-diffusion system for v^n (of Burgers equation type) and a scalar Neumann problem for $\tilde{p}^n$. Still, it was argued that the pressure $\tilde{p}^n$ were a mere fictitious quantity without any physical relevance. It remained the question: How can such a method work at all? A challenging problem for mathematical analysis!

The first convergence results for the projection method was already given by Chorin, but concerned only cases with absent rigid boundaries (all-space or spatially periodic problems). Later on, qualitative convergence was shown even for the pressure, but in a measure theoretical sense, too weak for practical purposes. Only recently, stronger results on the error behavior of this method have been obtained (see, e.g., Shen [84, 85] as well as [73], and the literature cited therein). The best known error estimate is

$$\|v^n - v(t_n)\|_\Omega + \|\tilde{p}^n - p(t_n)\|_{\Omega'} = \mathcal{O}(k), \tag{4.16}$$

where $\Omega' \subset\subset \Omega$ is a subdomain with positive distance to the boundary $\partial\Omega$. This shows that the quantities $\tilde{p}^n$ are indeed reasonable approximations to the pressure $p(t_n)$, and finally confirms that Chorin's original method is a first-order time stepping scheme for the incompressible Navier-Stokes problem. The key to this result is the re-interpretation of the projection method in the context of the "pressure stabilization methods". To this end, one inserts the quantity $v^{n-1} = \tilde{v}^{n-1} - k\nabla\tilde{p}^{n-1}$ into the momentum equation, obtaining

$$k^{-1}(\tilde{v}^n - \tilde{v}^{n-1}) - \nu\Delta\tilde{v}^n + (\tilde{v}^n\cdot\nabla)\tilde{v}^n + \nabla\tilde{p}^{n-1} = f^n, \quad \tilde{v}^n_{|\partial\Omega} = 0, \tag{4.17}$$

$$\nabla\cdot\tilde{v}^n - k\Delta\tilde{p}^n = 0, \quad \partial_n\tilde{p}^n_{|\partial\Omega} = 0. \tag{4.18}$$

This looks like an approximation of the Navier-Stokes equations involving a first-order (in time) "pressure stabilization" term, i.e., the projection method can be viewed as a pressure stabilization method with a global stabilization parameter $\varepsilon = k$, and an explicit treatment of the pressure term. Moreover, it appears that the pressure error is actually confined to a small boundary strip of width $\delta \approx \sqrt{\nu k}$

and decays exponentially into the interior of Ω. In fact, it was conjectured that, setting $d(x) = \mathrm{dist}(x, \partial\Omega)$,

$$|\tilde{p}^n(x) - p(x, t_n)| \le c \exp\left(-\alpha \frac{d(x)}{\sqrt{\nu k}}\right) \sqrt{k} + O(k). \tag{4.19}$$

This conjecture is supported by numerical experiments for the pressure stabilization method applied to the stationary Stokes problem and by some model situation analysis in E/Liu [25]. The analysis of this boundary layer phenomenon requires the study of the singularly perturbed Neumann problem

$$\left(\nu^{-1}\nabla\cdot\Delta_D^{-1}\nabla - \epsilon\Delta_N\right)q = \epsilon\Delta p, \quad \text{in } \Omega, \quad \partial_n q_{|\partial\Omega} = \partial_n p_{|\partial\Omega}, \tag{4.20}$$

where Δ_D denotes the Laplacian operator corresponding to Dirichlet boundary conditions. Clearly, $\nabla\cdot\Delta_D^{-1}\nabla$ is a zero-order operator mapping $L^2(\Omega)$ into $L^2(\Omega)$. For this problem, one would like to know a decay estimate of the form

$$\|q\|_{\Omega_\delta} \le c \exp\left(-\alpha \frac{\delta}{\sqrt{\epsilon}}\right) \|\nabla p\| + c\nu\epsilon\, \|\Delta p\|, \tag{4.21}$$

for interior subdomains $\Omega_\delta := \{x \in \Omega,\, \mathrm{dist}(x, \partial\Omega) > \delta\}$, from which a pointwise result like (4.19) could be inferred. Such an estimate could be proven in [73] only for the case that the "global" operator $\nabla\cdot\Delta_D^{-1}\nabla$ is replaced by the "local" identity operator. In the general case the corresponding result is still an open problem.

Open Problem 4.2: *Prove an analogue of the a priori decay estimate (4.21) for the non-local operator* $\nabla\cdot\Delta_D^{-1}\nabla$.

The occurrence of the pressure boundary layer is demonstrated in Figure 22 for a simple model problem on the unit square with known polynomial solution. It is even possible to recover the optimal-order accuracy of the pressure, $\mathcal{O}(h^2)$, at the boundary by postprocessing, e.g. by linear or quadratic extrapolation of pressure values from the interior of the domain; see Figure 23 and Blum [15] for more details on this matter.

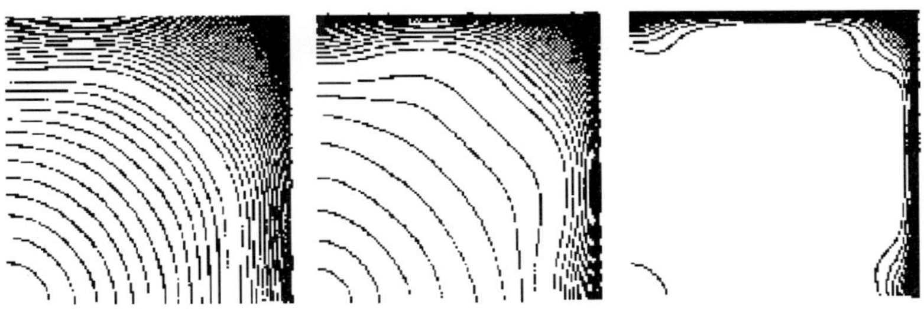

FIGURE 22. Sequence of pressure-error isolines obtained by the Chorin scheme with $k = 2.5\cdot10^{-2}$, $6.25\cdot10^{-3}$, $1.56\cdot10^{-3}$ (model problem with $\nu = 1$ on the square); from Prohl [69].

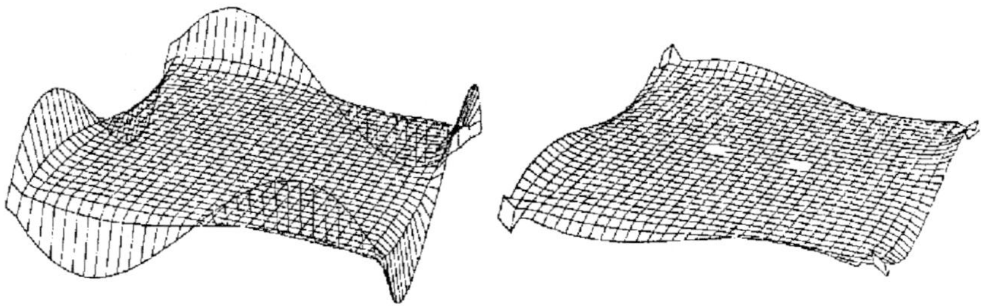

FIGURE 23. Pressure error plots for a polynomial Stokes solution before (left) and after (right) correction by extrapolation to the boundary; from Blum [15].

An important step towards the solution of the "boundary layer problem" has been made in Prohl [69, 70] by introducing the "Chorin-Uzawa scheme", which reads as follows:

(i) Implicit "Burgers step":

$$k^{-1}(\tilde{v}^n - v^{n-1}) - \Delta\tilde{v}^n + \tilde{v}^n\cdot\nabla\tilde{v}^n + \nabla(\tilde{p}^{n-1} - p^{n-1}) = f^n, \quad \tilde{v}^n_{|\partial\Omega} = 0.$$

(ii) Pressure Poisson problem:

$$\Delta\tilde{p}^n = k^{-1}\nabla\cdot\tilde{v}^n, \quad \partial_n\tilde{p}^n_{|\partial\Omega} = 0.$$

(iii) Pressure and velocity update:

$$v^n = \tilde{v}^n - k\nabla\tilde{p}^n, \quad p^n = p^{n-1} - \alpha\nabla\cdot\tilde{v}^n, \quad \alpha < 1.$$

The reference to the name "Uzawa" is due to the fact that this scheme partially resembles the structure of the well-known Uzawa algorithm for solving stationary saddle-point problems; see Girault/Raviart [29]. It corresponds to a quasi-compressibility method using the regularization

$$\nabla\cdot\tilde{v}^n + \alpha^{-1}k\partial_t p^n = 0. \tag{4.22}$$

This splitting scheme does not introduce a singular perturbation in the pressure equation and is therefore supposed to be free of any spatial boundary layer. However, it suffers from a "boundary layer" at time $t = 0$ in case of natural initial data not satisfying unrealistic global compatibility conditions; recall Section 2 for a discussion of such conditions. The conjectured suppression of the spatial pressure boundary layer by the Chorin-Uzawa scheme is confirmed by computational tests; see the example presented in Figure 24. A supporting analysis has been given in Prohl [70] for a modification of the Chorin-Uzawa method to a "multi-component scheme" which allows for the convergence estimate

$$\|p^n - p(t_n)\| \le ck, \quad t_n \ge 1. \tag{4.23}$$

Figures 24 show pressure error plots obtained for a given polynomial solution on the unit-square with viscosity $\nu = 1$; the time step is $k = 1/100$ and the

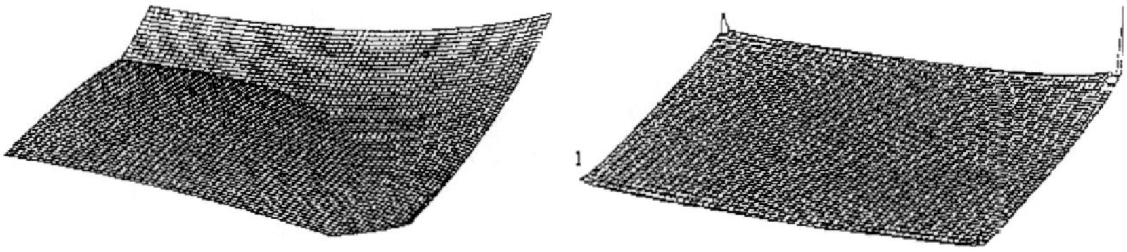

FIGURE 24. Pressure error plots for a polynomial solution produced by the standard Chorin scheme (left) and the Chorin-Uzawa scheme (right); from Prohl [69, 70].

spatial discretization uses the Q_1/Q_1 Stokes element with pressure stabilization on a uniform mesh with mesh-size $h = 1/64$.

The projection approach can be extended to formally higher order projection methods. The most popular example is Van Kan's Method [100]: For admissible starting values v^0 and p^0 compute, for $n \geq 1$ and some $\alpha \geq \frac{1}{2}$:

(i) $\tilde{v}^n \in \mathbf{H}$ (second order implicit Burgers step), satisfying

$$k^{-1}(\tilde{v}^n - v^{n-1}) - \tfrac{1}{2}\nu\Delta(\tilde{v}^n + v^{n-1}) + \tilde{v}^n\cdot\nabla\tilde{v}^n + \nabla p^{n-1} = f^{n-1/2};$$

(ii) $p^n \in H^1(\Omega)$: $\qquad v^n = \tilde{v}^n - \alpha k\nabla(p^n - p^{n-1})$.

A careful examination of this scheme shows that it can also be interpreted as a certain pressure stabilization method using a stabilization of the form

$$\nabla\cdot v - \alpha k^2\partial_t\Delta p = 0, \quad \text{in } \Omega, \quad \partial_n p_{|\partial\Omega} = 0, \tag{4.24}$$

i.e., this method may be viewed as an (implicit) quasi-compressibility method of the form (4.11) with $\varepsilon \approx k^2$; see [74] and Shen [86].

The projection method may be combined with any of the spatial discretizations described in Section 3. It should be remarked that the simple first-order Chorin scheme is not suitable for computing stationary limits since it has not the from of a fixed-point iteration. In contrast to that, the second-order scheme of Van Kan is designed as a defect-correction iteration and may therefore lead to convergence towards steady state solutions. However, in this case it requires extra pressure stabilization when used together with the conforming Q_1/Q_1 Stokes element; in fact the stabilizing effect of the projection step disappears as $\alpha k^2\partial_t\Delta p \to 0$.

Open Problem 4.3: *The efficient use of projection methods requires an automatic time-step-size control which should monitor deviation from the fully coupled solution. Design such a method for high-order schemes.*

5. Solution of the algebraic systems

In this section, we describe solution algorithms for the finite-dimensional problems arising from the discretization presented in the previous sections. These problems form huge and highly nonlinear algebraic systems with a characteristic structure which is exploited by the algorithms. The solution procedure consists of several nested loops. Usually the outermost loop is an implicit time-iteration. In each time step, the arising nonlinear system is solved by a quasi-Newton or defect-correction iteration. The discretization by finite elements leads to a sparse structure in the system matrices which is exploited by the iterative solution method. Even in the case of the Laplace operator (which is always a part of the system), the inversion by a direct solver or a simple iterative scheme like the "conjugate gradient" (CG) method is prohibitive due to the bad conditioning of the matrix with decreasing mesh size. Therefore, the use of multigrid methods is mandatory, either directly as solvers or as preconditioners for a robust iterative schemes like the "generalized minimal residual" (GMRES) algorithm. Since the systems to be solved are in general non-symmetric and indefinite, the construction of "good" multigrid algorithms requires special care.

5.1. Linearization

The time stepping schemes described above require in each time step the solution of nonlinear systems of the form

$$[\sigma M + \nu A + N(v)]v + Bp = g, \tag{5.1}$$

$$-B^T v + \epsilon C p = c, \tag{5.2}$$

where $\sigma = (\theta k)^{-1}$ and (on a quasi-uniform mesh) $\epsilon \sim h^2$. The operators involved correspond to differential operators as follows:

$$M \sim id., \quad A \sim -\mathrm{diag}(\Delta_D), \quad N(v) \sim v\cdot\nabla,$$
$$B \sim \nabla, \quad -B^T \sim \div, \quad C \sim -\Delta_N,$$

where Δ_D and Δ_N denote the Laplacian operator combined with (homogeneous) Dirichlet or Neumann boundary conditions, respectively. The right-hand sides g and c contain information from the preceding time level. Here and below, the same notation is used for the (discrete) velocity v and pressure p and the corresponding nodal vectors. The following iteration schemes are formulated on the continuous level without incorporating stabilization, i.e., we set $\epsilon = 0$ and $c = 0$.

a) Newton method:
Starting from some initial values $v^0, p^0 \in \mathbf{H} \times L$ (for example, taken from the preceding time level), one iterates:

1. Defect: $d^l = g - \big(\sigma M + \nu A + N(v^l)\big)v^l - Bp^l$.
2. Correction: $[\sigma M + \nu A + N'(v^l)]w^l + Bq^l = d^l, \quad B^T w^l = 0.$
3. Update: $v^{l+1} = v^l + \lambda_l w^l, \quad p^{l+1} = p^l + \lambda_l q^l \quad$ (λ_l damping factor).

This iteration has been observed to converge very fast, provided that it converges at all. The problem is its lack of robustness particularly in the case of larger Reynolds numbers. This is due to the structure of the operator to be inverted in each iteration step:

$$N'(v)w = v\cdot\nabla w + w\cdot\nabla v.$$

It contains a reaction term $w\cdot\nabla v$ which effects the main diagonal of the system matrix in an uncontrolled manner and may cause divergence of the iteration. This problem may be avoided by simply dropping the reaction term in the Jacobian which results in the following fixed-point defect correction iteration.

b) Fixed-point defect correction:
Starting from some initial values $v^0, p^0 \in \mathbf{H} \times L$ (taken again from the preceding time level), one iterates:

1. Defect: $d^l = g - \big(\sigma M + \nu A + N(v^l)\big)v^l - Bp^l$.
2. Correction: $[\sigma M + \nu A + N(v^l)]w^l + Bq^l = d^l, \quad B^T w^l = 0$.
3. Update: $v^{l+1} = v^l + \lambda_l w^l, \quad p^{l+1} = p^l + \lambda_l q^l \quad$ (λ_l damping factor).

In this scheme the preconditioning operator $\tilde{A}'(v^l) = v^l\cdot\nabla$ only contains a transport term which can be stabilized by any of the methods described above: upwinding, streamline diffusion, etc. Normally, within the time stepping scheme, only a few (usually 3–5) steps of the defect correction iteration are necessary for reducing the initial residual down to the level of the discretization error. This is our method of choice used in the codes mentioned in the Introduction.

c) Nonlinear multigrid iteration:
The multigrid method can be applied directly to the nonlinear system; see Hackbusch [36]. This may lead to faster convergence but its optimization is difficult and depends very much on the particular problem. Because of this lack of robustness, we do not advocate "nonlinear" multigrid for solving the Navier-Stokes equations.

d) Nonlinear least-squares cg method:
The (nonlinear) *least squares cg-method* for solving systems like (5.1) has been proposed by Glowinski/Periaux [31]. Starting from an initial guess x^0, a sequence of approximate solutions $(x^l)_{l\geq 0}$ is obtained by minimizing the least squares functional

$$\|\nabla w\|^2 \to \min! \tag{5.3}$$

where w is determined by $\{v, p\}$ through the equation

$$[\sigma M + \nu A]w - Bq = \text{"defect" of } \{v, p\}, \quad B^T w = 0. \tag{5.4}$$

It can be seen that each nonlinear cg-step actually requires only the solution of three linear Stokes problems which can be efficiently done by linear multigrid techniques. This method is very robust as it is based on the minimization of a positive functional, but the speed of convergence drastically slows down for larger Reynolds numbers. For example, the 3-D driven cavity problem can be solved by the stationary version of the least squares cg method up to about Re = 2000; for further details, see [38].

5.2. Solution of the linearized problems

The problem to be solved has the form

$$\begin{bmatrix} S & B \\ -B^T & \epsilon C \end{bmatrix} \begin{bmatrix} v \\ p \end{bmatrix} = \begin{bmatrix} g \\ c \end{bmatrix}, \qquad \mathcal{A} = \begin{bmatrix} S & B \\ -B^T & \epsilon C \end{bmatrix}, \tag{5.5}$$

where, with some initial guess $\bar{v}$,

$$S = \sigma M + \nu A + N(\bar{v}).$$

The difficulty with this system is that the matrix $\mathcal{A}$ is neither symmetric nor definite. It is usually too large for the application of direct solvers (like the LU decomposition by Gaussian elimination) and also the traditional iterative methods (like SOR iteration or Krylov space schemes) do not work sufficiently well. This suggests the use of multigrid methods which are particularly suited on very fine meshes. However, the construction of efficient multigrid algorithms for solving the indefinite system (5.5) is not at all straightforward. Therefore, as a simpler alternative the Schur complement approach has become popular which will be described in the following subsection.

5.2.1. SCHUR-COMPLEMENT ITERATION In the system matrix $\mathcal{A}$ the main block S is regular and usually robust to be inverted. Hence, the velocity unknowns may be eliminated from the system by inverting S which leads to:

$$[B^T S^{-1} B + \epsilon C]p = B^T S^{-1} g + c, \quad v = S^{-1}(g - Bp). \tag{5.6}$$

The "Schur-complement" matrix $\Sigma = B^T S^{-1} B + \epsilon C$ is regular. Neglecting the influence of the nonlinear term $N(\bar{v})$, its condition number behaves like

$$\text{cond}(\Sigma) = \mathcal{O}(h^{-2}) \text{ for } \nu k \ll h^2, \quad \text{cond}(\Sigma) = \mathcal{O}(1) \text{ for } \nu k \gg h^2.$$

This suggests the use of iterative methods for its inversion, e.g., Krylov space methods like the GMRES or the bi-cg-stab method. In the essentially nonstationary case, $\nu k \geq h^2$, only a few iteration steps suffice. In nonstationary computations where $\nu k \ll h^2$, preconditioning by an approximation of the Neumann-type operator $B^T M^{-1} B$ is necessary. In each iteration step the operator S^{-1} has to be evaluated which amounts to solving a linear transport-diffusion problem. In the case of the nonconforming $\tilde{Q}_1/P_0$ Stokes elements combined with upwind stabilization of advection S becomes an M-matrix. This facilitates the iterative inversion of Σ, particularly by multigrid methods. The non-exact inversion of Σ makes the step $q \to \tilde{\Sigma}^{-1} q$ a preconditioning step within the iteration for inverting Σ. Hence, the number of inner iteration steps should be kept fixed during the whole solution process. Another strategy for compensating for the error in the evaluation of S^{-1} is to embed the outer iteration (5.6) into a defect correction process; see Bank, et al., [6]. The convergence usually deteriorates for increasing Reynolds number, because of loss of "symmetry", and for decreasing time step k, because of the bad conditioning of the operator $B^T B \sim \Delta$. For larger Reynolds number the convergence of the Schur complement iteration becomes slow and special preconditioning is necessary. The construction of effective preconditioners is not easy since the operator Σ is not available as an explicit matrix. Another stability problem occurs on meshes containing cells with large aspect ratio. Because of this lack of robustness, the Schur complement method has less potential than the direct multigrid approach which will be described below.

Open Problem 5.1: *Derive a formula for the dependence of the conditioning of the Schur complement operator* $\Sigma = B^T S^{-1} B + \epsilon C$ *on the Reynolds number and on the mesh aspect ratio* σ_h .

5.3. Linear multigrid solution

The main idea underlying a multigrid method is the fast reduction of high-frequency error components by "cheap" relaxation ("smoothing") on the fine meshes and the reduction of the remaining low-frequency error components by defect correction on coarser meshes ("coarse-grid correction"); see Hackbusch [36] and Wesseling [104], for an introduction to multigrid methods.

5.3.1. MULTIGRID AS A PRECONDITIONER Let $\mathcal{A}$ be the finite element system matrix of the linearized equation (5.5) or an appropriate approximation. While the theory of multigrid is well developed for scalar elliptic equations, the situation is less clear for complicated systems as considered in this paper. From mathematical analysis, we know that the use of the multigrid method as a preconditioner in an outer iteration (e.g., a Krylov space method such as GMRES) requires less restrictive assumptions than using the multigrid method directly as a solver. In the first case, denoting by $\mathcal{M}$ the action of a multigrid step, it is sufficient to have an upper bound for the condition of the product $\mathcal{MA}$, whereas in the second case, the eigenvalues of the iteration matrix $\mathcal{B} = I - \mathcal{MA}$ have to be uniformly bounded away from one. Therefore, we choose the first option to construct a robust iteration scheme for the system (5.5). As basic solver, one may use the "generalized minimal residual method" (GMRES) for the preconditioned matrix $\mathcal{MA}$. Here, the multigrid operator $\mathcal{M}$ can be interpreted as a certain approximate inverse $\mathcal{M} \approx \mathcal{A}^{-1}$. It is not necessary to calculate this matrix explicitly; it is sufficient to evaluate the matrix-vector product $\mathcal{M}\xi$, i.e., to apply the multigrid iteration for a fixed right-hand side.

5.3.2. MULTIGRID AS A SOLVER The multigrid iteration makes use of the hierarchy of finite element spaces

$$V_0 \subset V_1 \subset \ldots \subset V_L ,$$

obtained, for example, in the course of a systematic mesh refinement process; strategies for an automatic adaptive mesh refinement will be discussed below in Section 7. The connection between these spaces is given by "prolongation operators" $P^l_{l-1} : V_{l-1} \to V_l$ and "restriction operators" $R^{l-1}_l : V_l \to V_{l-1}$. In the finite element context, these operators are given naturally as

$$P^l_{l-1} \text{ injection}, \quad R^{l-1}_l \ L^2 \text{ Projection}.$$

The main ingredients of a multigrid scheme are the smoothing operators S_l on each grid level $0 \le l \le L$ ($l=0$ corresponding to the coarse initial mesh and $l=L$ to the finest mesh). The explicit form of these operators will be described below. The multigrid iteration

$$\mathcal{M}\xi = \mathcal{M}(l, z_0, \xi), \tag{5.7}$$

on level l with initial guess z_0 and with m_1 pre- and m_2 post-smoothing steps is recursively defined as follows:

Multigrid Algorithm $\mathcal{M}(l, z_0, \xi)$ for $l \geq 0$:
For $l=0$, the multigrid algorithm is given by an exact solver $\mathcal{M}(l, z_0, \xi) := \mathcal{A}_0^{-1}\xi$.
For $l>0$, the following recursive iteration is performed:

1. Pre-smoothing m_1 times: $z_1 := S_l^{m_1} z_0$.
2. Residual on level l: $r_l := \xi - \mathcal{A}_k z_1$.
3. Restriction to level $l-1$: $r_{l-1} := R_l r_l$.
4. Coarse grid correction starting with $q_0 = 0$: $q := \mathcal{M}(l-1, q_0, r_{l-1})$.
5. Prolongation to level l: $z_2 := z_1 + P_{l-1} q$.
6. Post-smoothing m_2 times: $\mathcal{M}(l, z_0, \xi) := S_l^{m_2} z_2$.

If the multigrid recursion is applied γ-times on each mesh-level, one speaks of a V-cycle for $\gamma = 1$ and of a W-cycle for $\gamma = 2$. If the multigrid iteration is used only as a preconditioner for a robust outer iteration scheme, usually the V-cycle suffices. If multigrid is used as the primary solver, particularly in the case of nonsymmetric problems, the W-cycle is more robust and therefore to be preferred. In this case, the F-cycle as indicated in the figure below is a compromise between V- and W-cycle. The multigrid cycle with $\gamma > 2$ becomes too expensive and is not used.

FIGURE 25. Scheme of the multigrid V-cycle (left) and the F-cycle (right)

The design of a multigrid algorithm for solving the system (5.5) requires special care. In particular, the choice of the smoother is a delicate matter since the standard fixed-point iterations do not work for the indefinite matrix $\mathcal{A}$. This problem can be tackled in various ways.

(1) Damped Jacobi smoother: In the case $\epsilon > 0$, the matrix $\mathcal{A}$ is weakly definite which makes it possible to apply even standard methods like the damped Jacobi iteration. However, the resulting algorithm is not very robust and parameter tuning requires care; see [38] for an application to 3-dimensional model problems. For larger Reynolds number the method slows down and multigrid convergence may get lost.

(2) Block-Gauss-Seidel smoother: A simple and successful smoother for the matrix $\mathcal{A}$ can be obtained by a cell-wise blocking of the physical variables within

a global Gauss-Seidel iteration. This was originally proposed by Vanka [101] for a finite difference discretization of the Navier-Stokes problem. We briefly discuss its analogue for the nonconforming "rotated" $\tilde{Q}_1/P_0$ Stokes element. The velocity and pressure unknowns corresponding to a cell K or a patch of cells are grouped together. Indicating the corresponding element system matrices by index "*loc*", these blocks of local velocity and pressure unknowns are simultaneously updated according to

$$S_{loc}v_{loc}^{t+1} + B_{loc}p_{loc}^{t+1} = \text{"known"}, \quad B_{loc}^T v_{loc}^{t+1} = \text{"known"},$$

where $S_{loc} = \sigma M_{loc} + \nu A_{loc} + N_{loc}(\bar{v}_h)$. This iteration sweeps over all cell-blocks. The local Stokes problems have the dimension $d_{loc} = 9$ (in 2D) or $d_{loc} = 19$ (in 3D), respectively. The corresponding matrices (in 2D) are described in the following figure.

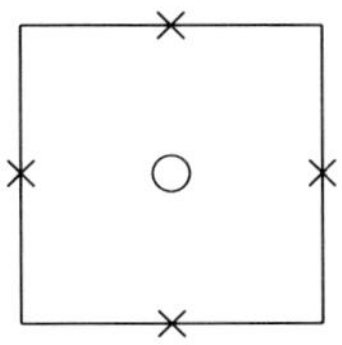

$\times$: node for v
$\bigcirc$: node for p

$$\mathcal{A}_{loc} = \begin{bmatrix} S_{loc,1} & O & B_{loc,1} \\ O & S_{loc,2} & B_{loc,2} \\ -B_{loc,1}^T & -B_{loc,2}^T & 0 \end{bmatrix}.$$

For cost reduction, the main diagonal blocks $S_{loc,i}$ may be "lumped", $S_{loc,i} \approx D_{loc,i}$. Furthermore, for increasing robustness, the iteration is damped, $v_h^{t+1} = v_h^t + \omega(\tilde{v}_h^{t+1} - v_h^t)$ with $\omega \approx 0.9$. The good performance of this smoother for the $\tilde{Q}_1/P_0$-element has been demonstrated in Schreiber/Turek [83], Schieweck [82], and Turek [97], and for the Q_1/Q_1-element in Becker [7]. We illustrate the performance of the multigrid algorithm described above for the $\tilde{Q}_1/P_0$-element by results obtained for solving the driven-cavity problem on grids as shown below.

FIGURE 26. Driven cavity mesh (left) and computed results: pressure isolines (middle), velocity plot (right).

We note that a similar block-iteration can also be used in the context of a *incomplete* block-LU-decomposition for generating a multigrid smoother; for a detailed discussion of this approach see Braack [17].

TABLE 2. Multigrid convergence rates (2 pre- and 1 post-smoothing step by the "Vanka smoother") and number of outer fixed-point iterations on uniformly refined meshes.

#cells	1600	6400	25600	#iter
Re = 1	0.081	0.096	0.121	4
Re = 100	0.098	0.099	0.130	6
Re = 1000	0.227	0.245	0.168	9
Re = 5000	0.285	0.368	0.370	18

From the common multigrid theory for elliptic equations we know that point iterations loose the smoothing property for large mesh-aspect ratios σ_h. The remedy is the use of a smoother which becomes a direct solver in the limit $\sigma_h \to \infty$. Consequently, since our smoother acts like a point-Gauss-Seidel iteration on the velocity unknowns, we expect problems in the case of strongly stretched grids. Our strategy to overcome this difficulty is as follows: Since we expect the cell aspect ratio σ_K to be large only in a small part of the computational domain, we should use an *adaptive* smoother. This means that we will combine the point smoother with a more robust version just where we need it, for instance on elements with large σ_K. In this approach the nodes are grouped in the direction of the anisotropic mesh refinement and iterated implicitly leading to a process which may be termed "stringwise" block-Gauss-Seidel method.

Let us finally mention a critical problem especially in the use of non-uniform grids. The use of iterative solvers makes it necessary to define a stopping criterion. To this end, we need to measure the residual in the right norm. Clearly, the common weighting by the number of unknowns is not appropriate on non-uniform grids. For an approach towards a solution of this problem based on the Galerkin orthogonality inherent to the multigrid process, we refer to [11] and Becker [8].

(3) Discrete projection smoother: Finally, we present an approach to constructing multigrid solvers for the indefinite system (5.5) which uses the idea of operator splitting as introduced above in Section 4 on time-discretization schemes; see Turek [95, 97]. This method is particularly efficient in the nonstationary case when $\sigma = 1/k$ balances ν/h^2. In the following, we consider the linearized problem arising within a time-stepping scheme as described above in combination with spatial discretization by a Stokes element which does not need pressure stabilization. This problem has the form

$$Sv^n + Bp^n = g^n, \quad B^T v^n = 0, \tag{5.8}$$

with the (momentum) matrix $S = \sigma M + \nu A + N(\bar{v}^n)$. The right-hand side g^n and the approximation $\bar{v}^n$ are given from the preceding time level. Elimination of the velocity unknown yields again the Schur complement formulation

$$B^T S^{-1} Bp^n = B^T S^{-1} g^n, \quad v^n = S^{-1}(g^n - Bp^n). \tag{5.9}$$

We have already mentioned that the solution of this problem by Krylov space methods with evaluation of S^{-1} by multigrid iteration becomes increasingly inefficient for small time step k, larger Reynolds number, and on strongly anisotropic meshes. This problem can be overcome by using instead a simple Richardson iteration for the Schur complement equation (5.9) with a preconditioner of the form $B^TC^{-1}B$. Popular choices for the preconditioning operator C are:

- $C^{-1} = I$ (corresponds to the SIMPLE algorithm).
- $C^{-1} = \bar{M}^{-1}$ (lumped mass preconditioning).
- $C^{-1} = \bar{M}^{-1} + \alpha^{-1}B^TB$ (Turek's preconditioner)

The resulting iteration is termed "discrete projection method" (see Turek [95]):

$$p^{n,l+1} = p^{n,l} - (B^TC^{-1}B)^{-1}(B^TS^{-1}Bp^{n,l} - B^TS^{-1}g^n). \qquad (5.10)$$

After L iteration steps, on sets $p^n := p^{n,L}$ and computes the corresponding velocity component by solving:

$$Sv^n = g^n - Bp^n + \alpha^{-1}(\alpha I - Sc^{-1})B(p^{n,L} - p^{n,L-1}),$$

with some relaxation parameter $\alpha \in (0,1)$. This construction of v^n ensures that the resulting velocity is in the discrete sense divergence-free, $B^Tv^n = 0$, and suggests the name "projection method" for the whole scheme. The discrete projection method is then used as a smoother within an outer multigrid iteration.

In the special case $L = 1$, this scheme corresponds to a discrete version of the classical projection methods of Chorin (for the choice $p^{n,0} := 0$) and of Van Kan (for the choice $p^{n,0} := p^{n-1}$, see Gresho [32]. This operator-splitting time-stepping scheme has the form:

1. $S\tilde{v}^n = g^n - kBp^{n-1}$ (Burgers step),
2. $B^T\bar{M}^{-1}Bq^n = k^{-1}B^T\tilde{v}^n$ (Pressure Poisson equation),
3. $v^n = \tilde{v}^n - k\bar{M}^{-1}Bq^n$ (Velocity update),
4. $p^n = p^{n-1} + \alpha q^n$ (Pressure update).

All these schemes are variants of the "segregated" solution approach containing the schemes of SIMPLE-type and other pressure correction schemes as special cases; for a survey see [97] and [98].

The multigrid method with smoothing by the discrete projection iteration (5.10) has proven to be a very efficient solution method for the fully coupled problem (5.8); it is robust for all relevant Reynolds number (laminar flows) and time steps. The whole solution process is based on efficient and robust "inner" multigrid solvers for the subproblems "Burgers equation" and "pressure Poisson equation". The concrete implementation of this algorithm (as described in Turek [97]) requires about 1 KByte memory per mesh cell and shows almost meshsize-independent convergence behavior. As the result, 3D simulations with more than 10^7 unknowns requiring about 1 GByte of memory can be done on modern workstations.

Open Problem 5.2: *Derive a good preconditioner (smoother) for the Schur complement iteration (5.10) in the transport-dominant case.*

6. A review of theoretical analysis

In this section, we give an account of the available theoretical analysis for the discretization described in the previous sections. We concentrate on the practical impact of these theoretical results; the main topics are:

- Problem of regularity at "$t = 0$".
- Problem of global convergence up to "$t = \infty$".
- Problem of realistic error constants.

We will identify some critical shortcomings of the available theory which lead to challenging questions for further analysis.

We assume that the stationary or nonstationary Navier-Stokes equations are discretized by the finite element method as specified in Section 3 combined with one of the time-stepping schemes described in Section 4.

(I) For the spatial discretization, we recall the following two representative examples of (quadrilateral) Stokes elements:

a) the nonconforming "rotated" d-linear $\widetilde{Q}_1/P_0$ element;
b) the conforming d-linear Q_1/Q_1 element with pressure stabilization.

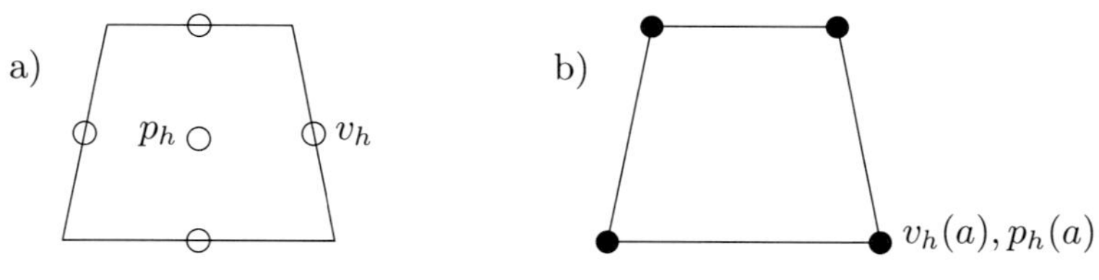

These discretizations are of second order expressed in terms of local approximation properties of the finite element functions used:

$$\inf_{\phi_h \in \mathbf{H}_h} \|v - \phi_h\| \le c\,h^2 \|\nabla^2 \phi\|, \quad \phi \in \mathbf{H} \cap \mathbf{H}^2(\Omega).$$

(II) For the time discretization, we think of the Crank-Nicolson scheme or the Fractional-Step-θ scheme which are both of second order in terms of local truncation error, e.g., for the Crank-Nicolson scheme applied to the homogeneous heat equation, there holds

$$\|k^{-1}(v^n - v^{n-1}) + \tfrac{1}{2}(A^n v^n + A^{n-1} v^{n-1})\| \le c\,k^2 \max_{[t_{n-1},t_n]} \|\partial_t^2 v\|_{-1}\,.$$

In view of these *local* approximation properties and the stability of the schemes, we expect a *global* a priori estimate for the errors $e_v^n := u(\cdot, t_n) - u_h^n$ and $e_p^n := p(\cdot, t_n) - p_h^n$ of the form

$$\max_{0<t_n\le T} \{\|e_v^n\| + \|e_p^n\|_{-1}\} \le C(\nu, T, \text{data})\{h^2 + k^2\}, \tag{6.1}$$

with an "error constant" $C(\nu, T, \text{data})$ depending on the viscosity parameter ν, the time-interval length $T > 0$, and assumed bounds M for the data of the problem, e.g.,

$$M := \|\nabla^2 v^0\| + \sup_{[0,T]} \{\|f\| + \|\partial_t f\|\} < \infty.$$

If additionally the domain Ω is sufficiently regular (say, convex or with C^2-boundary), it is guaranteed that the solution $\{v, p\}$ satisfies at least the a priori estimate

$$\sup_{(0,T]} \{\|\nabla^2 v\| + \|\partial_t v\| + \|\nabla p\|\} < \infty. \tag{6.2}$$

Clearly, the size of the error constant $C(\nu, T, \text{data})$ is of crucial importance for the practical value of the error estimate; we will come to this point in more detail, below. At first, we have to consider the question of whether an error estimate of the form (6.1) can be expected to hold at all. In general, the answer is "no", unless certain additional conditions are satisfied. This leads us to the following discussion of the "smoothing property".

6.1. The problem of regularity at "$t = 0$"

The second-order convergence of the time-stepping scheme expressed in the estimate (6.1) requires an a priori bound of the form $\sup_{(0,T]} \|\partial_t^2 v\| < \infty$. We have seen in Section 2 that there is a principle problem with assuming this degree of regularity in general. Even for arbitrarily smooth data the solution of the Navier-Stokes problem may suffer from

$$\lim_{t\to 0} \{\|\nabla^3 v(t)\| + \|\nabla \partial_t v(t)\|\} = \infty, \tag{6.3}$$

unless certain *non-local* (and non-verifiable) compatibility conditions are satisfied for the initial data. We recall from Section 2 the natural regularity assumption for the (nonstationary) Navier-Stokes equations (without additional compatibility condition):

$$v^0 \in \mathbf{J}_1(\Omega) \cap \mathbf{H}^2(\Omega) \quad \Rightarrow \quad \sup_{t\in(0,T]} \{\|\nabla^2 v(t)\| + \|\partial_t v(t)\|\} < \infty. \tag{6.4}$$

Accordingly, the best possible error estimate for the velocity which can be obtained under these "realistic" assumptions is

$$\sup_{t_n\in(0,T]} \|e_v^n\| = \mathcal{O}\{h^2 + k\}. \tag{6.5}$$

This estimate is only of first order in time, in contrast to the postulated second-order error estimate (6.1). As a result of the foregoing discussion we obtain the following:

Conclusion: *For any discretization of the nonstationary Navier-Stokes equations which requires more than the natural regularity inherent to the problem, meaningful higher-order error estimates must be of "smoothing type".*

We call an error estimate of type (6.1) a "smoothing error estimate" if it is of the form

$$\sup_{t_n\in(0,T]} \{t_n\|e_v^n\| + t_n^{3/2}\|e_p^n\|_{-1}\} \le C(\nu,T,\text{data})\{h^2+k^2\}. \tag{6.6}$$

This estimate reflects the well-known "smoothing behavior" of the exact solution $\{v,p\}$ as $t\to 0$ in the (realistic) situation (6.4):

$$\sup_{t\in(0,T]} \{t^{r/2-1}\|\nabla^r v(t)\| + t^{r-1}\|\partial_t^r v(t)\|\} \le c\{\|\nabla^2 v^0\| + \text{data}\}. \tag{6.7}$$

Smoothing error estimates of the form (6.6) have been established earlier for standard parabolic problems like the heat equation in the case of rough initial data; see, e.g., Thomée [89], as well as [62] and [71]. Corresponding results for the Navier-Stokes equation have been given in [43] for higher-order spatial semi-discretization and in [44] for the Crank-Nicolson time-stepping scheme. It turns out that due to the nonlinearity of the problem, the maximal achievable orders of smoothing error estimates under assumption (6.4) is $\mathcal{O}(h^6)$ for the spatial discretization and accordingly $\mathcal{O}(k^3)$ for the time stepping (provided that the scheme is strongly A-stable). This particularly implies the result (6.6) stated above. The existence of a natural order-barrier for the smoothing property of finite element Galerkin schemes applied to nonlinear problems has been established by Johnson, et al. [53]. We adapt the following example from [53] for the situation of H^2-regular initial data as relevant for the case of the nonlinear Navier-Stokes equations.

Example: *Example of limited smoothing property*
For $x\in(-\pi,\pi)$ and $t>0$, we consider the system of equations

$$\begin{aligned} \partial_t u - \partial_x^2 u &= 4\min\{v^2,1\}, \quad u(x,0)=u^0(x):=0,\\ \partial_t v - \partial_x^2 v &= 0, \quad v(x,0)=v^0(x):=m^{-r}\cos(mx), \end{aligned}$$

with periodic boundary conditions. For any fixed $m\in\mathbb{N}$ and $r\in\mathbb{N}\cup\{0\}$, the exact solution is

$$\begin{aligned} u(x,t) &= m^{-2r-2}\big(1-e^{2m^2t}\big)\big(1+e^{2m^2t}\cos(2mx)\big),\\ v(x,t) &= m^{-r}e^{-m^2t}\cos(mx)\,. \end{aligned}$$

For spatial semi-discretization of this problem, let the Galerkin method be used with the trial spaces

$$S_m := \operatorname{span}\big\{1,\cos(x),\sin(x),\dots,\cos((m-1)x),\sin((m-1)x)\big\},$$

and let P_m denote the L^2 projection onto S_m. Since $P_m v^0=0$, taking as usual $P_m u^0$ and $P_m v^0$ as initial values for the Galerkin approximation results in the Galerkin solutions $v_m(t)=0$ and $u_m(t)=0$. Consequently, for fixed $t>0$, there holds

$$\|(u_m-u)(t)\| = \|u(t)\| \sim \sqrt{2\pi}m^{-2r-2} = \sqrt{2}\|v^0\|_r h^{2r+2},$$

if we set $h:=m^{-1}$. This demonstrates that, for $v^0\in H^2(-\pi,\pi)$, i.e., for $r=2$, the best achievable order of approximation for $t>0$ is indeed $\mathcal{O}(h^6)$.

There is another remarkable aspect of the estimate (6.6) which concerns the Crank-Nicolson scheme. This scheme, due to its absent damping properties (not *strongly* A-stable), possesses only a reduced smoothing property. In consequence, even in the case of the linear heat equation, for initial data $v^0 \in L^2(\Omega)$ only qualitative convergence $\|e_v^n\| \to 0\,(h,k\to 0)$ can be guaranteed at fixed $t_n = t > 0$. For even stronger initial irregularity (e.g., $v^0 = \delta_x$ a Dirac measure) divergence $\|e^n\| \to \infty\ (h,k\to 0)$ occurs. However, the optimal smoothing behavior is recovered if one keeps the relation $k \sim h^2$. This undesirable step-size restriction can be avoided simply by starting the computation with a few (two or three) backward Euler steps; for examples and an analysis, see [61], [71], and the literature cited therein. Surprisingly, such a modification is not necessary for more regular initial data (half way up to the maximum regularity), $v^0 \in H_0^1(\Omega)\cap H^2(\Omega)$. In this case the Crank-Nicolson scheme admits an optimal-order smoothing error estimate of the form

$$\|e_v^n\| \le C\big(\nu, T, \|\nabla^2 v^0\|, \text{data}\big)\big\{h^2 + t_n^{-1}k^2\big\}, \quad t_n \in (0,T]. \tag{6.8}$$

For the heat equation this is easily seen by a standard spectral argument; see Chen/Thomée [23] and [71]. The extension of the smoothing error estimate (6.8) to the nonlinear (nonsymmetric and nonautonomous) Navier-Stokes equations is one of the main results in [44].

Finally, we mention some further results on smoothing error estimates relevant for the Navier-Stokes equations together with some open problems:

- Second-order projection schemes, particularly the Van Kan scheme, have been analyzed by Prohl [69, 70] and optimal-order smoothing error estimates have been established.
- The second-order smoothing property of the Fractional-Step-θ scheme has been proven by Müller [64].

Open Problem 6.1: *The construction of compatible initial data from experimental data has already been formulated as a problem. If this is not possible, it would be interesting to estimate the time length over which the error pollution effect of incompatible initial data persists.*

Open Problem 6.2: *Establish the optimal smoothing property of any higher-order ($q \ge 3$) time discretization schemes for the Navier-Stokes equations.*

6.2. The problem of convergence up to "$t = \infty$"

The error constants in the a priori error estimates (6.1) usually grow exponentially in time,

$$C(\nu, T, \text{data}) \sim Ke^{\kappa T},$$

unless the data of the problem is very small, actually of size ν^2, such that nonlinear perturbation terms can be absorbed into the linear main part. This exponential growth is unavoidable in general, due to the use of Gronwall's inequality in the proof. In fact, the solution to be computed may be exponentially unstable, so that a better error behavior cannot be expected. To improve on this situation, one has to make additional assumptions on the stability of the solution. Instead of requiring the data of the problem to be unrealistically small, the solution $\{v, p\}$ itself is supposed to be stable. This assumption may not be verifiable theoretically; nevertheless it may be justified in many situations in view of experimental evidence. A discussion of various types of stability concepts for nonstationary solutions of the Navier-Stokes equations in view of numerical approximation can be found in [42, 45].

Exponential Stability: A solenoidal solution v is called (conditionally) "exponentially stable", if for each sufficiently small initial perturbation $w^0 \in \mathbf{J}_1(\Omega)$, $\|w^0\| < \delta$, at any time $t_0 \ge 0$, the solution $\tilde{v}(t)$ of the perturbed problem

$$\partial_t \tilde{v} - \nu\Delta\tilde{v} + \tilde{v}\cdot\nabla\tilde{v} + \nabla q = 0, \quad t \ge t_0,$$

starting from $\tilde{v}(t_0) = v(t_0) + w^0$, satisfies

$$\|(v - \tilde{v})(t)\| \le Ae^{-\alpha(t-t_0)}\|w^0\|, \quad t \ge t_0, \tag{6.9}$$

with certain constants $A > 0$ and $\alpha > 0$. In this assumption it is essential that the decay of the perturbation is proportional to the size of the initial perturbation $\|w^0\|$. For global strong solutions this concept of exponential "L^2-stability" is equivalent to corresponding stability concepts expressed in terms of stronger norms, e.g. the H^1 norm; see [45]. It has been proved in a series of papers [42, 43, 44] that exponentially stable solutions can be approximated uniformly in time, i.e.,

$$\sup_{t_n \ge 1} \|v_h^n - v(\cdot, t_n)\| \le C\{h^2 + k^2\}. \tag{6.10}$$

In this estimate the error constant $C = C(A, \alpha)$ depends on the stability parameters of the solution. The proof uses a continuation argument. We sketch its essential steps for semi-discretization in time by the backward Euler scheme.

Proof of the global error estimate (6.10):
(i) We recall the *local* bound for the error $e^n = v(t_n) - v_k^n$,

$$\|e^n\| \le E(t_n)k, \quad t_n \ge 0, \tag{6.11}$$

involving the exponentially growing error constant $E(t) := Ke^{\kappa t}$. By continuity, it can be assumed that this error estimate holds with the same constant for all solutions neighboring the true solution v. The proof of this statement is technical and uses the particular properties of the discretization scheme considered. Further, let $T = Nk$ be a fixed time length such that with the stability parameters of the solution v, there holds

$$Ae^{-\alpha T} \le \tfrac{1}{2}. \tag{6.12}$$

(ii) Suppose now that the desired error estimate is already known to hold on some time interval $(0, t_m]$, $t_m \ge T$, with error constant $K_* := 2E(T)$. Let $\tilde{v}(t)$ be the solution of the perturbed problem

$$\partial_t \tilde{v} - \nu\Delta\tilde{v} + \tilde{v}\cdot\nabla\tilde{v} + \nabla\tilde{p} = 0, \quad t \ge t_m,$$

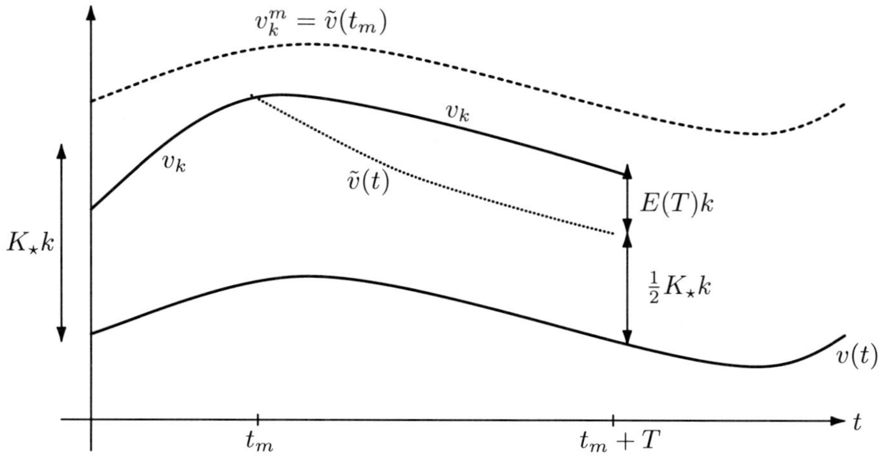

FIGURE 27. Scheme of induction proof (following [42])

starting at t_m with initial value $\tilde{v}(t_m) = v_k^m$. In virtue of the assumed exponential stability of the solution, there holds

$$\|(v - \tilde{v})(t)\| \leq Ae^{-\alpha(t-t_m)}\|e^m\|, \quad t > t_m,$$

for sufficiently small k guaranteeing $\|e^m\| < \delta$. Then, stepping forward by time length T , we obtain

$$\|e^{m+n}\| \leq \|(v - \tilde{v})(t_m + T)\| + \|\tilde{v}(t_m + T) - v_k^{m+n}\|.$$

Here, the first term is bounded by $K_* k$, in view of (6.11) and the induction assumption (6.12), while the second one can be controlled by $E(T)k$, using the local error estimate (6.11) for $\tilde{v} - v_k$. Hence, it follows that

$$\|e^{m+n}\| \leq K_* k.$$

The assertion then follows by induction with respect to multiples of T .

The argument presented for the global error estimate (6.10) appears simple and general; however, in concrete situations involving simultaneous discretization in space and time there are several technical difficulties. The initial value for the perturbed solution $\tilde{v}(t_n)$ in the induction step may not be admissible (i.e., not exactly divergence-free or even nonconforming). Further, the use of the local error bound (6.11) for the perturbed error $\tilde{v} - v_h$ requires control on higher-order regularity of the corresponding initial value $\tilde{v}(t_n)$. These and some other complications can be overcome as shown in [42, 43, 44], for different types of spatial as well as time discretization.

6.3. The problem of realistic error constants

In the preceding sections, we have discussed the derivation of *qualitative* a priori error estimates, local as well as global in time. Now, we turn to the more *quantitative* aspect of the size of error constants relating to the question of practical relevance of the a priori results. To this end, let us briefly summarize the results of a priori error analysis presented so far:

a) In the *stationary* case, we can guarantee convergence behavior like

$$\|e_v\| \le Ch^p$$

provided that the solution v is sufficiently smooth and locally unique (i.e., the linearization of the nonlinear Navier-Stokes operator at v is regular). Then, the error constant $C(\nu, v)$ depends on bounds on the regularity of v as well as its "stability", on the viscosity ν, and of course on the characteristics of the discretization.

b) In the *nonstationary* case, we can guarantee convergence behavior like

$$\|e_v^n\| \le C\{h^p + k^q\} \quad 0 \le t_n \le T,$$

provided again that the solution v is sufficiently smooth. The error constant $C(\nu, T, v)$ depends on bounds on the regularity of v, on the viscosity ν, and additionally on the length of the time interval T.

A question naturally arises: How large is C? In "normal" situations as, for example, for the Poisson problem or the heat equation, the error constant may be shown to be of moderate size $C \sim 1-10^4$, depending on the situation and the care spent in the estimation. The qualitative conclusion from the estimates may then be that the the error bound is reduced by a factor of $2^{-\min\{p,q\}}$ if the mesh size is halved in space and time. The (not unrealistic) hope is that this carries over to the true discretization error. Unfortunately, the Navier-Stokes equations do not at all a "normal" problem; it is of mixed elliptic-hyperbolic or parabolic-hyperbolic type with degenerating ellipticity. This has decisive consequences for the size of the error constants C.

Normalizing the flow configuration as usual to characteristic length $L = 1$ and velocity $U = 1$, the Reynolds number for common cases is $Re = \nu^{-1} \approx 1 - 10^5$, which relates to "laminar" flow, and the characteristic time length is $T \sim 1/\nu$. This means that it takes the time $T \approx \nu^{-1}$ for the flow to reach a characteristic limit behavior, e.g. stationary or time periodic. The question can now be made more precise: *How do the error constants C depend on Re?* This dependence has several sources:

- the explicit occurrence of ν in the differential operator,
- the dependence of the solution's regularity on ν (boundary layers),
- the dependence on the length of the time interval $T \sim 1/\nu$,
- the dependence of the solution's stability on ν.

Let us discuss the mechanisms of these dependencies separately.

(i) Structure of the differential operator:
The standard procedure in the stationary case is to absorb the lower-order terms into the linear main part $-\nu\Delta v$ which leads to the dependence

$$C \sim \nu^{-1}.$$

In the nonstationary case the lower-order terms are absorbed into the term $\partial_t v$ by the use of Gronwall's inequality resulting in

$$C \sim e^{KT/\nu}, \quad K \sim \sup_{(0,T]}\|\nabla v\| \sim \nu^{-1/2}.$$

This dependence on ν can be formally removed by using streamline-diffusion damping for the transport term, but it leaves the T-dependence.
(ii) Regularity of solution:
For small ν boundary layers of width $\delta \sim \sqrt{\nu}$ occur. This implies that

$$\sup_\Omega|\nabla v| \sim \nu^{-1/2}, \quad C \sim \|\nabla^p v\| \sim \nu^{-\alpha(d,p)}.$$

This problem can be solved by proper mesh refinement in the boundary layer.
(iii) Length of the time interval:
It was demonstrated above that the local "worst case" error constant

$$C \sim e^{KT/\nu}$$

becomes independent of the time interval-length T,

$$C \sim e^{KT_*/\nu},$$

if the solution can be assumed to be exponentially stable. Here, T_* is sufficiently large but fixed. However, tracing constants in the proof, we see that $T_* \sim \nu^{-1}$ rendering this formally global error bound practically meaningless for small ν.
(iv) Stability of the solution:
The argument for proving error estimates for nonlinear problems rely on assumptions on the stability of certain linearized tangent operators. The resulting error constants can in general not be assumed to behave better than

$$C \sim e^{KT/\nu}.$$

This exponential dependence seems unavoidable unless something different is shown in particular situations. The observability of laminar flows even for higher Reynolds numbers indicates that these flows may possess better stability properties than expressed by the "worst case" scenario addressed above. An analogous conclusion may be true even for turbulent flows with respect to certain averaged quantities.

Conclusion: *In general, one has to admit that the error constants depend exponentially on ν^{-1}, unless something different is proven. Realizing that even in the range of laminar flows, $20 \le Re \le 10^4$,*

$$e^{20} \approx 5{\cdot}10^8, \quad e^{100} \approx 10^{43}, \quad e^{1000} \approx \infty,$$

the practical meaning of available a priori error estimates seems rather questionable!

The above observation seems to indicate that there is a conceptual crisis in the theoretical support of CFD as far as it concerns the computational solution of the Navier-Stokes equations. This is contrasted by the abundant body of research papers reporting successful computations of viscous flows and the good agreement of the obtained results with experimental data. Hence, we reformulate the question: *Is there any theoretical support that certain flows (i.e., solutions of the Navier-Stokes equations) can actually be computed numerically.* If the answer were "no", everybody should be worried. We again emphasize that the presence of an asymptotic error estimate of the form

$$\|v - v_h\| \le C\{h^p + k^q\}$$

cannot be taken as justification for the meaningful performance of a numerical scheme, unless the error constant C is shown to be of moderate size at least for certain model situations of practical interest. Reliable flow simulation requires computable error bounds in terms of the approximate solution; the elements of such an *a posteriori* error analysis will be described in Section 7 below.

In proving useful error estimates, we have to deal with the question of proper concepts for describing the stability of solutions relevant for numerical approximation. Qualitatively, all stability concepts may be equivalent but this strongly depends on the viscosity parameter ν. The choice of the wrong norm may lead to unfavorable dependence on ν, like $\mathcal{O}(\nu^{-2})$ rather than the generic behavior $\mathcal{O}(\nu^{-1})$. Actually, the fundamental question whether there are practically interesting situations in which the solution of the Navier-Stokes equations are stable, with stability constant $c_S \sim \nu^{-1}$ seems open. Results in this direction appear necessary for a rigorous error analysis of discretization schemes. However, until now, practically meaningful a priori error bounds are not even available for such basic situations as Couette flow (constant sheer flow) and Poiseuille flow (constant pipe flow); we will address this question in more detail in the following section.

6.4. Towards a "quantitative" a priori error analysis

The following discussion is of conceptual nature. In order to abstract from the nonessential technicalities of finite element discretization, we consider the idealized situation of an "exactly" divergence-free approximation, using subspaces $\mathbf{V}_h \subset \mathbf{V} := \mathbf{J}_1(\Omega)$. Accordingly, the discretization delivers only approximations $v_h \in \mathbf{V}_h$ to the velocity $v \in \mathbf{V}$. The associated pressures p_h are then to be determined by post-processing. Further, we restrict us to the very basic case of homogeneous Dirichlet boundary conditions $v_{|\partial\Omega} = 0$.

6.4.1. THE STATIONARY CASE We begin with the stationary Navier-Stokes problem

$$-\nu\Delta v + v\cdot\nabla v + \nabla p = f, \quad \nabla\cdot v = 0, \quad \text{in } \Omega, \quad v_{|\partial\Omega} = 0. \tag{6.13}$$

Using again the notation

$$a(v,\psi) = \nu(\nabla v, \nabla\psi), \quad n(v,v,\psi) = (v\cdot\nabla v, \psi),$$

the *"pressure-free"* variational formulation seeks $v \in \mathbf{V}$, such that

$$A(v;\phi) := a(v,\phi) + n(v,v,\phi) = (f,\phi) \quad \forall \phi \in \mathbf{V}. \tag{6.14}$$

The corresponding finite element discretization seeks $v_h \in \mathbf{V}_h$, such that

$$A(v_h;\phi_h) = (f,\phi_h) \quad \forall \phi_h \in \mathbf{V}_h. \tag{6.15}$$

All error analysis of this discretization is based upon the (nonlinear) Galerkin orthogonality:

$$A(v;\phi_h) - A(v_h;\phi_h) = 0, \quad \phi_h \in \mathbf{V}_h. \tag{6.16}$$

In the following, we denote the error by $e := v - v_h$.

a) The "small data case", $\|\nabla v\| \sim \nu$:
The Fréchet derivative taken at v of the semi-linear Form $A(\cdot;\cdot)$ is given by

$$L(v;\phi,\psi) = a(\phi,\psi) + n(v,\phi,\psi) + n(\phi,v,\psi).$$

Under the "small data" assumption, this bilinear form is coercive on $\mathbf{V}$ with "stability constant" $c_S = c_S(\nu) \sim \nu^{-1}$:

$$\|\nabla\phi\| \leq c_S L(v;\phi,\phi).$$

Linearization and Galerkin orthogonality then leads to the relation

$$\|\nabla e\| \leq c_S L(v,e,e) = c_S\big\{n(e,e,e) + A(v;v-\phi_h) - A(v_h;v-\phi_h)\big\},$$

with an arbitrary approximation $\phi_h \in \mathbf{V}_h$ to v. From this we infer that, for sufficiently small h,

$$\|\nabla e\| \leq c_S C h, \tag{6.17}$$

with an error constant $C = C(\|\nabla^2 v\|, \text{data})$.

b) The general case of an "isolated" solution:
Now, the solution v is assumed to be stable in the sense that

$$\|\nabla\phi\| \leq c_S \sup_{\psi\in\mathbf{V}} \frac{L(v,\phi,\psi)}{\|\nabla\psi\|}, \tag{6.18}$$

with some "stability constant" $c_S = c_S(\nu, v)$. Again, by linearization and Galerkin orthogonality, it follows that

$$\|\nabla e\| \leq c_S C h. \tag{6.19}$$

Further, assuming stability of the Fréchet derivative in the form

$$\|\phi\| \leq c_S \sup_{\psi\in\mathbf{V}\cap\mathbf{H}^2(\Omega)} \frac{L(v,\phi,\psi)}{\|\nabla^2\psi\|}, \tag{6.20}$$

one may apply the usual duality argument to obtain the L^2-error bound

$$\|e\| \leq c_S C h^2. \tag{6.21}$$

These results rely on the assumption of stability of the problem expressed in terms of the stability constant c_S; see [56]. In order to use the resulting estimates, on has to determine these constants either analytically, which may rarely be possible, or computationally by solving dual problems. Numerical experiments for the driven-cavity problem reported by Boman [16] show a dependence like $c_S(\nu) \sim \nu^{-1}$ in the ("laminar") range $1 \le Re \le 10^3$. This investigation should be extended to other elementary flows in order to see whether linear growth $c_S \sim Re$ is generic for laminar flow. The answer is not clear yet, as indicated by the following simple example.

An example of "bad" stability (from Tobiska/Verfürth [91]*):* For the worst-case scenario, we quote the one-dimensional Burgers equation

$$-\nu v_{xx} + v v_x = 0, \quad x \in (-1,1), \quad v(-1) = 1,\ v(1) = -1.$$

The exact solution is $v(x) = -2\alpha_\nu \nu \tanh(\alpha_\nu x)$, where α_ν is the unique positive solution of $2\nu\alpha_\nu \tanh(\alpha_\nu) = 1$. Linearization at this solution results in the boundary value problem

$$-\nu z_{xx} + v z_x + z v_x = f \quad x \in (-1,1), \quad z(-1) = z(1) = 0,$$

which has the solution

$$z(x) = -\nu^{-1} e^{U(x)/\nu} \int_{-1}^{x} e^{-U(t)/\nu} \left(\int_0^t f(s)\, ds + c \right) dt,$$

where U is a primitive of v and the constant c is determined by imposing the boundary condition $z(1) = 0$. We ask for the best possible bound in the stability estimate

$$\nu \|z\|_{H^1} \le C(\nu) \|f\|_{H^{-1}}.$$

For the particular choice $f(x) = \cosh(\alpha_\nu x)$, there holds

$$z(x) = \frac{\cosh^3(\alpha_\nu) - \cosh^3(\alpha_\nu x)}{3\alpha_\nu^2 \nu \cosh^2(\alpha_\nu x)}.$$

Since

$$\|z\|_\infty \le \sqrt{2}\|z\|_1, \quad \|f\|_{-1} \le 2\sqrt{2}\|f\|_\infty,$$

it follows that, for $\nu \le 1$,

$$\|z\|_\infty \ge c\nu e^{1/\nu} \|f\|_\infty,$$

and consequently, $C(\nu) \sim e^{1/\nu}$. This seems to indicate that Burgers equation is not numerically solvable which, however, contradicts practical evidence. The explanation may be that for the performance of discretization stability is essential in other more local measures then those considered above.

Open Problem 6.3: *Explain the success of discretization methods in computing solutions to the Burgers equation despite its bad conditioning with respect to the "energy norm".*

6.4.2. THE NONSTATIONARY CASE We now turn to the nonstationary Navier-Stokes problem posed on a time interval $I=[0,T]$,

$$\partial_t v - \nu\Delta v + v\cdot\nabla v + \nabla p = f \;\text{ in } \Omega\times I, \quad v_{|\partial\Omega}=0, \;\; v_{|t=0}=v^0. \tag{6.22}$$

The corresponding (pressure-free) space-time variational formulation uses the function space $\mathbf{V}(I) := H^1(I;\mathbf{V})$ and the space-time forms

$$(\phi,\psi)_I = \int_I (\phi,\psi)\,dt, \quad a_I(\phi,\psi) = \int_I a(\phi,\psi)\,dt,$$

$$n_I(v,\phi,\psi) = \int_I n(v,\phi,\psi)\,dt.$$

Then, a solution $v\in\mathbf{V}(I)$ is sought satisfying

$$A(v;\phi) = F(\phi) \quad \forall\phi\in\mathbf{V}(I), \tag{6.23}$$

where $F(\phi) := (f,\phi)_I + (v^0,\phi(0))$ and

$$A(v;\phi) := (\phi,\psi)_I + a_I(\phi,\psi) + n_I(v,\phi,\psi) + (v^0,\phi(0)).$$

Here, the initial condition $v(0)=v^0$ is incorporated weakly.

In the following conceptual discussion, we restrict ourselves to the semi-discretization in time leaving the spatial variable "continuous". The discretization is by the "discontinuous" Galerkin method of degree $r=0$ ("dG(0) method") which is a variant of the backward Euler scheme. The time interval $I=[0,T]$ is decomposed like $0=t_0<t_1<\cdots<t_{M+1}=T$, and we set

$$I_m = [t_{m-1},t_m), \quad k(t)\equiv k_{|I_m} = t_m - t_{m-1}.$$

For piecewise continuous functions on this decomposition, we write

$$v_\pm^m = \lim_{s\to 0_+} v(t_m \pm s), \quad [v]^m = v_+^m - v_-^m.$$

Accordingly, we define the discrete semi-linear form

$$\begin{aligned} A_k(v;\phi) \;\; := \;\; & \sum_{m=0}^{M} \Big\{ (\partial_t v,\phi)_m + a_m(v,\phi) + n_m(v,v,\phi) \Big\} \\ & + \sum_{m=1}^{M} ([v]^m,\phi_+^m) + (v_+^0,\phi_+^0), \end{aligned}$$

and the time-discrete spaces

$$\mathbf{V}_k(I) = \{ v_k \in L^2(I;\mathbf{V}),\; v_{|I_m}\in P_0(I_m),\; m=1,\dots,M \}.$$

The time-discrete approximation then seeks a $v_k\in\mathbf{V}_k(I)$, such that

$$A_k(v_k;\phi_k) = F(\phi_k) \quad \forall\phi_k\in\mathbf{V}_k(I). \tag{6.24}$$

We note that this formulation contains the initial condition $v(0)=v^0$ in the weak sense. The essential feature of the dG(r) schemes are their Galerkin orthogonality property. Using the fact that the continuous solution v also satisfies the discrete equation (6.24), we have

$$A_k(v;\phi_k) - A_k(v_k;\phi_k) = 0, \quad \phi_k \in \mathbf{V}(I). \tag{6.25}$$

In order to estimate the error $e := v - v_k$, we may employ a "parabolic" duality argument. The adjoint of the Fréchet derivative of the governing semi-linear form taken at the solution v is given by

$$L^*(v;\phi,z) = (\phi, -\partial_t z)_I + a_I(\phi,z) + n_I(v,\phi,z) + n_I(\phi,v,z) - (\phi(T),z(T)).$$

Then, we introduce the "dual solution" $z \in \mathbf{V}(I)$ as solution of the space-time "dual problem"

$$L^*(v;\phi,z) = (e(T),\phi(T)) \quad \forall \phi \in \mathbf{V}(I). \tag{6.26}$$

Taking now $\phi = e$ in (6.26), we obtain the error representation

$$\begin{aligned} \|e(T)\| &= L^*(v;e,z) \\ &= (e,-\partial_t z)_I + a_I(e,z) + n_I(v,e,z) + n_I(e,v,z). \end{aligned}$$

Using the Galerkin orthogonality (6.25) and interpolation estimates on each subinterval I_m, we conclude the estimate

$$\begin{aligned} \|e(T)\| &= \sum_{m=0}^{M} \Big\{ (f, z - z_k)_m - ([v]^m, z_+^m - z_{k,+}^m) \Big\} \\ &\le c_S \big\{ \max_I \|[v]^m\| + \max_I \|kf\| \big\}, \end{aligned}$$

with the "stability constant" $c_S = c_S(\nu, T, v)$ given by

$$|\log(k_M)|^{-1/2} \Big\{ \|z\|_I + |\log(k_M)|^{-1/2} \int_0^{T-k_M} \|\partial_t z\|\,dt \Big\} \le c_S \|z(T)\|. \tag{6.27}$$

The result is the "final time" a priori error estimate

$$\|e(T)\| \le c L_k c_S \max_I \|k \partial_t v\|, \tag{6.28}$$

where $L_k := \max_I \big(1 + \log(k)\big)^{1/2}$.

Conclusion: The foregoing discussion shows that proving a priori error estimates for numerical approximations is closely connected with the study of stability properties of linearizations of the Navier-Stokes equations, i.e. with hydrodynamic stability; for more details see [55]. The goal is to derive a priori estimates of the type (6.18), (6.20) and (6.27) with quantified constants $c_S = c_S(\nu, T, \text{data})$. The dependence of these constants on the Reynolds number may be *linear* in "good" cases or may deteriorate to *exponential* in "bad" cases. Such exponentially unstable flows are not computable over relevant intervals of time. The same question will also be crucial for the derivation of *a posteriori* error estimates discussed in Section 7, below.

6.5. The problem of stability constants (A critical review of hydrodynamic stability theory)

Most of the traditional stability theory for fluid flow is of qualitative nature being based on eigenvalue criteria through a linearized stability argument. The theoretical results are in good agreement with experiments concerning the *critical* Reynolds number at which the first bifurcation occurs; well-studied examples are the Bénard problem and the Taylor-Couette problems. But they do not fit with experiments for the other fundamental case of parallel flow. There are several paradoxes observed:

- Poiseuille flow (between two parallel plates) is predicted to turn turbulent at $Re \sim 5772$ through $2d$ Tollmien-Schlichting waves, while experiments show instability with essential $3d$ features somewhere in the range $Re \sim 1000 - 10000$ depending on the experimental setup.
- Couette flow (parallel shear flow) is supposed to be stable for all $Re > 0$, but experiments show instability for $Re \sim 300 - 1500$.

This failure of theory was blamed on the deficiency of *linearized* stability theory being valid only for *small* perturbations. However, *linearized stability theory* is okay, but was only wrongly interpreted. In dynamic systems governed by *non-normal* matrices, one has to look at the size of the total amplification factors for the initial perturbation and not only at the sign of the eigenvalue's real parts. It is observed that, for example in Poiseuille flow there occurs amplification by a factor of 10^4 for $Re \geq 549$. Some of the relevant references on this subject are Landahl [59] and Trefethen et al. [92], to mention only a few. More references can be found in [54, 55] where this new concept in hydrodynamic stability theory is discussed in view of numerical approximation. In fact, the question of quantitative hydrodynamic instability and that of numerical computability of laminar flows are closely related. In transition to turbulence one seeks to establish *lower bounds* on the growth of perturbations in order to understand how a laminar flow may develop into a turbulent flow. In error control in CFD for laminar flows one seeks *upper bounds* of the growth of perturbations related to discretizations of the Navier-Stokes equations.

We want to illustrate the phenomenon of error amplification by two simple examples taken from [55].

a) An ODE model:
At first, we consider the simple ODE system

$$\dot{w}_1 + \nu w_1 + w_2 = 0, \tag{6.29}$$

$$\dot{w}_2 + \nu w_2 = 0. \tag{6.30}$$

Here, νw_i stand for the diffusion terms and w_2 in the first equation for the coupling in the transport term of the linearized perturbation equation of the Navier-Stokes equations. The corresponding coefficient matrix

$$A = \begin{pmatrix} \nu & 1 \\ 0 & \nu \end{pmatrix}$$

is non-normal; the only eigenvalue $\lambda=\nu$ has algebraic multiplicity two. This is just the situation described above. For this linear system the solution corresponding to the initial values $w(0)=w^0$ is given by

$$w_1(t)=e^{-\nu t}w_1^0-te^{-\nu t}w_2^0,\quad w_2(t)=e^{-\nu t}w_2^0.$$

We see the exponential decay of the second component and the linear growth over the interval $[0,\nu^{-1}]$ to size $w_1(\nu^{-1})=\nu^{-1}e^{-1}w_2^0$ of the first component before the exponential decay sets in. The component w_2 acts like a catalyst in the first equation. Although exponentially decaying it first causes w_1 to grow; the later exponential decay is irrelevant when by the growth to size $\nu^{-1}w_1^0$ the linearization is no longer valid.

b) A simple flow model:
Next, we consider a very simple configuration: the flow in an infinite pipe $\Omega=\mathbb{R}\times\omega$ extending in the x_1-axis with cross section ω in the (x_2,x_3)-plane. The flow is driven by a volume force $f=(f_1(x_2,x_3,t),0,0)^T$ (gravitation) in x_1-direction. The solution is supposed to have the form (like x_1-independent Poiseuille flow):

$$v=(v_1(x_2,x_3,t),0,0)^T,\quad p=p(x,t).$$

Then the Navier-Stokes equations take the form

$$\partial_t v_1-\nu\Delta v_1+\partial_1 p=f_1\quad \text{in}\,\omega\,,\quad v_{1|\partial\omega}=0\,.$$

The corresponding linearized perturbation equation is

$$\begin{aligned}
\partial_t w_1-\nu\Delta w_1+v_1\partial_1 w_1+\partial_2 v_1 w_2+\partial_3 v_1 w_3+\partial_1 q &= 0\,,\\
\partial_t w_2-\nu\Delta w_2+v_1\partial_1 w_2 \qquad\qquad\qquad\qquad +\partial_2 q &= 0\,,\\
\partial_t w_3-\nu\Delta w_3+v_1\partial_1 w_3 \qquad\qquad\qquad\qquad +\partial_3 q &= 0\,,
\end{aligned}$$

with the incompressibility condition $\partial_2 w_2+\partial_3 w_3=0$, and the initial and boundary conditions $w_{|t=0}=w^0$ and $w_{|\partial\omega}=0$. Even this simple problem is still too complex for an explicit solution. Therefore, we simplify it further by assuming that the perturbed solution $\{w,q\}$ is also independent of x_1. This corresponds to looking at a fluid in a long vertical tube under gravity or in a long rotating tube with varying speed of rotation. Under this assumption the perturbation equation reduces to

$$\begin{aligned}
\partial_t w_1-\nu\Delta w_1+\partial_2 v_1 w_2+\partial_3 v_1 w_3 &= 0\,,\\
\partial_t w_2-\nu\Delta w_2+ \qquad\qquad\qquad +\partial_2 q &= 0\,,\\
\partial_t w_3-\nu\Delta w_3+ \qquad\qquad\qquad +\partial_3 q &= 0\,.
\end{aligned}$$

In this system the equations for the components $\bar{w}:=\{w_2,w_3\}$ together with the constraint $\partial_2 w_2+\partial_3 w_3=0$ form a two-dimensional Stokes problem which can be solved independently of the first equation. Hence, we are in a similar situation as in the above ODE example. For the Stokes subsystem we have the standard a priori estimate

$$\|\bar{w}(t)\|\le e^{-\kappa\nu t}\|\bar{w}^0\|,\quad t\ge 0,$$

with $\kappa = \text{diam}(\omega)$. The first equation does not contain the pressure. Using the result for $\bar{w}$ we obtain for the first component w_1 the bound

$$\|w_1(t)\| \leq ce^{-\kappa\nu t}\{t\|w_1^0\| + \|\bar{w}^0\|\}, \quad t \geq 0. \tag{6.31}$$

Hence, we see that for this model problem, one can show that the error constant in the a priori error estimate (6.1) grows at most linearly with the Reynolds number:

$$C(\nu, T, \text{data}) \sim \max\{T, Re\}.$$

It is an open question whether this linear dependence on Re is generic for a larger class of flow problems. Numerical experiments for the lid-driven cavity flow show such a dependence.

Open Problem 6.4: *Prove a posteriori stability estimates like (6.31) for more practical problems (e.g. Poiseuille flow) possibly with respect to different norms. Is there any indication that linear growth in time of perturbations may be generic to the Navier-Stokes equations?*

7. Error control and mesh adaptation

This section is devoted to concepts of error estimation and mesh optimization. The goal is to develop techniques for reliable estimation of the discretization error in quantities of physical interest as well as economical mesh adaptation. The use of a finite element Galerkin discretization provides the appropriate framework for a mathematically rigorous error analysis. On the basis of computable a posteriori error bounds the mesh is locally refined within a feed-back process yielding economical mesh-size distributions for prescribed error tolerance or maximum number of cells. On the resulting sequence of refined meshes the discrete problems are solved by multi-level techniques.

The general concept of residual-based error control for finite element methods is described in the survey article by Eriksson/Estep/Hansbo/Johnson [26]; this technique has then been further developed for various situations in [12, 14]. The application to incompressible flows is extensively discussed in Becker [7, 9]. Extensions to compressible flow including chemical reactions are given in Braack [17]; see also [18]. A survey of applications of this approach to a variety of other problems can be found in [75].

7.1. Principles of error estimation

The discretization error in a cell K splits into two components, the *locally* produced error (truncation error) and the *transported* error (pollution error)

$$e_K^{tot} = e_K^{loc} + e_K^{trans}. \tag{7.1}$$

The effect of the cell residual ρ_K on the local error $e_{K'}$, at another cell K', is governed by the Green function of the continuous problem. This is the general philosophy underlying our approach to error control.

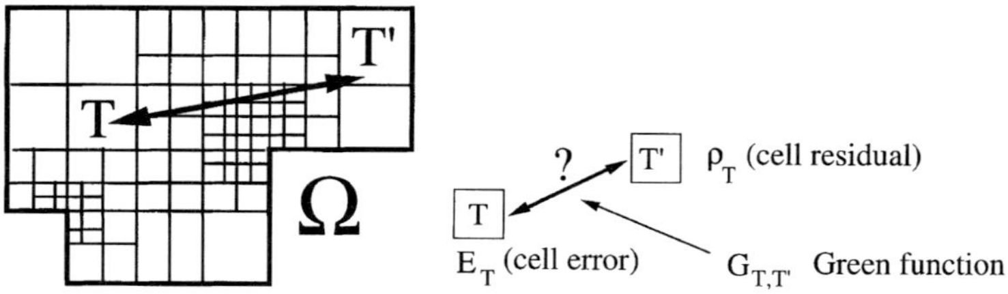

(I) A priori error analysis: The classical *a priori* error estimation aims at estimating the error to be expected in a computation which is still to be done. These bounds are expressed in terms of powers of a mesh size h and involve constants which depend on the (unknown) exact solution. In this way, only asymptotic (as $h \to 0$) information about the error behavior is provided but no quantitatively useful error bound. In particular, no criterion for local mesh adaptation is obtained.

(II) A posteriori error analysis: The *a posteriori* error analysis generates error estimates in the course of the computation. Accordingly, these bounds are in terms of computable local residuals of the approximate solution and do not require information about the exact solution. However, a posteriori error analysis usually does not provide a priori information about the convergence of the discretization process as $h \to 0$.

We illustrate the basic principles underlying error estimation by considering perturbations of linear algebraic systems. Let $A, \tilde{A} \in \mathbb{R}^{n\times n}$, $b, \tilde{b} \in \mathbb{R}^n$ be given and solve

$$Ax = b\,, \qquad \tilde{A}\tilde{x} = \tilde{b} \text{ (perturbed problem).} \tag{7.2}$$

For estimating the error $e := x - \tilde{x}$, there are several approaches. The *a priori* method uses the "truncation error" $\tau := \tilde{A}x - \tilde{b} = \tilde{A}(x - \tilde{x})$,

$$e = \tilde{A}^{-1}\tau \quad \Rightarrow \quad \|e\| \le \tilde{c}_S\|\tau\|, \tag{7.3}$$

with the "discrete" stability constant $\tilde{c}_S := \|\tilde{A}^{-1}\|$. The *a posteriori* method uses the "residual" $\rho := b - A\tilde{x} = A(x - \tilde{x})$,

$$e = A^{-1}\rho \quad \Rightarrow \quad \|e\| \le c_S\|\rho\|, \tag{7.4}$$

with the "continuous" stability constant $c_S := \|A^{-1}\|$. Alternatively, we may use the solution z of the "dual problem" $A^*z = \|e\|^{-1}e$, to obtain

$$\|e\| = (e, A^*z) = (b - A\tilde{x}, z) = (\rho, z) \le \|\rho\|\,\|z\| \le c_S^*\|\rho\|, \tag{7.5}$$

with the "dual" stability constant $c_S^* := \|A^{*-1}\|$. Of course, this approach does not yield a new result in estimating the error in the l_2-norm. But it shows the way to bound other error quantities as for example single components $|e_i|$.

An analogous argument can also be applied in the case of nonlinear equations. Let $F, \tilde{F} : \mathbb{R}^n \to \mathbb{R}^n$ be (differentiable) vector functions and solve

$$F(x) = b\,, \qquad \tilde{F}(\tilde{x}) = \tilde{b} \text{ (perturbed problem).} \tag{7.6}$$

Then, the residual $\rho := b - F(\tilde{x})$ satisfies

$$\rho = F(x) - F(\tilde{x}) = \left(\int_0^1 F'(\tilde{x} + se)\, ds \right) e =: L(x, \tilde{x})e, \tag{7.7}$$

with the Jacobian F'. The term in parentheses defines a linear operator $L(x, \tilde{x}) : \mathbb{R}^n \to \mathbb{R}^n$ which depends on the (unknown) solution x. It follows that $\|e\| \le c_S \|\rho\|$, with the (nonlinear) stability constant $c_S := \|L(x, \tilde{x})^{-1}\|$. Below, we will use this duality technique for generating a posteriori error estimates in Galerkin finite element methods for differential equations.

7.1.1. A DIFFUSION MODEL PROBLEM For illustrating our concept, we start with the (scalar) model diffusion problem

$$-\Delta u = f \ \text{ in } \Omega, \qquad u = 0 \ \text{ on } \partial\Omega, \tag{7.8}$$

posed on a polygonal domain $\Omega \subset \mathbb{R}^2$. In its variational formulation one seeks $u \in V := H_0^1(\Omega)$ satisfying

$$(\nabla u, \nabla \phi) = (f, \phi) \quad \forall \phi \in V. \tag{7.9}$$

We consider a finite element approximation using piecewise (isoparametric) bilinear shape functions (see Section 3). The corresponding finite element spaces $V_h \subset V$ are defined on decompositions $\mathbb{T}_h$ of $\bar{\Omega}$ into quadrilaterals ("cells") K of width $h_K := diam(K)$. We write again $h := \max_{K \in \mathbb{T}} h_K$ for the maximal *global* mesh width. Simultaneously, the notation $h = h(x)$ is used for the continuously distributed mesh-size function defined by $h_{|K} = h_K$. For ease of mesh refinement and coarsening we allow "hanging nodes", but at most one per edge. The shape of the corresponding modified basis function is shown in Figure 28.

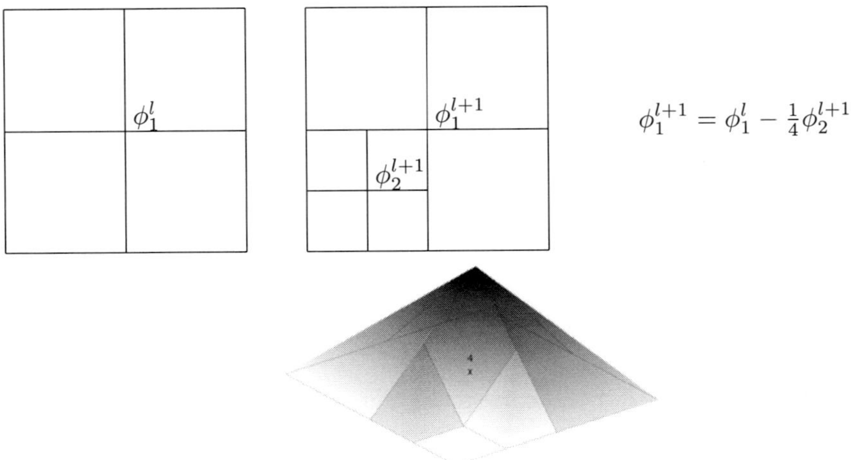

FIGURE 28. Q_1 nodal basis function on a patch of cells with a hanging node

The discrete problem determines $u_h \in V_h$ by

$$(\nabla u_h, \nabla\phi_h) = (f, \phi_h) \quad \forall\phi_h \in V_h. \tag{7.10}$$

We recall the "Galerkin orthogonality" of the error $e := u - u_h$,

$$(\nabla e, \nabla\phi_h) = 0, \quad \phi_h \in V_h. \tag{7.11}$$

We seek to derive a posteriori error estimates. Let $J(\cdot)$ be an arbitrary "error functional" defined on V and $z \in V$ the solution of the corresponding dual problem

$$(\nabla\phi, \nabla z) = J(\phi) \quad \forall\phi \in V. \tag{7.12}$$

Setting $\phi = e$ in (7.12) results in the error representation

$$\begin{aligned} J(e) &= (\nabla e, \nabla z) = (\nabla e, \nabla(z - I_h z)) \\ &= \sum_{K\in\mathbb{T}} \Big\{(-\Delta u + \Delta u_h, z - I_h z)_K - (\partial_n u_h, z - I_h z)_{\partial K}\Big\} \\ &= \sum_{K\in\mathbb{T}} \Big\{(f + \Delta u_h, z - I_h z)_K - \tfrac{1}{2}(n\cdot[\nabla u_h], z - I_h z)_{\partial K}\Big\}, \end{aligned} \tag{7.13}$$

where $[\nabla u_h]$ is the jump of ∇u_h across the interelement boundary. In the second equation, we have used galerkin orthogonality. This gives us the a posteriori error estimate

$$|J(e)| \le \eta(u_h) := \sum_{K\in\mathbb{T}_h} h_K^4 \Big\{\rho_K(u_h)\omega_K(z) + \rho_{\partial K}(u_h)\omega_{\partial K}(z)\Big\}, \tag{7.14}$$

with the cell residuals

$$\rho_K(u_h) := h_K^{-1}\|f + \Delta u_h\|_K, \quad \rho_{\partial K}(u_h) := h_K^{-3/2}\|n\cdot[\nabla u_h]\|_{\partial K},$$

and the weights

$$\omega_K(z) := h_K^{-3}\|z - I_h z\|_K, \quad \omega_{\partial K}(z) := \tfrac{1}{2}h_K^{-5/2}\|z - I_h z\|_{\partial K}.$$

These quantities are normalized, such that they can be expected to approach certain mesh-independent limits as $h \to 0$. The interpretation of the relation (7.14) is that the weights $\omega_K(z)$ describe the dependence of $J(e)$ on variations of the cell residuals $\rho_K(u_h)$,

$$\frac{\partial J(e)}{\partial \rho_K} \approx h_K^4\omega_K(z) \approx h_K^4 \max_K|\nabla^2 z|.$$

We remark that in a finite difference discretization of the model problem (7.8) the corresponding "influence factors" behave like $\omega_K(z) \approx h_K^2 \max_K |z|$.

In practice the weights $\omega_K(z)$ have to be determined computationally. Let $z_h \in V_h$ be the finite element approximation of z,

$$(\nabla\phi_h, \nabla z_h) = J(\phi_h) \quad \forall\phi_h \in V_h. \tag{7.15}$$

We can estimate

$$\omega_K(z) \le c_I h_K^{-1} \|\nabla^2 z\|_K \approx c_I \max_K |\nabla_h^2 z_h|, \tag{7.16}$$

where $\nabla_h^2 z_h$ is a suitable difference quotient approximating $\nabla^2 z$. The interpolation constant is usually in the range $c_I \approx 0.1 - 1$ and can be determined by calibration. Alternatively, we may construct from $z_h \in V_h$ a patchwise biquadratic interpolation $I_h^{(2)} z_h$ and replace $z - I_h z$ in the weight $\omega_K(z)$ by $I_h^{(2)} z_h - z_h$. This gives an approximation to $\omega_K(z)$ which is free of interpolation constants.

One may try to further improve the quality of the error estimate by solving local defect equations, either Dirichlet problems (à la Babuska/Miller) or Neumann problems (à la Bank/Weiser); see Backes [4]. References for these approaches are Verfürth [102] and Ainsworth/Oden [1]. Comparison with simpler mesh adaptation techniques, e.g. refinement criteria based on difference quotients of the computed solution, local gradient recovery "ZZ technique" (à la Zienkiewicz/Zhu [106]), or other local "ad hoc" criteria have been reported in Braack [17] and in [75].

By the same type of argument, one can also derive the traditional global error estimates in the energy and the L^2 norm.

(i) Energy-norm error bound: Using the functional

$$J(\phi) := \|\nabla e\|^{-1}(\nabla e, \nabla\phi)$$

in the dual problem, we obtain the estimate

$$\|\nabla e\| \le \sum_{K\in\mathbb{T}} h_K^4\, \rho_K(u_h)\, \omega_K(z) \le c_I \sum_{K\in\mathbb{T}} h_K^2\, \rho_K(u_h)\, \|\nabla z\|_{\tilde K},$$

where $\tilde K$ is the union of all cells neighboring K. In view of the a priori bound $\|\nabla z\| \le c_S = 1$, this implies the a posteriori error estimate

$$\|\nabla e\| \le \eta_E(u_h) := c_I\Big(\sum_{K\in\mathbb{T}} h_K^4\, \rho_K(u_h)^2\Big)^{1/2}. \tag{7.17}$$

(ii) L^2-norm error bounds: Using the functional

$$J(\phi) := \|e\|^{-1}(e, \phi)$$

in the dual problem, we obtain the estimate

$$\|e\| \le \sum_{K\in\mathbb{T}} h_K^4\, \rho_K(u_h)\, \omega_K(z) \le c_I \sum_{K\in\mathbb{T}} h_K^3\, \rho_K(u_h)\, \|\nabla^2 z\|_K.$$

In view of the a priori bound $\|\nabla^2 z\| \le c_S$ ($c_S = 1$ if Ω is convex), this implies the a posteriori error bound

$$\|e\| \le \eta_{L^2}(u_h) := c_I c_S\Big(\sum_{K\in\mathbb{T}} h_K^6\, \rho_K(u_h)^2\Big)^{1/2}. \tag{7.18}$$

7.1.2. A TRANSPORT MODEL PROBLEM As a simple model, we consider the scalar transport equation

$$\beta\cdot\nabla u = f\,, \tag{7.19}$$

on a domain $\Omega \subset \mathbb{R}^2$ with inflow boundary condition $u = g$ along the "inflow boundary" $\partial\Omega_- = \{x \in \partial\Omega,\, n\cdot\beta < 0\}$. Accordingly, $\partial\Omega_+ = \partial\Omega \setminus \partial\Omega-$ is the "outflow boundary". The transport vector β is assumed as constant for simplicity; therefore, the natural solution space is

$$V := \{v \in L^2(\Omega),\, \beta\cdot\nabla v \in L^2(\Omega)\}.$$

This problem is discretized using the Galerkin finite element method with streamline diffusion stabilization as described above. On quadrilateral meshes $\mathbb{T}_h$, we define again subspaces $V_h = \{v \in H^1(\Omega),\, v_{|K} \in \widetilde{Q}_1(K),\, K \in \mathbb{T}_h\}$, where $\widetilde{Q}_1$ is the space of "isoparametric" bilinear functions on cell K. The discrete solution $u_h \in V_h$ is defined by

$$(\beta\cdot\nabla u_h - f, \phi + \delta\beta\cdot\nabla\phi) + (n\cdot\beta(g-u_h), \phi)_{\partial\Omega_-} = 0 \qquad \forall\phi \in V_h, \tag{7.20}$$

where the stabilization parameter is determined locally by $\delta_K = h_K$. In this formulation the inflow boundary condition is imposed in the weak sense. This facilitates the use of a duality argument in generating a posteriori error estimates. Let $J(\cdot)$ be a given functional with respect to which the error $e = u - u_h$ is to be controlled. Following our general approach, we consider the corresponding dual problem

$$(\beta\cdot\nabla\phi, z + \delta\beta\cdot\nabla z) - (n\cdot\beta\phi, z)_{\partial\Omega_-} = J(\phi) \qquad \forall\phi \in V, \tag{7.21}$$

which is a transport problem with transport in the negative β-direction. We note that the stabilized bilinear form $A_h(\cdot,\cdot)$ is used in the duality argument, in order to achieve an optimal treatment of the stabilization terms; for a detailed discussion of this point see [75] and [47]. The error representation reads

$$J(e) = (\beta\cdot\nabla e, z - z_h + \delta\beta\cdot\nabla(z - z_h)) - (n\cdot\beta e, z - z_h)_{\partial\Omega_-},$$

for arbitrary $z_h \in V_h$. This results in the a posteriori error estimate

$$|J(e)| \le \eta(u_h) := c_I \sum_{K\in\mathbb{T}} h_K^4 \Big\{\rho_K(u_h)\omega_K(z) + \rho_{\partial K}(u_h)\omega_{\partial K}(z)\Big\}, \tag{7.22}$$

with the cell residuals

$$\rho_K(u_h) := h_K^{-1}\|f - \beta\cdot\nabla u_h\|_K, \quad \rho_{\partial K}(u_h) := h_K^{-3/2}\|n\cdot\beta(u_h - g)\|_{\partial K\cap\partial\Omega_-},$$

and cell weights (setting $\xi := z - z_h$)

$$\omega_K(z) := h_K^{-3}\{\|\xi\|_K + \delta_K\|\beta\cdot\nabla\xi\|_K\}, \quad \omega_{\partial K}(z) := h_K^{-5/2}\|\xi\|_{\partial K\cap\partial\Omega_-}.$$

We note that this a posteriori error bound explicitly contains the mesh size h_K and the stabilization parameter δ_K as well. This gives us the possibility to simultaneously adapt both parameters, which may be particularly advantageous in capturing sharp layers in the solution.

We want to illustrate the features of the error estimate (7.22) by a simple thought experiment. Let $\Omega = (0,1)\times(0,1)$ and $f = 0$. We take the functional

$$J(u) := (1, n\cdot\beta u)_{\partial\Omega_+}.$$

The corresponding dual solution is $z \equiv 1$, so that $J(e) = 0$. This implies

$$(1, n\cdot\beta u_h)_{\partial\Omega_+} = (1, n\cdot\beta u)_{\partial\Omega_+} = -(1, n\cdot\beta g)_{\partial\Omega_-}\,.$$

recovering the well-known global conservation property of the scheme.

7.1.3. EVALUATION OF THE ERROR ESTIMATES To evaluate the error estimates (7.14) or (7.22), one may solve the corresponding perturbed dual problem numerically by the same method as used in computing u_h, yielding an approximation $z_h \in V_h$ to the exact dual solution z. However, the use of the same meshes for computing primal and dual solution is by no means obligatory. In fact, in the case of dominant transport it may be advisable to compute the dual solution on a different mesh; see [47] for examples. Then, the weights ω_K can be determined numerically in different ways:

1. We may take $z_h = I_h z \in V_h$ as the nodal interpolation of z and use the local interpolation properties of finite elements to obtain

$$\omega_K = h_K^{-3}\|z - I_h z\|_K \le c_I h_K^{-1}\|\nabla^2 z\|_K,$$

with an interpolation constant $c_I \approx 0.1 - 1$. Here, $\nabla^2 z$ is the tensor of second derivatives of z. Then, approximation by second-order difference quotients of the computed discrete dual solution $z_h \in V_h$ yields

$$\omega_K \approx c_I|\nabla_h^2 z_h(x_K)|\,, \tag{7.23}$$

x_K being the center point of K.

2. Computation of a discrete dual solution $z_{h'} \in V_{h'}$ in a richer space $V_{h'} \supset V_h$ (e.g., on a finer mesh or by higher-order elements) and setting

$$\omega_K \approx h_K^{-3}\|z_{h'} - I_h z_{h'}\|_K\,, \tag{7.24}$$

where $I_h z_{h'} \in V_h$ denotes the generic nodal interpolation.

3. Interpolation of the discrete dual solution $z_h \in V_h$ by higher order polynomials on certain cell-patches, e.g., biquadratic interpolation $I_h^{(2)} z_h$:

$$\omega_K \approx h_K^{-3}\|I_h^{(2)} z_h - z_h\|_K\,. \tag{7.25}$$

Analogous approximations can be used for the weights $\omega_{\partial K}$. Option (2) is quite expensive and rarely used. Since we normally do not want to spend more time in evaluating the error estimate than for solving the primal problem, we recommend option (1) or (3). Notice that option (3) does not involve an interpolation constant which needs to be specified. The computational results reported in [14] indicate that the use of biquadratic interpolation on patches of four quadrilaterals is more accurate than using the finite difference approximation (7.23).

7.2. Strategies for mesh adaptation

We use the notation introduced above: u is the solution of the variational problem posed on a 2-dimensional domain Ω, u_h is its piecewise linear (or bilinear) finite element approximation. Further, $e = u - u_h$ is the discretization error and $J(\cdot)$ a linear error functional for measuring e. We suppose that there is an a posteriori error estimate of the form

$$|J(e)| \le \eta := \sum_{K\in\mathcal{T}_h} h_K^4\, \rho_K(u_h)\, \omega_K(z), \tag{7.26}$$

with the cell residuals $\rho_K(u_h)$ and weights $\omega_K(z)$. Accordingly, we define the local "error indicators"

$$\eta_K := h_K^4\, \rho_K(u_h)\, \omega_K(z).$$

The mesh design strategies are oriented towards a prescribed tolerance TOL for the error quantity $J(e)$ and the number of mesh cells N which measures the complexity of the computational model. Usually the admissible complexity is constrained by some maximum value $N_{\max}$.

There are various strategies for organizing a mesh adaptation process on the basis of the a posteriori error estimate (7.26).

- *Error balancing strategy:* Cycle through the mesh and equilibrate the local error indicators,

$$\eta_K \approx \frac{TOL}{N} \quad \Rightarrow \quad \eta \approx TOL. \tag{7.27}$$

  This process requires iteration with respect to the number of cells N.
- *Fixed fraction strategy:* Order cells according to the size of η_K and refine a certain percentage (say 30%) of cells with largest η_K (or those which make up 30% of the estimate value η) and coarsen those cells with smallest η_K. By this strategy, we may achieve a prescribed rate of increase of N (or keep it constant as may be desirable in nonstationary computations).
- *Mesh optimization strategy:* Use the representation

$$\eta := \sum_{K\in\mathcal{T}_h} h_K^4\, \rho_K(u_h)\, \omega_K(z) \approx \int_\Omega h(x)^2 A(x)\, dx \tag{7.28}$$

  for generating a formula for an optimal mesh-size distribution $h_{opt}(x)$.

We want to discuss the strategy for deriving an optimal mesh-size distribution in more detail. As a side-product, we will also obtain the justification of the error equilibration strategy. Let $N_{\max}$ and TOL be prescribed. We assume that for $TOL \to 0$, the cell residuals and the weights approach certain limits,

$$\rho_K(u_h) \approx h_K^{-3/2} \|n\cdot[\nabla u_h]\|_{\partial K} \;\to\; \rho(x_K) \approx |D^2 u(x_K)|,$$
$$\omega_K(z) \approx h_K^{-5/2} \|z - I_h z\|_{\partial K} \;\to\; \omega(x_K) \approx |D^2 z(x_K)|.$$

These properties can be proven on uniformly refined meshes by exploiting superconvergence effects, but still need theoretical justification on locally refined meshes. This suggests to assume that

$$\eta \approx \tilde{\eta} := \int_\Omega h(x)^2 A(x)\,dx, \qquad N = \sum_{K\in\mathcal{T}_h} h_K^2 h_K^{-2} \approx \int_\Omega h(x)^{-2}\,dx, \tag{7.29}$$

with the weighting function $A(x) = \rho(x)\omega(x)$. Now, let us consider the mesh optimization problem

$$\eta \to min!, \quad N \le N_{\max}.$$

Applying the usual Lagrange approach yields the necessary optimality conditions

$$\frac{d}{dt}\Big[\int_\Omega (h+t\phi)^2 A\,dx + (\lambda+t\mu)\Big((h+t\mu)^{-2}\,dx - N_{\max}\Big)\Big]_{t=0} = 0,$$

for any variations ϕ and μ. From this, we infer that

$$2h(x)A(x) - 2\lambda h(x)^{-3} = 0, \qquad \int_\Omega h(x)^{-2}\,dx - N_{\max} = 0.$$

Hence, we obtain

$$h(x) = \big(\lambda^{-1}A(x)\big)^{-1/4} \quad \Rightarrow \quad \tilde{\eta} = h^4 A \equiv \lambda^{-1},$$

and

$$\lambda^{-1/2}\int_\Omega A(x)^{1/2}\,dx = N_{\max}, \quad W := \int_\Omega A(x)^{1/2}\,dx.$$

This gives us a formula for the "optimal" mesh-size distribution:

$$\lambda = \Big(\frac{W}{N_{\max}}\Big)^2 \quad \Rightarrow \quad h_{opt}(x) = \Big(\frac{W}{N_{\max}}\Big)^{1/2} A(x)^{-1/4}. \tag{7.30}$$

In an analogous way, we can also treat the adjoint optimization problem $N \to min!$, $\eta \le TOL$. We note that even for a rather "singular" error functional $J(\cdot)$ the quantity W is bounded, e.g.,

$$J(e) = \nabla e(0) \quad \Rightarrow \quad A(x) \approx |x|^{-3} \quad \Rightarrow \quad W = \int_\Omega |x|^{-3/2}\,dx < \infty.$$

Open Problem 7.1: *Make the "mesh optimization strategy" rigorous, i.e., prove the proposed convergence of cell weights and residuals under (local) mesh refinement. This could be accomplished by proving that for piecewise linear or d-linear approximation, there holds*

$$\overline{\lim}_{h\to 0}\{\max_{\bar{\Omega}} |\nabla_h^2 u_h|\} \le c(u),$$

where $\nabla_h^2 u_h$ is a suitable second-order difference quotient.

7.2.1. COMPUTATIONAL TESTS *(I) The diffusion model problem:* We begin with the model diffusion problem (7.8) posed on the rectangular domain $\Omega = (-1,1)\times(-1,3)$ with slit at $(0,0)$. In the presence of a reentrant corner, here a slit, with angle $\omega = 2\pi$, the solution involves a "corner singularity". It can be written in the form $u = \psi r^{1/2} + \tilde{u}$, with r being the distance to the corner point and $\tilde{u} \in H^2(\Omega)$. We want to illustrate how the singularity introduced by the weights interacts with the pollution effect caused by the slit singularity. Let the goal be the accurate computation of the a derivative value $J(u) = \partial_1 u(P)$ at the point $P = (0.75, 2.25)$. In this case the dual solution z behaves like

$$|\nabla^2 z(x)| \approx d(x)^{-3} + r(x)^{-3/2},$$

where $d(x)$ and $r(x)$ are the distance functions with respect to the points P and $(0,0)$, respectively. Notice that in this case, the dual solution does not exist in the sense of $H_0^1(\Omega)$, such that for practical use, we have to regularize the functional $J(u) = \partial_1 u(P)$ appropriately. It follows that

$$|\partial_1 e(P)| \approx c_I \sum_{K\in\mathbb{T}_h} h_K^4 \rho_K(u_h)\Big\{d_K^{-3} + r_K^{-3/2}\Big\}. \tag{7.31}$$

Equilibrating the local error indicators yields

$$\eta_K \approx \frac{h_K^4}{d_K^3} \approx \frac{TOL}{N} \quad\Rightarrow\quad h_K^2 \approx d_K^{3/2}\Big(\frac{TOL}{N}\Big)^{1/2},$$

and, consequently,

$$N = \sum_{K\in\mathbb{T}_h} h_K^2 h_K^{-2} = \Big(\frac{N}{TOL}\Big)^{1/2} \sum_{K\in\mathbb{T}_h} h_K^2 d_K^{-3/2} \approx \Big(\frac{N}{TOL}\Big)^{1/2}.$$

This implies that $N_{opt} \approx TOL^{-1}$ which is better than what could be achieved on a uniformly refined mesh. In fact, the *global* energy-error estimate leads to a mesh efficiency like $J(e) \sim N^{-1/2}$, i.e., $N_{opt} \approx TOL^{-2}$. This predicted asymptotic behavior is well confirmed by the results of our computational test shown in Figures 29 and 30 (for more details, we refer to [14]).

(II) The transport model problem: Next, we consider the model problem (7.19) on the unit square $\Omega = (0,1)\times(0,1) \subset \mathbb{R}^2$ with the right-hand side $f \equiv 0$, the (constant) transport coefficient $\beta = (1, 0.5)^T$, and the inflow data

$$g(x,0) = 0, \quad g(0,y) = 1\,.$$

The quantity to be computed is part of the outflow as indicated in Figure 31:

$$J(u) := \int_\Gamma \beta{\cdot}n u\, ds\,.$$

The mesh refinement is organized according to the "fixed fraction strategy" described above. In Table 3, we show results for this test computation. The corresponding meshes and the primal as well as the dual solution are presented in Figure 31. Notice that there is no mesh refinement enforced along the upper line

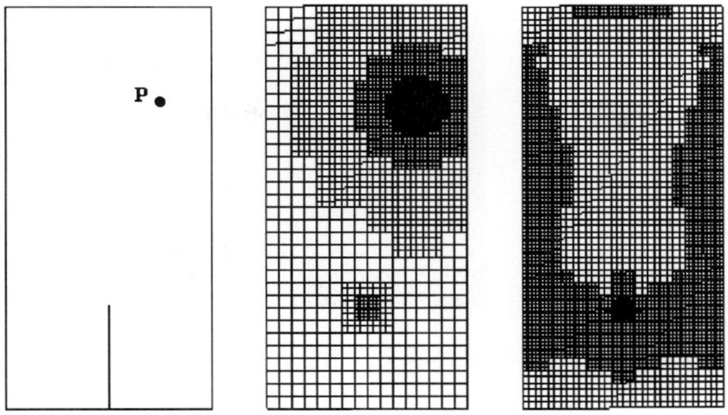

FIGURE 29. Refined meshes with about 5,000 cells for computing $\partial_1 u(P)$ using the weighted error estimate η_{weight} (middle) and the energy error estimate η_E (right); from Backes [4].

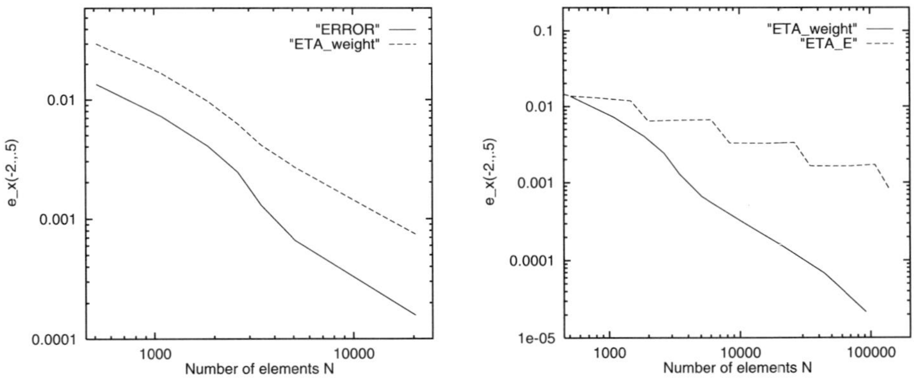

FIGURE 30. Test results on the slit domain obtained by the weighted error estimator η_{weight}: comparison with the true error (left) and comparison of efficiency with that of the energy error estimator η_E (right); from Backes [4].

of discontinuity of the dual solution since here the residual of the primal solution is almost zero.

7.3. A general paradigm for a posteriori error estimation

The approach to residual-based error estimation described above for the linear model problem can be extended to general nonlinear systems. We outline the underlying concept in an abstract setting following the general paradigm introduced by Johnson, et al. [26].

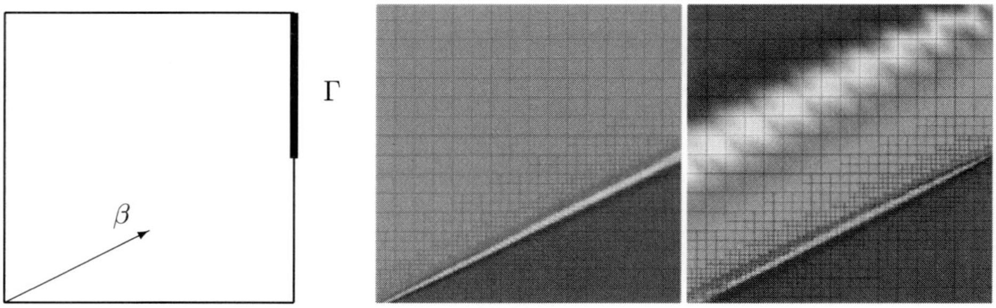

FIGURE 31. Configuration and grids of the test computation for the model transport problem (7.19): primal solution (left) and dual solution (right) on an adaptively refined mesh.

TABLE 3. Convergence results of the test computation for the model transport problem (7.19).

Level	N	$J(e)$	η	$\eta/J(e)$
0	256	2.01e-2	2.38e-2	1.18
1	310	1.82e-2	1.96e-2	1.08
2	634	1.09e-2	1.21e-2	1.11
3	964	7.02e-3	8.23e-3	1.17
4	1315	6.25e-3	7.88e-3	1.26
5	1540	5.37e-3	6.94e-3	1.29
6	2050	4.21e-3	5.37e-3	1.27
7	2128	4.11e-3	5.21e-3	1.27

Let V be a Hilbert space with inner product $(\cdot,\cdot)$ and corresponding norm $\|\cdot\|$, $A(\cdot;\cdot)$ a continuous semi-linear form and $F(\cdot)$ a linear form defined on V. We seek a solution $u \in V$ to the abstract variational problem

$$A(u;\phi) = F(\phi) \qquad \forall\, \phi \in V. \tag{7.32}$$

This problem is approximated by a Galerkin method using a sequence of finite dimensional subspaces $V_h \subset V$ parameterized by a discretization parameter h. The discrete problems seek $u_h \in V_h$ satisfying

$$A(u_h;\phi_h) = F(\phi_h) \qquad \forall\, \phi_h \in V_h. \tag{7.33}$$

The key feature of this approximation is the "Galerkin orthogonality" which in this nonlinear case reads as

$$A(u;\phi_h) - A(u_h;\phi_h) = 0, \quad \phi_h \in V_h. \tag{7.34}$$

By elementary calculus, there holds

$$A(u;\phi) - A(u_h;\phi) = \int_0^1 A'(su + (1-s)u_h; e, \phi)\, ds,$$

with $A'(v;\cdot,\cdot)$ denoting the tangent form of $A(\cdot;\cdot)$ at some $v \in V$. This leads us to introduce the bilinear form

$$L(u, u_h; \phi, \psi) := \int_0^1 A'(su + (1-s)u_h; \phi, \psi)\, ds\,,$$

which depends on the solutions u as well as u_h. Then, denoting the error by $e = u - u_h$, there holds

$$\begin{aligned} L(u, u_h; e, \phi_h) &= \int_0^1 A'(su + (1-s)u_h; e, \phi_h)\, ds \\ &= A(u;\phi_h) - A(u_h;\phi_h) = 0\,, \qquad \phi_h \in V_h. \end{aligned}$$

Suppose that the quantity $J(u)$ has to be computed, where $J(\cdot)$ is a linear functional defined on V. For representing the error $J(e)$, we use the solution $z \in V$ of the *dual problem*

$$L(u, u_h; \phi, z) = J(\phi) \qquad \forall\, \phi \in V. \tag{7.35}$$

Assuming that this problem is solvable and using the Galerkin orthogonality (7.34), we obtain the error representation

$$J(e) = L(u, u_h; e, z - z_h) = F(z - z_h) - A(u_h; z - z_h), \tag{7.36}$$

with any approximation $z_h \in V_h$. Since the bilinear form $L(u, u_h; \cdot, \cdot)$ contains the unknown solution u in its coefficient, the evaluation of (7.36) requires approximation. The simplest way is to replace u by u_h, yielding a perturbed dual problem

$$L(u_h, u_h; \phi, \tilde{z}) = J(\phi) \qquad \forall\, \phi \in V. \tag{7.37}$$

We remark that the bilinear form $L(u_h, u_h; \cdot, \cdot)$ used in (7.37) is identical to the tangent form $A'(u_h; \cdot, \cdot)$. Controlling the effect of this perturbation on the accuracy of the resulting error estimate may be a delicate task and depends strongly on the particular problem under consideration. Our own experience with several different types of problems (including the Navier-Stokes equations) indicates that this problem is less critical as long as the continuous solution is stable. The crucial problem is the numerical computation of the perturbed dual solution $\tilde{z}$ by solving a discretized dual problem

$$L(u_h, u_h; \phi, \tilde{z}_h) = J(\phi) \qquad \forall\, \phi \in V_h. \tag{7.38}$$

This then results in a practically useful error estimate $J(e) \approx \tilde{\eta}(u_h)$.

Open Problem 7.2: *Analyze the effect of the error introduced by the linearization steps (7.37) and (7.38) on the quality of the a posteriori error bound and design a reliable strategy for controlling this error.*

7.3.1. THE NESTED SOLUTION APPROACH For solving the nonlinear problem (7.32) by the adaptive Galerkin finite element method (7.33), we employ the following iterative scheme. Starting from a coarse initial mesh $\mathbb{T}_0$, a hierarchy of refined meshes $\mathbb{T}_i$, $i \geq 1$, and corresponding finite element spaces V_i is generated by a nested solution process.

(0) *Initialization* $i = 0$: Start on coarse mesh $\mathbb{T}_0$ with

$$v_0^{(0)} \in V_0.$$

(1) *Defect correction iteration:* For $i \geq 1$, start with

$$v_i^{(0)} = v_{i-1} \in V_i.$$

(2) *Iteration step:* Evaluate the defect

$$(d_i^{(j)}, \phi) := F(\phi) - A(v_i^{(j)}; \phi), \quad \phi \in V_i.$$

Choose a suitable approximation $\tilde{A}'(v_i^{(j)}; \cdot, \cdot)$ to the derivative $A'(v_i^{(j)}; \cdot, \cdot)$ (with good stability and solubility properties) and solve the correction equation

$$\delta v_i^{(j)} \in V_i : \quad \tilde{A}'(v_i^{(j)}; \delta v_i^{(j)}, \phi) = (d_i^{(j)}, \phi) \quad \forall \phi \in V_i.$$

For this, Krylov-space or multigrid iterations are employed using the hierarchy of already constructed meshes $\{\mathbb{T}_i, \dots, \mathbb{T}_0\}$. Then, update $v_i^{(j+1)} = v_i^{(j)} + \lambda_i \delta v_i^{(j)}$ ($\lambda_i \in (0, 1]$ a relaxation parameter), set $j = j+1$ and go back to (2). This process is repeated until a limit $v_i \in V_i$, is reached with a certain required accuracy.

(3) *Error estimation:* Solve the (linearized) discrete dual problem

$$z_i \in V_i : \quad A'(v_i; \phi, z_i) = J(\phi) \quad \forall \phi \in V_i,$$

and evaluate the a posteriori error estimate

$$|J(e_i)| \approx \tilde{\eta}(v_i).$$

For controlling the reliability of this bound, i.e. the accuracy in the determination of the dual solution z, one may check whether $\|z_i - z_{i-1}\|$ is sufficiently small; if this is not the case, additional global mesh refinement is advisable. If $\tilde{\eta}(v_i) \leq TOL$ or $N_i \geq N_{\max}$, then stop. Otherwise cell-wise mesh adaptation yields the new mesh $\mathbb{T}_{i+1}$. Then, set $i = i + 1$ and go back to (1).

This nested solution process is employed in the application presented below. Notice that the derivation of the a posteriori error estimate (3) involves only the solution of *linearized* problems. Hence, the whole error estimation may amount only to a relatively small fraction of the total cost for the solution process.

7.4. Application to the Navier-Stokes equations

The results in this section are collected from Becker [7, 9]; see also [12]. We consider the stationary Navier-Stokes equations

$$-\nu\Delta v + v\cdot\nabla v + \nabla p = 0, \qquad \nabla\cdot v = 0\,, \tag{7.39}$$

in a bounded domain $\Omega \subset \mathbb{R}^2$, with boundary conditions as described in Section 2,

$$v|_{\Gamma_{rigid}} = 0, \quad v|_{\Gamma_{in}} = v^{in}, \quad \nu\partial_n v - pn|_{\Gamma_{out}} = 0.$$

As an example, we consider the flow around the cross section of a cylinder in a channel shown in Figure 32. This is part of a set of benchmark problems discussed in Schäfer/Turek [81].

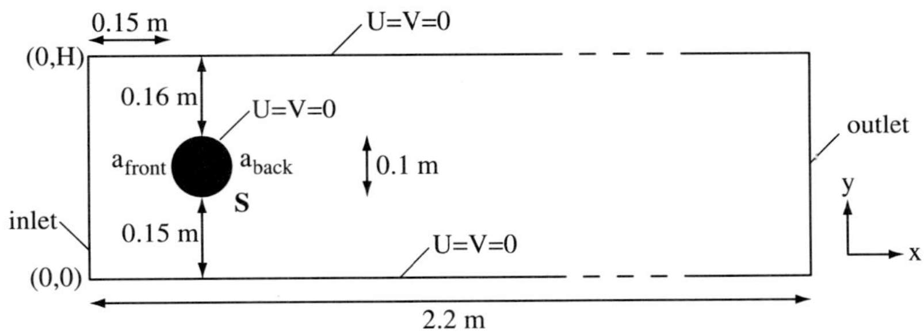

FIGURE 32. Configuration of the benchmark "flow around a cylinder"

Quantities of physical interest are, for example,

$$\begin{aligned}
\text{pressure drop:} \qquad J_{\Delta p}(v,p) &= p(a_{front}) - p(a_{back}),\\
\text{drag coefficient:} \qquad J_{drag}(v,p) &= \frac{2}{\bar{U}^2 D}\int_S n\cdot\sigma(v,p)e_x\,ds,\\
\text{lift coefficient:} \qquad J_{lift}(v,p) &= \frac{2}{\bar{U}^2 D}\int_S n\cdot\sigma(v,p)e_y\,ds,
\end{aligned}$$

where S is the surface of the cylinder, D its diameter, e_x and e_y the cartesian unit vectors, $\bar{U}$ the reference velocity, and $\sigma(v,p) = \frac{1}{2}\nu(\nabla v + \nabla v^T) + pI$ the stress force acting on S. In our example, the Reynolds number is $Re = \bar{U}^2 D/\nu = 20$, such that the flow is stationary. For evaluating the drag and lift coefficients, one may use another representation obtained by the Stokes formula, e.g., for the drag:

$$J_{drag} := \frac{2}{\bar{U}^2 D}\int_S n\cdot\sigma(v,p)e_x\,ds = \frac{2}{\bar{U}^2 D}\int_\Omega \{\sigma(v,p)\nabla\bar{e}_x + \nabla\sigma(v,p)\cdot\bar{e}_x\}\,dx,$$

where $\bar{e}_x$ is an extension of e_x to the interior of Ω with support along S; see Giles, et al. [28] and Becker [9]. This representation in terms of a domain integral is more robust and accurate than the original one involving a contour integral.

The discretization is by the finite element Galerkin method using the conforming Q_1/Q_1 Stokes element described in Section 3 with least-squares pressure stabilization and streamline diffusion stabilization for the transport. In order to incorporate this scheme in the abstract framework described above, we rewrite it in a more compact form. To this end, we introduce the Hilbert-spaces $\mathbf{V} := \mathbf{H}\times L$ of pairs $u := \{v,p\}$ and their discrete analogues $\mathbf{V}_h := \mathbf{H}_h \times L_h$ of pairs $u_h := \{v_h, p_h\}$. Accordingly the Navier-Stokes equations can be written in vector form as follows:

$$Lu := \begin{bmatrix} -\nu\Delta v + v\cdot\nabla v + \nabla p \\ \nabla\cdot v \end{bmatrix} = \begin{bmatrix} f \\ 0 \end{bmatrix}.$$

Further, for $u := \{v,p\}$ and $\phi = \{\psi,\chi\}$, we define the semi-linear form

$$A(u;\phi) := \nu(\nabla v, \nabla\psi) + (v\cdot\nabla v, \psi) - (p, \nabla\cdot\psi) + (\nabla\cdot v, \chi),$$

and the linear functional $F(\phi) := (f,\psi)$. Then, the stationary version of the variational formulation (2.8) is written in the following compact form: Find $u \in \mathbf{V} + (v_h^{in}, 0)^T$, such that

$$A(u;\phi) = F(\phi) \quad \forall\phi \in \mathbf{V}. \tag{7.40}$$

Using the weighted L^2-bilinear form

$$(v,w)_h := \sum_{K\in\mathbb{T}_h} \delta_K(\nabla v, \nabla w)_K,$$

the stabilized finite element approximation reads as follows: Find $u_h \in \mathbf{V}_h + (v^{in}, 0)^T$ such that

$$A(u_h;\phi_h) + (Lu_h, S\phi_h)_h = (F,\phi_h) + (F, S\phi_h)_h \quad \forall\phi_h \in \mathbf{V}_h, \tag{7.41}$$

where the stabilization operator S is defined by

$$S\phi := \begin{bmatrix} \nu\Delta\psi + v\cdot\nabla\psi + \nabla\chi \\ 0 \end{bmatrix},$$

with the parameter δ specified by (3.36). This formulation contains the pressure and transport stabilization as described in Section 3.

The question is now how to construct a mesh as economical as possible on which the quantities $J_{\Delta p}(v,p)$, $J_{drag}(v,p)$ and $J_{lift}(v,p)$ can be computed to the required accuracy, say, of 1%. The a priori design of such a mesh is a difficult task as will be demonstrated by the results of numerical tests below; Figure 33 shows a collection of possible *a priori* meshes.

Now, we will discuss the use of *a posteriori* techniques for constructing economical meshes. We denote the discretization error for the pressure by $e_p := p - p_h$ and that for the velocity by $e_v := v - v_h$. By standard arguments relying on the coerciveness properties of the Fréchet derivative of the operator L, one derives

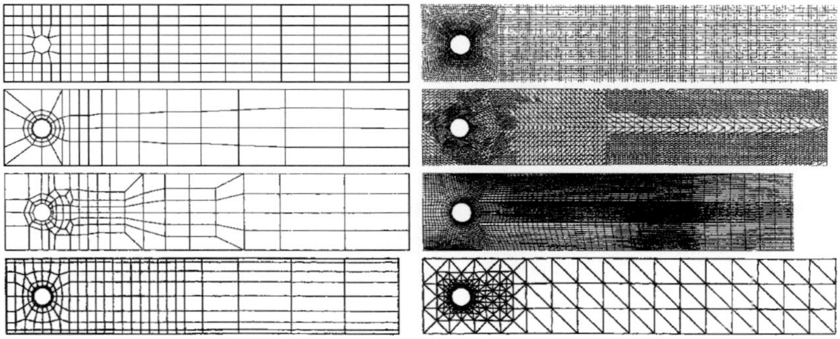

FIGURE 33. Examples of meshes designed for the benchmark problem "flow around a cylinder"; the first three meshes on the left, "Grid 1", "Grid 2", and "Grid 3", are coarse initial meshes which are to be uniformly refined.

the following energy-norm *a posteriori* error estimate

$$\|\nabla e_v\| + \|e_p\| \;\le\; c_I c_S\Big(\sum_{K\in\mathcal{T}_h} \Big\{(h_K^2+\delta_K)\|R(u_h))\|_K^2 + \qquad (7.42)$$

$$+\|\nabla\cdot v_h\|_K^2 + \nu h_K\|n\cdot[\nabla v_h]\|_{\partial K}^2\Big\} + \cdots\Big)^{1/2},$$

with the residual $R(u_h) := \nu\Delta v_h - v_h\cdot\nabla v_h - \nabla p_h$. In this estimate the " ... " stand for additional terms representing the errors in approximating the inflow data and the curved boundary S; they can be expected to be small compared to the other residual terms and are usually neglected. In this estimate the *interpolation constant* c_I can be determined and is of moderate size $c_I \sim 0.2$. The most critical point is the *stability constant* c_S which is completely unknown. It is related to the constant in the coerciveness estimate of the tangent form $A'(v;\cdot,\cdot)$ of $A(\cdot;\cdot)$ taken at the solution v,

$$\|\nabla w\| + \|q\| \;\le\; c_S \sup_{\phi\in V}\left\{\frac{A'(v;z,\phi)}{\|\nabla\psi\|+\|\chi\|}\right\},$$

where $z=\{w,q\}$ and $\phi=\{\psi,\chi\}$. In order to use this error bound for mesh-size control, we have set it to $c_S=1$.

The error estimate (7.42) is not appropriate for controlling the error in local quantities like drag and lift since it measures the residual uniformly over the whole computational domain. One way of introducing more a priori information into the mesh refinement process based on (7.42) is to start from an initial mesh which is already refined towards the contour S. Alternatively, one may also use (on heuristic grounds) additional weighting factors which enforce stronger mesh refinement in the neighborhood of S. The resulting global error indicator reads as follows:

$$\|\nabla e_v\| + \|e_p\| \;\le\; c_I c_S\Big(\sum_{K\in\mathbb{T}_h} \sigma_K\Big\{(h_K^2+\delta_K)\|R(u_h))\|_K^2 \\ +\|\nabla\cdot v_h\|_K^2 + \nu h_K\|n\cdot[\nabla v_h]\|_{\partial K}^2\Big\} + \cdots\Big)^{1/2}, \tag{7.43}$$

where the weights σ_K are chosen large along S.

Correctly *weighted a posteriori* error estimates can be obtained following the general line of argument described above. The approximate dual problem seeks $z := \{w, q\} \in V$ satisfying

$$A'(u_h;\varphi,z) + (L'(u_h)^*\varphi, Sz)_\delta = J(\varphi) \qquad \forall \varphi \in V\,, \tag{7.44}$$

where $A'(u_h;\cdot,\cdot)$ and $L'(u_h)^*$ are the tangent form and adjoint tangent operator of $A(\cdot;\cdot)$ and $L(\cdot)$, respectively. The resulting weighted a posteriori estimate for the error $e := u - u_h$ becomes

$$|J(e)| \le \sum_{K\in\mathbb{T}_h} \Big\{\rho_K\,\omega_K + \rho_{\partial K}\,\omega_{\partial K} + \rho_K^{div}\,\omega_K^{div} + \cdots\Big\}, \tag{7.45}$$

with the local residual terms and weights defined by

$$\begin{aligned}
\rho_K &= \|R(u_h)\|_K, \quad \omega_K = \|w-w_h\|_K + \delta_K\|v_h\cdot\nabla(w-w_h) + \nabla(q-q_h)\|_K,\\
\rho_{\partial K} &= \tfrac12\nu\|n\cdot[\nabla v_h]\|_{\partial K}, \quad \omega_{\partial K} = \|w-w_h\|_{\partial K},\\
\rho_K^{div} &= \|\nabla\cdot v_h\|_K, \quad \omega_K^{div} = \|q-q_h\|_K.
\end{aligned}$$

The dots " ... " stand again for additional terms measuring the errors in approximating the inflow and the curved cylinder boundary. For more details on this aspect, we refer to [14] and Becker [9]. The bounds for the dual solution $z = \{w, q\}$ are obtained computationally by replacing the unknown solution u in the convection term by its approximation u_h and solving the resulting linearized problem on the same mesh. From this approximate dual solution $\tilde z_h$, patchwise biquadratic interpolations are taken to approximate z in evaluating the weights $\omega_K^{(i)}$, $I_h^{(2)}\tilde z_h - z_h \approx z - z_h$. This avoids the occurrence of interpolation constants. Table 4 shows the corresponding results for the pressure drop computed on four different types of meshes:

(i) Hierarchically refined meshes starting from coarse meshes of type "Grid 1" and "Grid 2" as shown in Figure 33.

(ii) Adapted meshes using the global energy-norm error estimate (7.42) with enforced refinement along the contour S; see Figure 34.

(iii) Adapted meshes using the weighted error estimate (7.45) for the pressure drop; see Figure 35.

These results demonstrate clearly the superiority of the weighted error estimate (7.45) in computing local quantities. It produces an error of less than 1% already after 6 refinement cycles on a mesh with less than 1400 unknowns while the other algorithms use more than 21000 unknowns to achieve the same accuracy (the corresponding values are printed in boldface). Corresponding sequences of meshes generated by the weighted energy error estimate (7.42) and the energy-error estimate (7.43) are seen in Figures 34 and 35.

Table 5 contains some results of the computation of drag and lift coefficients using the corresponding weighted error estimates. The effectivity index is defined by $I_{eff} := \eta(u_h)/|J(e)|$. Finally, Figure 36 shows plots of the dual solutions occurring in the computation of pressure drop, drag and lift.

TABLE 4. Results of the pressure drop computation (ref. value $\Delta p = 0.11752016\ldots$); a) upper row: on uniformly refined meshes of type Grid1 and Grid2, b) lower row: on adaptively refined meshes starting from a coarse mesh Grid1; from Becker [7].

Uniform Refinement, Grid1		
L	N	Δp
1	2268	0.109389
2	8664	0.110513
3	33840	0.113617
4	133728	0.115488
5	**531648**	**0.116486**

Uniform Refinement, Grid2		
L	N	Δp
1	1296	0.106318
2	4896	0.112428
3	19008	0.115484
4	**74880**	**0.116651**
5	297216	0.117098

Adaptive Refinement, Grid1		
L	N	Δp
2	1362	0.105990
4	5334	0.113978
6	**21546**	**0.116915**
8	86259	0.117379
10	330930	0.117530

Weighted Adaptive Refinement		
L	N	Δp
4	650	0.115967
6	**1358**	**0.116732**
9	2858	0.117441
11	5510	0.117514
12	8810	0.117527

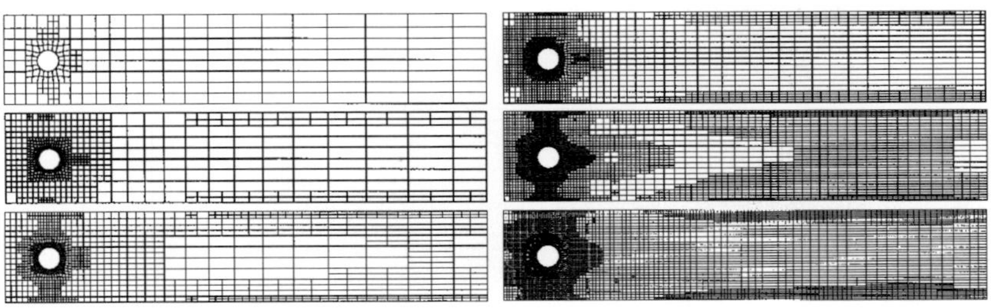

FIGURE 34. A sequence of refined meshes generated by the (heuristically) weighted global energy estimate; from Becker [7].

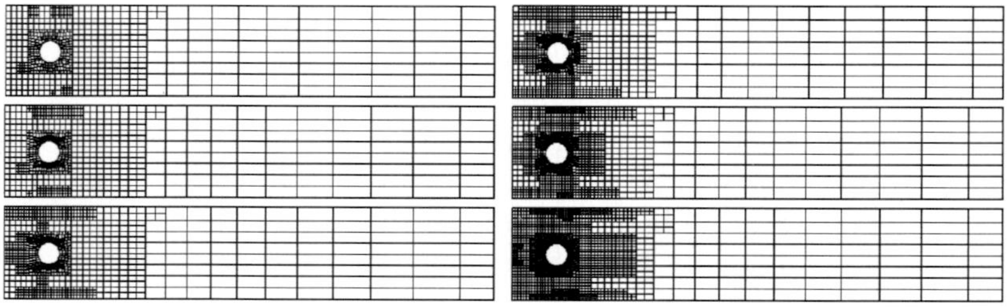

FIGURE 35. A sequence of refined meshes generated by the weighted error estimate for the pressure drop; from Becker [7].

TABLE 5. Results of the cylinder flow computations of drag and lift (ref. values $c_{drag} = 5.579535\dots$ and $c_{lift} = 0.0106189\dots$) on adaptively refined meshes starting from a coarse mesh of type Grid1; from Becker [9].

Computation of drag					Computation of lift				
L	N	c_{drag}	η_{drag}	I_{eff}	L	N	c_{lift}	η_{lift}	I_{eff}
3	251	5.780186	$2.0e{-}1$	0.5	3	296	0.007680	$2.9e{-}3$	5.0
4	587	5.637737	$5.8e{-}2$	0.6	4	764	0.009249	$1.4e{-}3$	5.0
5	1331	5.568844	$1.0e{-}2$	1.6	5	1622	0.009916	$7.3e{-}4$	5.0
6	3953	5.576580	$2.5e{-}3$	2.0	6	4466	0.010144	$5.0e{-}4$	2.5
7	8852	5.578224	$8.7e{-}4$	2.5	7	8624	0.010267	$3.8e{-}5$	2.0
8	16880	5.578451	$6.5e{-}4$	1.6	8	18093	0.010457	$1.9e{-}5$	2.0
9	34472	5.578883	$2.1e{-}4$	2.0	9	34010	0.010524	$1.2e{-}4$	1.6

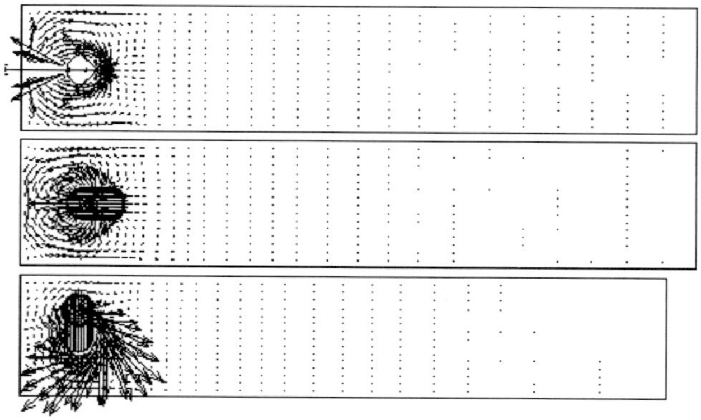

FIGURE 36. Velocity plots of the dual solution for pressure drop (top), drag (middle), and lift (bottom); from Becker [9].

7.5. The nonstationary case

The extension of the approach to mesh adaptivity described above to the *nonstationary* Navier-Stokes equations is presently under development.

(I) The traditional method for a posteriori time-step selection is based on the concept of controlling the *local* "truncation error" but neglecting the *global* error accumulation. In its simplest form this strategy uses the condition

$$\tfrac{1}{3}c_S k_n^2 (U_{k/2}^n - U_k^n) \approx TOL, \tag{7.46}$$

where U_k^n and $U_{k/2}^n$ are the solutions computed from the preceding approximation U^{n-1} at t_{n-1} by a second-order scheme (e.g. the Crank-Nicolson scheme) with time-step sizes k and $k/2$, respectively. For a more detailed description of techniques of this type, we refer to Turek [97].

(II) The extension of the residual-based error control described above to nonstationary problems is based on a time discretization which has also the features of a Galerkin method. These are for example the so-called "continuous" or "discontinuous" Galerkin methods of polynomial degree $r \geq 0$ ("cG(r)" or "dG(r)" methods). The lowest-order examples are the dG(0) method which (in the autonomous case) is equivalent to the backward Euler scheme, the dG(1) method which is similar to an implicit Runge-Kutta scheme of third order, and the cG(1) method which can be interpreted as a variant of the Crank-Nicolson scheme. In particular, the dG(1) method is attractive for solving the nonstationary Navier-Stokes problem because of its superior accuracy (compared to the dG(0) method).

The result of error estimation using a duality argument is an a posteriori error estimate of the form (see [55] and Hartmann [39])

$$\|U_k^n - u(\cdot, t_n)\| \approx \sum_{m=1}^{n} \{k_m^3 \omega_m \|d_t U_k^m\| + \cdots\} \tag{7.47}$$

where $d_t U_k^m = k_m^{-1}(U_k^m - U_k^{m-1})$ are the time-difference quotients of the computed solution and ω_m are weighting factors obtained by solving a "backward in time" space-time dual problem. The dots "..." refer to residual terms of the spatial discretization. The main problem with this approach is its huge computational work; in a nonlinear problem the "forward" solution $\{U_k^m\}_{m=0}^n$ enters the linearized dual problem as coefficient and needs therefore to be stored over the whole time interval. Moreover, in this way error control can be achieved only at single times t_n or for the time-averaged error. Controlling the error uniformly in time requires (theoretically) to solve a dual problem at each discrete time level resulting in prohibitively high cost. The economical realization of this concept for computing nonstationary flows involving global error control is still an open problem.

Open Problem 7.3: *Devise a strategy for adapting the stabilization parameters δ_K simultaneously with the mesh size h_K on the basis of the a posteriori error estimate (7.45).*

Open Problem 7.4: *Derive an a posteriori error estimate of the form (7.45) for the full space-time discretization of the Navier-Stokes equations and device a strategy for simultaneous adaptation of mesh sizes* h_K *and time steps* k_n .

8. Extension to weakly compressible flows

In this last section, we discuss the extension of the computational methodology described above to certain *compressible* flows. The flows of interest are those in which density changes are induced by temperature gradients resulting for example from heat release by chemical reactions. Such "weakly" compressible flows are characterized by low-Mach-number speed and hydrodynamically incompressible behavior. Here, the dominant problem is that of stiff velocity-pressure coupling while shocks or large pressure gradients do not develop. We recall the system of conservation equations for mass, momentum and energy, in the case of a stationary flow:

$$\begin{aligned} \nabla\cdot[\rho v] &= 0\,, && (8.1)\\ \rho v\cdot\nabla v - \nabla\cdot[\mu\nabla v + \tfrac{1}{3}\mu\nabla\cdot vI] + \nabla p_{tot} &= \rho f\,, && (8.2)\\ c_p\rho v\cdot\nabla T - \nabla\cdot[\lambda\nabla T] &= h\,. && (8.3) \end{aligned}$$

Here, again v is the velocity, ρ the density, p_{tot} the (total) pressure and T the temperature of the fluid occupying a two- or three-dimensional region Ω. The dynamic viscosity $\mu > 0$, the heat capacity $c_p > 0$, the heat conductivity λ, the external volume force f and the heat source h are given. Since we only consider low-speed flows, the influence of stress and hydrodynamic pressure in the energy equation can be neglected. In general, f as well as h implicitly depend on the temperature T and on further quantities describing the release of heat for example through chemical reactions. Here, we will simply consider the heat source h as given. The coupling between pressure and density is assumed as that of a perfect gas,

$$p_{tot} = R\rho T, \tag{8.4}$$

where R is the gas constant. As mentioned above, we consider hydrodynamically incompressible flows. Accordingly, the pressure is split into two parts,

$$p_{tot}(x,t) = \bar{p}(t) + p_{hyd}(x,t),$$

namely the spatial mean value

$$\bar{p} := |\Omega|^{-1}\int_\Omega p_{tot}(x,t)\,dx,$$

and the "hydrodynamic pressure" $p_{hyd}(x,t)$. In a weakly compressible flow, the pressure variation due to hydrodynamic mechanisms is assumed to be small compared to the mean value of the total pressure,

$$|p_{hyd}| \ll |\bar{p}|,$$

which is determined by thermodynamic effects. Accordingly, we call $P_{th}(t) = \bar{p}(t)$ the "thermodynamic pressure". In the "low-Mach-number approximation" the hydrodynamic pressure occurs in the momentum equation

$$\rho v{\cdot}\nabla v - \nabla{\cdot}[\mu\nabla v + \tfrac{1}{3}\mu\nabla{\cdot}vI] + \nabla p_{hyd} = \rho f, \tag{8.5}$$

while the pressure-density coupling in the equation of state (8.4) is expressed in terms of the "thermodynamic pressure"

$$\rho = \frac{P_{th}}{RT}. \tag{8.6}$$

In many applications, this set of equations has to be supplemented by further conservation equations for species concentrations and complicated nonlinear source terms representing the chemical reactions. Here, we restrict ourselves to the simple case of low-Mach-number flow, where temperature variations are induced by outer source terms. The thermodynamic pressure $P_{th}(t)$ is supposed to be determined by a priori considerations; for more details, see Braack [17] and also [18].

Open Problem 8.1: *Estimate the error caused by neglecting stress and hydrodynamic pressure in the energy equation. Prove corresponding error bounds for the low-Mach-number approximation in terms of the Mach number.*

Since in the above approximation the density occurs as a secondary variable determined by the temperature through the equation of state, it appears natural to use the pressure $p := p_{hyd}$ together with the velocity v and the temperature T as primal variables in the computational model. We use the equation of state to rewrite the continuity equation as an equation for velocity and temperature:

$$\nabla{\cdot}v - T^{-1}v{\cdot}\nabla T = 0\,. \tag{8.7}$$

Furthermore, introducing the modified pressure $p := p_{hyd} - \frac{1}{3}\mu\nabla{\cdot}v$, the momentum equation can be written as

$$\rho v{\cdot}\nabla v - \nabla{\cdot}[\mu\nabla v] + \nabla p = \rho f\,, \tag{8.8}$$

while the energy equation keeps the form

$$c_p\rho v{\cdot}\nabla T - \nabla{\cdot}[\lambda\nabla T] = h\;. \tag{8.9}$$

The temperature-dependent functions $\mu = \mu(T)$ and $c_p = c_p(T)$ are usually given in terms of polynomial fits from data bases. The density ρ is expressed by the algebraic relation (8.6) in terms of the temperature. The system is closed by imposing appropriate boundary conditions,

$$v_{|\Gamma_{rigid}} = 0, \quad v_{|\Gamma_{in}} = v^{in}, \quad \mu\partial_n v + pn_{|\Gamma_{out}} = 0, \quad T_{|\partial\Omega} = \hat{T}, \tag{8.10}$$

where again Γ_{rigid}, Γ_{in}, Γ_{out} are the rigid part, the inflow part and the outflow part of the boundary $\partial\Omega$, respectively. For questions of well-posedness of this type of problem, we refer to the relevant literature, e.g., Feistauer [27] and Lions [60].

The starting point for a finite element discretization of problem (8.7), (8.8), (8.9), and (8.6) is again its variational formulation. To formulate this, we introduce the natural function spaces as already used above,

$$L \subset L^2(\Omega), \quad \mathbf{H} \subset H^1(\Omega)^d, \quad R \subset H^1(\Omega).$$

for the pressure $p \in L$, the velocity $v \in \mathbf{H}$, and the temperature $T \in R$. For a compact notation, we set $V := L \times \mathbf{H} \times R$. Prescribed Dirichlet data $\hat{v}$ and $\hat{T}$ can be included by seeking the weak solutions in appropriate sub-manifolds,

$$p \in L, \quad v \in \hat{v} + \mathbf{H}, \quad T \in \hat{T} + R.$$

Then, the triple $u := \{p, v, T\}$ is determined by the variational equations

$$\begin{aligned} (\nabla\cdot v, \chi) - (T^{-1}v\cdot\nabla T, \chi) &= 0, \quad \forall \chi \in L, & (8.11)\\ (\rho v\cdot\nabla v, \psi) + (\mu\nabla v, \nabla\psi) - (p, \nabla\cdot\psi) &= (\rho f, \psi) \quad \forall \psi \in \mathbf{H}, & (8.12)\\ (\rho c_p v\cdot\nabla T, \pi) + (\lambda\nabla T, \nabla\pi) &= (h, \pi) \quad \forall \pi \in R. & (8.13) \end{aligned}$$

In the following analysis, we consider for simplicity only the case of pure Dirichlet boundary conditions. In this case the pressure is determined only modulo constants and the corresponding solution space is $L = L_0^2(\Omega)$.

Now, the finite element discretization replaces the (infinite dimensional) function spaces L, $\mathbf{H}$, and R by finite dimensional *discrete* spaces denoted by L_h, $\mathbf{H}_h$, and R_h. Here, we think of finite element spaces based for example on conforming Q_1 approximation for all physical quantities. The corresponding *discrete* solutions $p_h \in L_h$, $v_h \in \hat{v}_h + \mathbf{H}_h$, and $T_h \in \hat{T}_h + R_h$ are determined through the system

$$\begin{aligned} (\nabla\cdot v_h, \chi_h) - (T_h^{-1}v_h\cdot\nabla T_h, \chi_h) &= 0, \quad \forall \chi_h \in L_h, & (8.14)\\ (\rho v_h\cdot\nabla v_h, \psi_h) + (\mu\nabla v_h, \nabla\psi_h) - (p_h, \nabla\cdot\psi_h) &= (\rho f, \psi_h) \quad \forall \psi_h \in \mathbf{H}_h, & (8.15)\\ (\rho c_p v_h\cdot\nabla T_h, \pi_h) + (\lambda\nabla T_h, \nabla\pi_h) &= (h, \pi_h) \quad \forall \pi_h \in R_h, & (8.16) \end{aligned}$$

with coefficients $\mu = \mu(T_h)$ and $c_p = c_p(T_h)$. The compact formulation of the system (8.14)–(8.16) makes use of the semi-linear form

$$\begin{aligned} A(u;\phi) \;:=\; & (\nabla\cdot v, \chi) - (T^{-1}v\cdot\nabla T, \chi) + (\rho v\cdot\nabla v, \psi) + (\mu\nabla v, \nabla\psi)\\ & -(p, \nabla\cdot\psi) + (\rho c_p v\cdot\nabla T, \pi) + (\lambda\nabla T, \nabla\pi), \end{aligned}$$

and the linear form $F(\phi) = (\rho f, \psi) + (h, \pi)$, defined for triples $u = \{p, v, T\}$, $\phi = \{\chi, \psi, \pi\} \in V$. With this notation, the problem reads as follows: Find $u \in \hat{u} + V$, such that

$$A(u;\phi) = F(\phi) \qquad \forall \phi \in V. \tag{8.17}$$

where $\hat{u}$ represents Dirichlet boundary data for all components. The corresponding discrete problem reads: Find $u_h \in \hat{u}_h + V_h$, such that

$$A(u_h;\phi_h) = F(\phi_h) \qquad \forall \phi_h \in V_h. \tag{8.18}$$

In general this system is unstable and needs stabilization with respect to the stiff velocity-pressure coupling as well as the transport terms.

8.1. Least-squares stabilization

The stabilization is introduced into the system (8.18) by using pressure stabilization and streamline diffusion as discussed above in the context of the incompressible Navier-Stokes equations. The corresponding stabilization terms are listed below:

- Pressure stabilization:

$$\begin{aligned} s_h^p(u_h,\chi_h) &= \sum_{K\in\mathcal{T}_h} \alpha_K(\bar{v}_h\cdot\nabla v_h - \nabla\cdot[\mu\nabla v_h] + \nabla p_h, \nabla\chi_h)_K\,, \\ r_h^p(u_h,\chi_h) &= \sum_{K\in\mathcal{T}_h} \alpha_K(\rho f, \nabla\chi_h)_K\,. \end{aligned}$$

- Streamline diffusion for the velocities:

$$\begin{aligned} s_h^v(u_h,\psi_h) &= \sum_{K\in\mathcal{T}_h} \delta_K(\rho\bar{v}_h\cdot\nabla v_h - \nabla\cdot[\mu\nabla v_h] + \nabla p_h, \rho\bar{v}_h\cdot\nabla\psi_h)_K\,, \\ r_h^v(u_h,\psi_h) &= \sum_{K\in\mathcal{T}_h} \delta_K(\rho f, \rho\bar{v}_h\cdot\nabla\psi_h)_K\,. \end{aligned}$$

- Streamline diffusion for the temperature:

$$\begin{aligned} s_h^T(u_h,\pi_h,\chi_h) &= \sum_{K\in\mathcal{T}_h} \gamma_K(\rho c_p\bar{v}_h\cdot\nabla T_h - \nabla\cdot[\lambda\nabla T_h], \rho c_p\bar{v}_h\cdot\nabla\pi_h)_K\,, \\ r_h^T(u_h,\pi_h,\chi_h) &= \sum_{K\in\mathcal{T}_h} \gamma_K(h, \rho c_p\bar{v}_h\cdot\nabla\pi_h)_K\,. \end{aligned}$$

Here, $\bar{v}_h$ is a suitable approximation to the current velocity field v_h, taken for example from a preceding iteration step. We denote the sum over these h-dependent stabilization terms by $s_h(\cdot,\cdot)$ and $r_h(\cdot)$, respectively,

$$\begin{aligned} s_h(u_h,\phi) &:= s_h^p(u_h,\chi) + s_h^v(u_h,\psi) + s_h^T(u_h,\pi), \\ r_h(u_h,\phi) &:= r_h^p(u_h,\chi) + r_h^v(u_h,\psi) + r_h^T(u_h,\pi). \end{aligned}$$

Then, with $A_h(\cdot;\cdot) := A(\cdot;\cdot) + s_h(\cdot,\cdot)$ and $F_h(\cdot) := F(\cdot) + r_h(\cdot)$, the discrete equations can be written in compact form

$$A_h(u_h;\phi) = F_h(\phi_h) \qquad \forall\phi\in V_h. \tag{8.19}$$

In order to ensure symmetry for the resulting stabilized system, α_K should be taken equal to δ_K. The stability and consistence of this formulation can be analyzed by similar techniques as used in the case of the incompressible Navier-Stokes equations; see Braack [17]. One obtains the following condition for the parameters δ_K:

$$\delta_K = \left[\frac{\mu}{h_K^2} + \frac{|\rho\bar{v}_h|_\infty}{h_K}\right]^{-1}, \quad \gamma_K = \left[\frac{\lambda}{h_K^2} + \frac{|\rho c_p\bar{v}_h|_\infty}{h_K}\right]^{-1}. \tag{8.20}$$

Open Problem 8.2: *Derive a formula for the stabilization parameters δ_K and γ_K which leads to a robust scheme on general meshes with large aspect ratio σ_h.*

8.2. Computational approach

For solving the model for weakly compressible flow introduced above, we want to use the methodology discussed for incompressible flows.

(i) Explicit defect correction coupling:
The simplest way to use an "incompressible solver" for computing weakly compressible flows is by a defect correction iteration. The step $\{p_h^{l-1}, v_h^{l-1}, T_h^{l-1}\} \to \{p_h^l, v_h^l, T_h^l\}$ of this scheme proceeds as follows:

1. The nonlinear coefficients (density, transport vectors, etc.) are frozen at $\{p_h^{l-1}, v_h^{l-1}, T_h^{l-1}\}$. Corresponding corrections $\{\delta p_h^l, \delta v_h^l, \delta T_h^l\} \in V_h$ are determined by solving the linearized system:

$$\begin{aligned} (\nabla\cdot\delta v_h^l, \chi_h) &= (d_p^{l-1}, \chi_h), \quad \forall \chi_h \in L_h, \\ (\rho v_h^{l-1}\cdot\nabla\delta v_h^l, \phi_h) + (\mu\nabla\delta v_h^l, \nabla\phi_h) - (\delta p_h^l, \nabla\cdot\phi_h) &= (d_v^{l-1}, \phi_h) \quad \forall \phi_h \in \mathbf{H}_h, \\ (\rho c_p v_h^{l-1}\cdot\nabla\delta T_h^l, \pi_h) + (\lambda\nabla\delta T_h^l, \nabla\pi_h) &= (d_t^{l-1}, \pi_h) \quad \forall \pi_h \in R_h, \end{aligned}$$

where d_p^{l-1}, d_v^{l-1}, and d_T^{l-1} are the defects of the iteration $\{p_h^{l-1}, v_h^{l-1}, T_h^{l-1}\}$. For the sake of robustness, the pressure and transport stabilization described above has also to be applied to this problem.

2. The new solution vector is obtained by

$$p_h^l = p_h^{i-1} + \kappa_l \delta p_h^l, \quad v_h^l = v_h^{i-1} + \kappa_l \delta v_h^l, \quad T_h^l = T_h^{i-1} + \kappa_l \delta T_h^l,$$

with some relaxation parameter $\kappa_l \in (0,1]$, and the density is updated according to $\rho_h^l = P_{th}/(RT_h^l)$.

3. The iteration is continued until some stopping criterion is satisfied.

In each step of this iteration a linearized Navier-Stokes problem supplemented by a heat transfer equation is to be solved. This may be accomplished by using the methods described above for the incompressible Navier-Stokes equations. Hence, the "incompressible solver" is used for preconditioning the defect correction iteration for solving the full system (8.18). However, this simple defect correction process may converge very slowly in the case of large temperature gradients (e.g., caused by strong heat release in chemical reactions). This lack of robustness can be cured by making the iteration more implicit.

(ii) Semi-implicit defect correction coupling:
In order to achieve better control on the variation of temperature, one may use the following more implicit iteration:

1. The nonlinear coefficients (density, transport vectors, etc.) are frozen at $\{p_h^{l-1}, v_h^{l-1}, T_h^{l-1}\}$. Corresponding corrections $\{\delta p_h^l, \delta v_h^l, \delta T_h^l\} \in V_h$ are then determined by solving the linearized system:

$$\begin{aligned} (\nabla\cdot\delta v_h^l, \chi_h) + ((T_h^{l-1})^{-1} v_h^{l-1}\cdot\nabla\delta T_h^l, \chi) &= (d_p^{l-1}, \chi_h) \quad \forall \chi_h \in L_h, \\ (\rho v_h^{l-1}\cdot\nabla\delta v_h^l, \phi_h) + (\mu\nabla\delta v_h^l, \nabla\phi_h) - (\delta p_h^l, \nabla\cdot\phi_h) &= (d_v^{l-1}, \phi_h) \quad \forall \phi_h \in \mathbf{H}_h, \\ (\rho c_p v_h^{l-1}\cdot\nabla\delta T_h^l, \pi_h) + (\lambda\nabla\delta T_h^l, \nabla\pi_h) &= (d_t^{l-1}, \pi_h) \quad \forall \pi_h \in R_h, \end{aligned}$$

where d_p^{l-1}, d_v^{l-1}, and d_T^{l-1} are the defects of the iteration $\{p_h^{l-1}, v_h^{l-1}, T_h^{l-1}\}$. Again, the pressure and transport stabilization described above has to be applied.

2. The new solution vector is obtained by

$$p_h^l = p_h^{i-1} + \kappa_l \delta p_h^l, \quad v_h^l = v_h^{i-1} + \kappa_l \delta v_h^l, \quad T_h^l = T_h^{i-1} + \kappa_l \delta T_h^l,$$

with some relaxation parameter $\kappa_l \in (0,1]$, and the density is updated according to $\rho_h^l = P_{th}/(RT_h^l)$.

3. The iteration is continued until some stopping criterion is satisfied.

This solution method has been used in Braack [17] for the simulation of low-Mach-number combustion processes; see also [10] and [18].

8.3. The algebraic system

In each substep of the defect correction iterations described above, we have to solve linear problems for the coefficients $x_j = \{x_j^{(p)}, x_j^{(v)}, x_j^{(T)}\}$ including the components for pressure, velocity and temperature in the basis representations

$$p_h = \sum_{j=1}^N x_j^{(p)} \psi_j, \quad v_h = \sum_{j=1}^N x_j^{(v)} \psi_j, \quad T_h = \sum_{j=1}^N x_j^{(T)} \psi_j.$$

The system sub-matrices corresponding to the different components are obtained from the coupled system by taking first test functions of the form $\phi_h = \{\psi_h, 0, 0\}$:

$$\begin{aligned} A_h(u_h; \{\psi_h, 0, 0\}) &= (\nabla\cdot v_h, \psi_h) - (\bar{T}_h^{-1} \bar{v}_h \cdot \nabla T_h, \psi_h) \\ &\quad + \sum_{K\in\mathcal{T}_h} \delta_K (\rho \bar{v}_h \cdot \nabla v_h - \nabla\cdot(\mu \nabla v_h) + \nabla p_h, \nabla \psi_h)_K . \end{aligned}$$

Analogously, taking the test functions $\phi_h = \{0, \psi_h, 0\}$, we obtain the equation for the velocity components,

$$\begin{aligned} A_h(u_h; \{0, \psi_h, 0\}) &= (\rho \bar{v}_h \cdot \nabla v_h, \psi_h) + (\mu \nabla v_h, \nabla \psi_h) - (p_h, \nabla\cdot\psi_h) + \\ &\quad \sum_{K\in\mathcal{T}_h} \delta_K (\rho \bar{v}_h \cdot \nabla v_h - \nabla\cdot(\mu \nabla v_h) + \nabla p_h, \rho \bar{v}_h \cdot \nabla \psi_h)_K , \end{aligned}$$

and by taking the test functions $\phi_h = (0, 0, \pi_h)$ the equation for the temperature component,

$$\begin{aligned} A_h(u_h; \{0, 0, \psi_h\}) &= (\rho c_p \bar{v}_h \cdot \nabla T_h, \psi_h) + (\lambda \nabla T_h, \nabla \psi_h) + \\ &\quad \sum_{K\in\mathcal{T}_h} \gamma_K (\rho c_p \bar{v}_h \cdot \nabla T_h - \nabla\cdot[\lambda \nabla T_h], \rho c_p \bar{v}_h \cdot \nabla \psi_h) . \end{aligned}$$

Ordering the unknowns in a physically block-wise sense, i.e., marching through the set of nodal points and attaching to each node the corresponding submatrix containing the unknowns of all physical quantities, we obtain "nodal matrices" $\mathcal{A}_{ij}$ of the form

$$\mathcal{A}_{ij} = \begin{bmatrix} B_{pp} & B_{pv} & B_{pT} \\ B_{vp} & B_{vv} & B_{vT} \\ B_{Tp} & B_{Tv} & B_{TT} \end{bmatrix},$$

where the indices p, v, T indicate the corresponding contributions. Looking at the equations, we see that almost all physical components are coupled with each other; only the pressure does not appear in the temperature equation, i.e., $B_{Tp} = 0$. Several of the other couplings are of minor importance and may be neglected in building an approximating nodal matrix $\widetilde{\mathcal{A}}_{ij}$ to $\mathcal{A}_{ij}$. One could think of a complete decoupling of the flow variables $\{p, v\}$ from the temperature T (or other state variables describing for example chemical reactions) resulting in an approximation of the form

$$\widetilde{\mathcal{A}}_{ij} = \begin{bmatrix} B_{pp} & B_{pv} & 0 \\ B_{vp} & B_{vv} & 0 \\ 0 & 0 & B_{TT} \end{bmatrix}.$$

However, such a simplification is not appropriate in computing processes in which the temperature has a significant influence on the flow field and vice versa. For example, in combustion problems, density variations are mainly caused by changes of the temperature. A detailed discussion of this issue can be found in Braack [17] and in [18].

8.4. An example of chemically reactive flow

We close this section by presenting some results from Braack [17] on computations for low-Mach-number flows with chemical reactions. The configuration considered is the model of a methane burner with a complicated geometry and using a sophisticated reaction mechanism. A stoichiometric mixture of methane CH_4 and air O_2/N_2 flows from the bottom of the burner through a sample of slots of uniform width $2\,mm$ and three different heights (varying from $14\,mm$ to $11\,mm$). The columns have a uniform width of $1.5\,mm$. The inflow velocity is uniformly $0.2\,m/s$. The Reynolds number in this model is about $Re = 90$. The geometry is shown in Figure 37.

Due to the heating of the slots, the flow accelerates up to approximately $1\,m/s$. Since this is higher than the flame velocity of a stoichiometric methane flame the flame front is located above the slots. For lower inflow velocities, the flame moves downstream into the slots and extinguishes as a result of the heat loss by the cold walls.

If the solution is assumed to be spatially periodic, it is sufficient to restrict the computational domain Ω to only three slots, as shown in Figure 38. The boundaries at the left and right-hand of Ω are symmetry boundary conditions. The walls of the slots are described by Dirichlet conditions for the temperature and Neumann conditions for the species. The calculation on the coarsest mesh (with 1344 cells) uses a time-stepping procedure to provide a physically correct starting value. Then, on the finer meshes the stationary fixed-point defect correction iteration converges. In order to obtain ignition, the temperature for the initial solution is set to $2000\,K$ at the points above the slots. The reaction mechanism is that of Smooke [80] with 15 species and 84 elementary reactions (42 bidirectional),

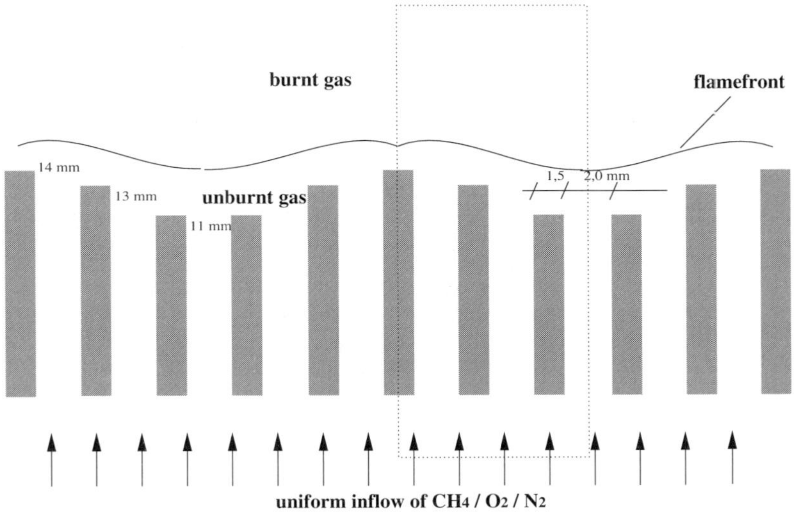

FIGURE 37. Geometry of a methane burner; from Braack [17].

supplemented by two further species, N and NO, and 4 additional reactions to describe their formation.

The solution is obtained on an adaptively refined mesh with refinement criterion based on the linear functional

$$J(u) = |\Omega|^{-1} \int_{\Omega} T \, dx \,,$$

in order to capture the temperature distribution accurately. The finest mesh is shown in Figure 38; we see local mesh refinement at the flame front and below the slots where the velocity field changes. The mesh is automatically adapted and no hand-fitting on the basis of a priori knowledge of the solution is necessary to find the appropriate balance of the mesh-size distribution. The CPU time required for such a simulation with about 5,000 cells ($\approx$ 100,000 unknowns) is approximately 6 hours on a Pentium II (233 Mhz) when the initial guess on the coarse grid with approximately 1300 cells is given.

The computed pressure and the main velocity component are shown in Figure 38. Due to the strong heat release the flow accelerates by a factor of 10 at the outflow of the slots. At the walls of the slots, Dirichlet conditions for the temperature are imposed, varying linearly from $298\,K$ at the bottom up to $393\,K$, $453\,K$ and $513\,K$ for the three different walls. This leads to a higher outflow velocity at the longer slot compared to the shorter ones. Therefore, the lift-off of the flame is substantially higher at the longer slot, leading to the common Bunsen cone formed by two neighboring longer slots.

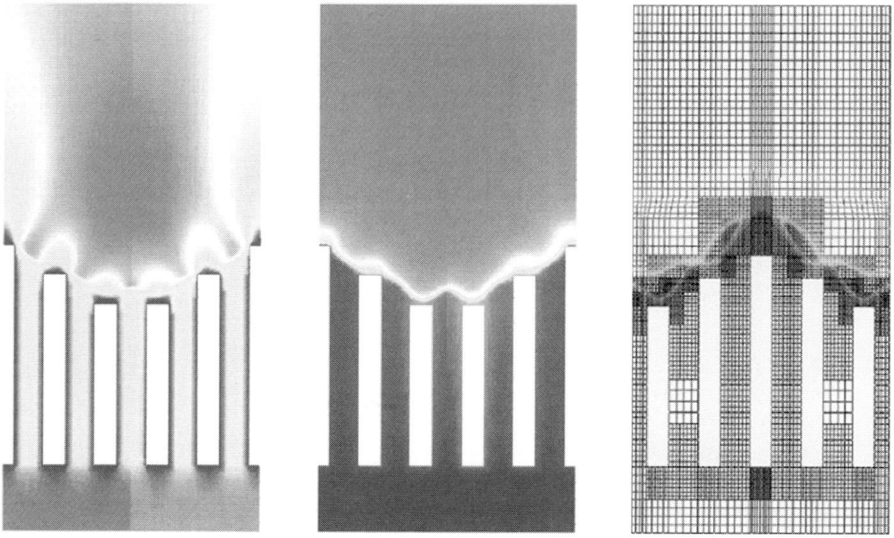

FIGURE 38. Results of the methane burner simulation: velocity and temperature profiles (left and middle), finest mesh with 5,000 cells (right); from Braack [17].

References

[1] M. Ainsworth and J. T. Oden (1997), *A posteriori error estimation in finite element analysis*, Comput. Meth. Appl. Mech. Engrg., 142, pp. 1–88.

[2] T. Apel and M. Dobrowolski (1992), *Anisotropic interpolation with applications to the finite element method*, Computing, 47, pp. 277–293.

[3] T. Apel (1999), *Anisotropic Finite Elements: Local Estimates and Applications*, Habilitation Thesis, Preprint 99-03, SFB 393, University of Magdeburg.

[4] E. Backes (1997), *Gewichtete a posteriori Fehleranalyse bei der adaptiven Finite-Elemente-Methode: Ein Vergleich zwischen Residuen- und Bank-Weiser-Schätzer*, Diploma Thesis, Institute of Applied Mathematics, University of Heidelberg.

[5] W. Bangerth and G. Kanschat (1999), *deal.II Homepage*, Technical Reference, Release 1.0, SFB 359, University of Heidelberg,
http://gaia.iwr.uni-heidelberg.de/~deal/.

[6] R. E. Bank, B. Weiser, and H. Yserentant (1990), *A class of iterative methods for solving saddle point problems*, Numer. Math., 56, 645–666.

[7] R. Becker (1995), *An Adaptive Finite Element Method for the Incompressible Navier-Stokes Equations on Time-Dependent Domains*, Doctor Thesis, Preprint 95-44, SFB 359, Nov. 1995, University of Heidelberg.

[8] R. Becker (1998), *An adaptive finite element method for the Stokes equations including control of the iteration error*, ENUMATH'95, Paris, Sept. 18–22. 1995, in: Proc. ENUMATH'97 (H. G. Bock, et al., eds.), pp. 609–620, World Scientific Publisher, Singapore.

[9] R. Becker (1998): *Weighted error estimators for finite element approximations of the incompressible Navier-Stokes equations*, Preprint 98-20, SFB 359, University of Heidelberg, submitted for publication.

[10] R. Becker, M. Braack, R. Rannacher, and C. Waguet (1998), *Fast and reliable solution of the Navier-Stokes equations including chemistry*, Proc. Conf. Applied Mathematics for Industrial Flow Problems (AMIF), San Feliu de Guixols (Spain), Oct. 1–3. 1998, Preprint 99-03 (SFB 359), University of Heidelberg, January 1999, to appear in Computing and Visualization in Sciences.

[11] R. Becker, C. Johnson, and R. Rannacher (1995), *Adaptive error control for multigrid finite element methods*, Computing, 55, pp. 271–288.

[12] R. Becker and R. Rannacher (1995), *Weighted a posteriori error control in FE methods*, ENUMATH'95, Paris, Sept. 18–22, 1995, Proc. ENUMATH'97 (H. G. Bock, et al., eds.), pp. 621–637, World Scientific Publishers, Singapore.

[13] R. Becker and R. Rannacher (1994), *Finite element solution of the incompressible Navier-Stokes equations on anisotropically refined meshes*, Proc. Workshop "Fast Solvers for Flow Problems", Kiel, Jan. 14–16, 1994 (W. Hackbusch and G. Wittum, eds.), pp. 52–61, NNFM, Vol. 49, Vieweg, Braunschweig.

[14] R. Becker and R. Rannacher (1996), *A feed-back approach to error control in finite element methods: Basic analysis and examples*, East-West J. Numer. Math., 4, pp. 237–264.

[15] H. Blum (1990), *Asymptotic Error Expansion and Defect Correction in the Finite Element Method*, Habilitation Thesis, University of Heidelberg.

[16] M. Boman (1995), *A Model Study of Hydrodynamic Stability*, Masters Thesis, Chalmers University of Technology, Gothenburg, Sweden.

[17] M. Braack (1998), *An Adaptive Finite Element Method for Reactive Flow Problems*, Doctor Thesis, Institute of Applied Mathematics, University of Heidelberg.

[18] M. Braack and R. Rannacher (1999), *Adaptive finite element methods for low-Mach-number flows with chemical reactions*, Lecture Series 1999-03, 30th Computational Fluid Dynamics, (H. Deconinck, ed.), von Karman Institute for Fluid Dynamics, Belgium.

[19] S. C. Brenner and R. L. Scott (1994), *The Mathematical Theory of Finite Element Methods*, Springer, Berlin-Heidelberg-New York.

[20] F. Brezzi and M. Fortin (1991), *Mixed and Hybrid Finite Element Methods*, Springer, Berlin-Heidelberg-New York.

[21] F. Brezzi and J. Pitkäranta (1984), *On the stabilization of finite element approximations of the Stokes equations*, Proc. Workshop Efficient Solution of Elliptic Systems (W. Hackbusch, ed.), Vieweg, Braunschweig.

[22] M. O. Bristeau, R. Glowinski, and J. Periaux (1987), *Numerical methods for the Navier-Stokes equations: Applications to the simulation of compressible and incompressible viscous flows*, Comput. Phys. Reports, 6, pp. 73–187.

[23] C. M. Chen and V. Thomée (1985), *The lumped mass finite element method for a parabolic problem*, J. Austral. Math. Soc., Ser. B, 26, pp. 329–354.

[24] A. J. Chorin (1968), *Numerical solution of the Navier-Stokes equations*, Math. Comp., 22, pp. 745–762.

[25] W. E and J. P. Liu (1998), *Projection method I: Convergence and numerical boundary layers*, SIAM J. Num. Anal., to appear.

[26] K. Eriksson, D. Estep, P. Hansbo, and C. Johnson (1995), *Introduction to adaptive methods for differential equations*, Acta Numerica 1995 (A. Iserles, ed.), pp. 105–158, Cambridge University Press.

[27] M. Feistauer (1993), *Mathematical Methods in Fluid Dynamics*, Longman Scientific & Technical, England.

[28] M. L. M. Giles, M. Larson, and E. Süli (1998), *Adaptive error control for finite element approximations of the lift and drag coefficients in viscous flow*, SIAM J. Numer. Anal, to appear.

[29] V. Girault and P.-A. Raviart (1986), *Finite Element Methods for the Navier-Stokes Equations*, Springer, Heidelberg.

[30] R. Glowinski (1985), *Viscous flow simulations by finite element methods and related numerical techniques*, in Progress in Supercomputing in Computational Fluid Dynamics (E.M. Murman and S.S. Abarbanel, eds.), pp. 173–210, Birkhäuser, Boston.

[31] R. Glowinski and J. Periaux (1987), *Numerical methods for nonlinear problems in fluid dynamics*, Proc. Int. Seminar on Scientific Supercomputers, Paris, North-Holland, Amsterdam.

[32] P. M. Gresho (1990), *On the theory of semi-implicit projection methods for viscous incompressible flow and its implementation via a finite element method that also introduces a nearly consistent mass matrix, Part 1: Theory*, Int. J. Numer. Meth. Fluids, 11, pp. 587–620.

[33] P. M. Gresho and S. T. Chan (1990), *On the theory of semi-implicit projection methods for viscous incompressible flow and its implementation via a finite element method that also introduces a nearly consistent mass matrix, Part 2: Implementation*, Int. J. Numer. Meth. Fluids, 11, pp. 621–660.

[34] P. M. Gresho (1991), *Incompressible fluid dynamics: Some fundamental formulation issues*, Ann. Rev. Fluid Mech., 23, pp. 413–453.

[35] P. M. Gresho and R. L. Sani (1998), *Incompressible Flow and the Finite Element Method*, John Wiley, Chichester.

[36] W. Hackbusch (1985), *Multi-Grid Methods and Applications*, Springer, Heidelberg-Berlin.

[37] P. Hanbo and C. Johnson (1995), *Streamline diffusion finite element methods for fluid flow*, in Finite Element Methods for Compressible and Incompressible Flow, Selected Topics from Previous VKI Lecture Series, (H. Deconinck, ed.), von Karman Institute for Fluid Dynamics, Belgium.

[38] J. Harig (1991), *Eine robuste und effiziente Finite-Elemente Methode zur Lösung der inkompressiblen 3D-Navier-Stokes Gleichungen auf Vektorrechnern*, Doctor Thesis, Institute of Applied Mathematics, University of Heidelberg.

[39] R. Hartmann (1998), *A posteriori Fehlerschätzung und adaptive Schrittweiten- und Ortsgittersteuerung bei Galerkin-Verfahren für die Wärmeleitungsgleichung*, Diploma Thesis, Institute of Applied Mathematics, University of Heidelberg.

[40] J. G. Heywood (1980), *The Navier-Stokes equations: On the existence, regularity and decay of solutions*, Indiana Univ. Math. J., 29, pp. 639–681.

[41] J. G. Heywood and R. Rannacher (1982), *Finite element approximation of the nonstationary Navier-Stokes Problem. I. Regularity of solutions and second order error estimates for spatial discretization*, SIAM J. Numer. Anal., 19, pp. 275–311.

[42] J. G. Heywood and R. Rannacher (1986), *Finite element approximation of the nonstationary Navier-Stokes Problem. II. Stability of solutions and error estimates uniform in time*, SIAM J. Numer. Anal., 23, pp. 750–777.

[43] J. G. Heywood and R. Rannacher (1988), *Finite element approximation of the nonstationary Navier-Stokes Problem. III. Smoothing property and higher order error estimates for spatial discretization*, SIAM J. Numer. Anal., 25, pp. 489–512.

[44] J. G. Heywood and R. Rannacher (1990), *Finite element approximation of the nonstationary Navier-Stokes Problem. IV. Error analysis for second-order time discretization*, SIAM J. Numer. Anal., 27, pp. 353–384.

[45] J. G. Heywood and R. Rannacher (1986), *An analysis of stability concepts for the Navier-Stokes equations*, J. Reine Angew. Math., 372, pp. 1–33.

[46] J. G. Heywood, R. Rannacher, and S. Turek (1992), *Artificial boundaries and flux and pressure conditions for the incompressible Navier-Stokes equations*, Int. J. Numer. Math. Fluids, 22, pp. 325–352.

[47] P. Houston, R. Rannacher, and E. Süli (1999), *A posteriori error analysis for stabilised finite element approximation of transport problems*, Report No. 99/04, Oxford University Computing Laboratory, Oxford, England OX1 3QD, submitted for publication.

[48] T. J. R. Hughes and A. N. Brooks (1982), *Streamline upwind/Petrov-Galerkin formulations for convection dominated flows with particular emphasis on the incompressible Navier-Stokes equation*, Comput. Meth. Appl. Mech. Engrg., 32, pp. 199–259.

[49] T. J. R. Hughes, L. P. Franca, and M. Balestra (1986), *A new finite element formulation for computational fluid mechanics: V. Circumventing the Babuska-Brezzi condition: A stable Petrov-Galerkin formulation of the Stokes problem accommodating equal order interpolation*, Comput. Meth. Appl. Mech. Engrg., 59, pp. 85–99.

[50] C. Johnson (1987), *Numerical Solution of Partial Differential Equations by the Finite Element Method*, Cambridge University Press, Cambridge-Lund.

[51] C. Johnson (1989), *The streamline diffusion finite element method for compressible and incompressible fluid flow*, Proc. Finite Element Method in Fluids VII, Huntsville.

[52] C. Johnson (1993), *A new paradigm for adaptive finite element methods*, Proc. MAFELAP'93 Conf., Brunel Univ., Uxbridge, UK, John Wiley, Chichester.

[53] C. Johnson, S. Larsson, V. Thomée, and L. B. Wahlbin (1987), *Error estimates for spatially discrete approximations of semilinear parabolic equations with nonsmooth initial data*, Math. Comp., 49, pp. 331–357.

[54] C. Johnson, R. Rannacher, and M. Boman (1995), *Numerics and hydrodynamic stability: Towards error control in CFD*, SIAM J. Numer. Anal., 32, pp. 1058–1079.

[55] C. Johnson, R. Rannacher, and M. Boman (1995), *On transition to turbulence and error control in CFD*, Preprint 95-06, SFB 359, University of Heidelberg.

[56] C. Johnson and R. Rannacher (1994), *On error control in CFD*, Proc. Int. Workshop "Numerical Methods for the Navier-Stokes Equations", Heidelberg, Oct. 25–28, 1993 (F.-K. Hebeker, R. Rannacher, and G. Wittum, eds.), pp. 25–28, NNFM, Vol. 47, Vieweg, Braunschweig.

[57] S. Kracmar and J. Neustupa (1994), *Modelling of flows of a viscous incompressible fluid through a channel by means of variational inequalities*, Z. Angew. Math. Mech., 74, pp. T637–T639.

[58] O. Ladyshenskaya (1969), *The Mathematical Theory of Viscous Incompressible Flow*, Gordon and Breach, London.

[59] M. Landahl (1980), *A note on an algebraic instability of viscous parallel shear flows*, J. Fluid Mech., 98, p. 243.

[60] P.-L. Lions (1998), *Mathematical Topics in Fluid Mechanics, Vol. 2, Compressible Models*, Clarendon Press, Oxford.

[61] M. Luskin and R. Rannacher (1982), *On the smoothing property of the Crank-Nicolson scheme*, Appl. Anal., 14, pp. 117–135.

[62] M. Luskin and R. Rannacher (1981), *On the smoothing property of the Galerkin method for parabolic equations*, SIAM J. Numer. Anal., 19, pp. 93–113.

[63] K. W. Morton (1996), *Numerical Solution of Convection-Diffusion Problems*, Applied Mathematics and Mathematical Computation, Vol. 12, Chapman&Hall.

[64] S. Müller-Urbaniak (1993), *Eine Analyse des Zwischenschritt-θ-Verfahrens zur Lösung der instationären Navier-Stokes-Gleichungen*, Doctor Thesis, Preprint 94-01, SFB 359, University of Heidelberg.

[65] S. Müller, A. Prohl, R. Rannacher, and S. Turek (1994), *Implicit time-discretization of the nonstationary incompressible Navier-Stokes equations*, Proc. Workshop "Fast

Solvers for Flow Problems", Kiel, Jan. 14–16, 1994 (W. Hackbusch and G. Wittum, eds.), pp. 175–191, NNFM, Vol. 49, Vieweg, Braunschweig.

[66] H. Oswald (1999), *Lösung der instationären Navier-Stokes-Gleichungen auf Parallelrechnern*, Doctor Thesis, Institute of Applied Mathematics, University of Heidelberg.

[67] O. Pironneau (1982), *On the transport-diffusion algorithm and its applications to the Navier-Stokes equations*, Numer. Math., 38, pp. 309–332.

[68] O. Pironneau (1983), *Finite Element Methods for Fluids*, John Wiley, Chichester.

[69] A. Prohl (1995), *Projektions- und Quasi-Kompressibilitätsmethoden zur Lösung der inkompressiblen Navier-Stokes-Gleichungen*, Doctor Thesis, Institute of Applied Mathematics, University of Heidelberg.

[70] A. Prohl (1998), *On Quasi-Compressibility Methods and Projection Methods for Solving the Incompressible Navier-Stokes Equations*, Teubner, Stuttgart.

[71] R. Rannacher (1984), *Finite element solution of diffusion problems with irregular data*, Numer. Math., 43, pp. 309–327.

[72] R. Rannacher (1998), *Numerical analysis of nonstationary fluid flow (a survey)*, in "Applications of Mathematics in Industry and Technology" (V.C.Boffi and H. Neunzert, eds.), pp. 34–53, B.G. Teubner, Stuttgart.

[73] R. Rannacher (1992), *On Chorin's projection method for the incompressible Navier-Stokes equations*, Proc. Oberwolfach Conf. Navier-Stokes Equations: Theory and Numerical Methods, September 1991 (J.G. Heywood, et al., eds.), pp. 167–183, LNM, Vol. 1530, Springer, Heidelberg.

[74] R. Rannacher (1993), *On the numerical solution of the incompressible Navier-Stokes equations*, Survey lecture at the annual GAMM-Conference 1992, Leipzig, March 24–27, Z. Angew. Math. Mech., 73, pp. 203–216.

[75] R. Rannacher (1998), *A posteriori error estimation in least-squares stabilized finite element schemes*, Comput. Meth. Appl. Mech. Engrg., 166, pp. 99–114.

[76] R. Rannacher (1999), *Error control in finite element computations*, Proc. NATO-Summer School "Error Control and Adaptivity in Scientific Computing" Antalya (Turkey), Aug. 1998 (H. Bulgak and C. Zenger, eds.), pp. 247–278, NATO Science Series, Series C, Vol. 536, Kluwer, Dordrecht.

[77] R. Rannacher and S. Turek (1992), *A simple nonconforming quadrilateral Stokes element*, Numer. Meth. Part. Diff. Equ., 8, pp. 97–111.

[78] R. Rautmann (1983), *On optimal regularity of Navier-Stokes solutions at time* $t = 0$, Math. Z., 184, pp. 141–149.

[79] H. Reichert and G. Wittum (1993), *On the construction of robust smoothers for incompressible flow problems*, Proc. Workshop "Numerical Methods for the Navier-Stokes Equations", Heidelberg, Oct. 25-28, 1993 (F.K. Hebeker, R. Rannacher, G. Wittum, eds.), pp. 207–216, Vieweg, Braunschweig.

[80] M. D. Smooke (1991), *Numerical Modeling of Laminar Diffusion Flames*, Progress in Astronautics and Aeronautics, 135.

[81] M. Schäfer and S. Turek (1996), *The benchmark problem "flow around a cylinde"*, Flow Simulation with High-Performance Computers (E.H. Hirschel, ed.), pp. 547–566, NNFM, Vol. 52, Vieweg, Braunschweig.

[82] F. Schieweck (1992), *A parallel multigrid algorithm for solving the Navier-Stokes equations on a transputer system*, Impact Comput. Sci. Engrg., 5, pp. 345–378.

[83] P. Schreiber and S. Turek (1993), *An efficient finite element solver for the nonstationary incompressible Navier-Stokes equations in two and three dimensions*, Proc. Workshop "Numerical Methods for the Navier-Stokes Equations", Heidelberg, Oct. 25–28, 1993 (F.K. Hebeker, R. Rannacher, G. Wittum, eds.), pp. 133–144, Vieweg, Braunschweig.

[84] J. Shen (1992), *On error estimates of projection methods for the Navier-Stokes Equations: First order schemes*, SIAM J. Numer. Anal., 29, pp. 57–77.

[85] J. Shen (1994), *Remarks on the pressure error estimates for the projection methods*, Numer. Math., 67, pp. 513–520.

[86] J. Shen (1996), *On error estimates of the projection methods for the Navier-Stokes Equations: 2nd order schemes*, Math. Comp., 65, pp. 1039–1065.

[87] G. Strang (1968), *On the construction and comparison of difference schemes*, SIAM J. Numer. Anal., 5, 506–517.

[88] R. Temam (1987), *Navier-Stokes Equations. Theory and numerical analysis*, 2nd ed., North Holland, Amsterdam.

[89] V. Thomée (1997), *Galerkin Finite Element Methods for Parabolic Problems*, Springer, Berlin-Heidelberg-New York.

[90] L. Tobiska and F. Schieweck (1989), *A nonconforming finite element method of upstream type applied to the stationary Navier-Stokes equation*, M^2AN, 23, pp. 627–647.

[91] L. Tobiska and R. Verfürth (1996), *Analysis of a streamline diffusion finite element method for the Stokes and Navier-Stokes equations*, SIAM J. Numer. Anal., 33, pp. 107–127.

[92] L. N. Trefethen, A. E. Trefethen, S. C. Reddy, and T. A. Driscoll (1992), *A new direction in hydrodynamical stability: Beyond eigenvalues*, Tech. Report CTC92TR115 12/92, Cornell Theory Center, Cornell University.

[93] S. Turek (1994), *Tools for simulating nonstationary incompressible flow via discretely divergence-free finite element models*, Int. J. Numer. Meth. Fluids, 18, pp. 71–105.

[94] S. Turek (1996), *A comparative study of some time-stepping techniques for the incompressible Navier-Stokes equations*, Int. J. Numer. Meth. Fluids, 22, pp. 987–1011.

[95] S. Turek (1997), *On discrete projection methods for the incompressible Navier-Stokes equations: An algorithmic approach*, Comput. Methods Appl. Mech. Engrg., 143, pp. 271–288.

[96] S. Turek (1998), *FEATFLOW: Finite Element Software for the Incompressible Navier-Stokes Equations*, User Manual, Release 1.3, SFB 359, University of Heidelberg, URL: http://gaia.iwr.uni-heidelberg.de/~featflow/.

[97] S. Turek (1999), *Efficient Solvers for Incompressible Flow Problems*, NNCSE, Vol. 6, Springer, Berlin-Heidelberg-New York.

[98] S. Turek, et al. (1999), *Virtual Album of Fluid Motion*, Preprint, Institute of Applied Mathematics, University of Heidelberg, in preparation.

[99] M. Van Dyke (1982), *An Album of Fluid Motion*, The Parabolic Press, Stanford.

[100] J. Van Kan (1986), *A second-order accurate pressure-correction scheme for viscous incompressible flow*, J. Sci. Stat. Comp., 7, pp. 870–891.

[101] S. P. Vanka (1986), *Block-implicit multigrid solution of Navier-Stokes equations in primitive variables*, J. Comp. Phys., 65, pp. 138–158.

[102] R. Verfürth (1996), *A Review of A Posteriori Error Estimation and Adaptive Mesh-Refinement Techniques*, Wiley/Teubner, New York-Stuttgart, 1996.

[103] C. Waguet (1999) *Adaptive Finite Element Computation of Chemical Flow Reactors*, Doctor Thesis, SFB 359, University of Heidelberg, in preparation.

[104] P. Wesseling (1992), *An Introduction to Multigrid Methods*, J. Wiley, Chichester.

[105] G. Wittum (1990), *The use of fast solvers in computational fluid dynamics*, in Proc. Eighth GAMM-Conference on Numerical Methods in Fluid Mechanics (P. Wesseling, ed.), pp. 574–581. LNFM, Vol. 29, Vieweg, Braunschweig.

[106] O. C. Zienkiewicz and J. Z. Zhu (1987), *A simple error estimator and adaptive procedure for practical engineering analysis*, Int. J. Numer. Meth. Engrg., 24, pp. 337–357.

[107] Guohui Zhou (1997), *How accurate is the streamline diffusion finite element method?*, Math. Comp., 66, pp. 31–44.

Rolf Rannacher
Institute of Applied Mathematics
University of Heidelberg
INF 293/294, D-69120 Heidelberg, Germany
E-Mail: `rannacher@iwr.uni-heidelberg.de`
URL: `http://gaia.iwr.uni-heidelberg.de`